LIVESTOCK

LIVESTOCK

Editor

Baldur Jaeger

scitus academics

Livestock

Edited by **Baldur Jaeger**

This Book is Printed on Acid-Free Paper. ♾

ISBN: 9781681179278

Library of Congress Control Number: 2017948956

SCITUS Academics LLC,

616, Corporate Way, Suite 2, 4766,

Valley Cottage, NY 10989

www.scitusacademics.com

Notice

Printed in United States of America.

Preface

Although food availability has augmented along with the rising human population over the last 30 years, there are still 800 million people suffering from malnutrition. This dilemma is not only the result of inadequate food production and scarce distribution, but also of the financial inability of the poor to purchase food of rational quality in adequate quantities to fulfill their needs. Livestock production comprises a very important component of the agricultural economy of developing countries, a contribution that goes beyond direct food production to include multipurpose uses, such as skins, fibre, fertilizer and fuel, as well as capital accumulation. Livestock are domesticated animals raised in an agricultural setting to produce commodities such as food, fiber, and labor. The livestock sector globally is highly dynamic in response to rapidly increasing demand for livestock products. In developed countries, demand for livestock products is stagnating, while many production systems are increasing their efficiency and environmental sustainability. Livestock systems have both positive and negative effects on the natural resource base, public health, social equity and economic growth. Currently, livestock is one of the fastest growing agricultural subsectors in developing countries. Its share of agricultural GDP is already 33 per cent and is quickly increasing. This growth is motivated by the swiftly growing demand for livestock products, this demand being driven by population growth, urbanization and rising incomes and the production response in different livestock systems has been associated with science and technology as well as increases in animal numbers.

This volume 'Livestock' presents selected topics in livestock science contributed by well-known experts in their respective areas. It consisting of topics in food-animal production systems, management of several animal products, health related issues in animals, and contaminants that may be found in animal foods. It presents the state of livestock production systems and the status of both intensive and extensive production systems globally in relation to recent trends, coupled with a brief assessment of whether these trends are likely to continue into the future. This book has been designed to meet the requirements of students, researchers who are working in different parts of the world in different environments.

Table of Contents

Chapter 1	Mapping the Global Distribution of Livestock 1
Chapter 2	Securing Sustainable Livestock Production Systems in an Uncertain Economic Climate: Nurturing Flexibility and Resilience 33
Chapter 3	Livelihood Activities and the Role of Livestock in Smallholder Farming Communities of Southern Zambia 51
Chapter 4	Livestock Demand, Global Land Use Changes, and Induced Greenhouse Gas Emissions 69
Chapter 5	Livestock metabolomics and the livestock metabolome: A systematic review 87
Chapter 6	Livestock feed resources, nutritional value and their implication on animal productivity in mixed farming system in Gasera and Ginnir Districts, Bale Zone, Ethiopia 121
Chapter 7	Livestock-Handling Related Injuries and Deaths 149

Chapter 8 Influence of breed, season and age on quality bovine semen used for artificial insemination 189

Chapter 9 Lactation Responses toward Milk Indigenous Enzymes 203

Chapter 10 Bacterial Tick-Borne Diseases of Livestock Animals 219

Chapter 11 Dual-Use Bioenergy-Livestock Feed Potential of Giant Miscanthus, Giant Reed, and Miscane 239

Chapter 12 The effect of feeding high fat diet to beef cattle on manure composition and gaseous emission from a feedlot pen surface 257

Chapter 13 Exploring Agritourism Experience and Perceptions of Pork Production 285

Chapter 14 Do Organic Livestock Farms Differ from Low-Input Conventional Ones? Insights Based on Beef Cattle in Southern Europe 305

Chapter 15 Molecular Characterization and Prevalence of Trypanosoma Species in Cattle from a Northern Livestock Area in Côte d' 327

Chapter 16 Livestock-associated risk factors for pneumonia in an area of intensive animal farming in the Netherlands 339

Chapter 17 Participatory Evaluation of the Relative Livestock Populations and the Assessment of the Status and Impacts of

Newcastle Disease in Rural Communities of Two Northeastern States, Nigeria 359

Chapter 18 Dynamics of Ruminant Livestock Management in the Context of the Nigerian Agricultural System 375

Index 397

CHAPTER 1

Mapping the Global Distribution of Livestock

Timothy P. Robinson[1,2*], G. R. William Wint[3], Giulia Conchedda[2], Thomas P. Van Boeckel,[4,5,6,7] Valentina Ercoli[2], Elisa Palamara[2], Giuseppina Cinardi[2], Laura D'Aietti[2], Simon I. Hay[8,9], Marius Gilbert

1 Livestock Systems and Environment Research Theme (LSE), International Livestock Research Institute (ILRI), Nairobi, Kenya

2 Animal Production and Health Division (AGA), Food and Agriculture Organization of the United Nations (FAO), Rome, Italy

3 Environmental Research Group Oxford (ERGO) - Department of Zoology, University of Oxford, Oxford, United Kingdom

4 Biological Control and Spatial Ecology, Université Libre de Bruxelles, Brussels, Belgium

5 Fonds National de la Recherche Scientifique, Brussels, Belgium

6 Department of Ecology and Evolutionary - Biology Department, Princeton University, Princeton, New Jersey, United States of America

7 Princeton Environmental Institute, Princeton, New Jersey, United States of America

8 Spatial Ecology and Epidemiology Group - Department of Zoology, University of Oxford, Oxford, United Kingdom

9 Fogarty International Center, National Institutes of Health, Bethesda, Maryland, United States of America

ABSTRACT

Livestock contributes directly to the livelihoods and food security of almost a billion people and affects the diet and health of many more. With estimated standing populations of 1.43 billion cattle, 1.87 billion sheep and goats, 0.98 billion pigs, and 19.60 billion chickens, reliable and accessible information on the distribution and abundance of livestock is needed for a many reasons. These include analyses of the social and economic aspects of the livestock sector; the environmental impacts of livestock such as the production and management of waste, greenhouse gas emissions and livestock-related land-use change; and large-scale public health and epidemiological investigations. The Gridded Livestock of the World (GLW) database, produced in 2007, provided modelled livestock densities of the world, adjusted to match official (FAOSTAT) national estimates for the reference year 2005, at a spatial resolution of 3 minutes of arc (about 5×5 km at the equator). Recent methodological improvements have significantly enhanced these distributions: more up-to date and detailed sub-national livestock statistics have been collected; a new, higher resolution set of predictor variables is used; and the analytical procedure has been revised and extended to include a more systematic assessment of model accuracy and the representation of uncertainties associated with the predictions. This paper describes the current approach in detail and presents new global distribution maps at 1 km resolution for cattle, pigs and chickens, and a partial distribution map for ducks. These digital layers are made publically available via the Livestock Geo-Wiki (http://www.livestock.geo-wiki.org), as will be the maps of other livestock types as they are produced.

1. INTRODUCTION

Livestock contributes directly to the livelihoods and food security of almost a billion people [1]–[2]. In recent decades, the world food economy has seen a shift towards increased consumption of animal-source foods. In developing countries, the meat and dairy sectors have grown at average yearly rates of 5.1 percent and 3.6 percent respectively since 1970 [3]. Most of this increase has been to supply expanding urban areas and has been primarily driven by population growth and increasing incomes, closely linked to urbanisation, making livestock one of the fastest-growing sectors in agriculture [1]. Globally, as human population growth slows and relatively high consumption levels of animal-source foods are reached in some of the countries that have seen the more rapid increases, Brazil and China in particular, growth in demand for meat, milk and eggs is also expected to slow down [4]. Notwithstanding regional reductions in growth, in many developing countries *per capita* consumption of animal-source foods is projected to continue rising and even to accelerate [3]. As well as the economic benefits that a growing livestock sector confers on the economies of these countries, and the potential improvements in food security and nutrition among the world's poor, some 766 million poor (<US$ 2 per day) livestock keepers could benefit directly [5]. This is particularly the case in mixed farming systems where livestock serve many

socio-economic functions and promote arable agriculture through manure and draft power [6].

Much of this demand-led growth, and particularly that stemming from the burgeoning urban centres, is likely to be supplied by intensification of production and will generally not be met by small-holders except in particular circumstances where small-holders are well-connected to markets and do not suffer from poor economies of scale, such as the small-holder dairy sector in India [7]. Whilst livestock sector growth presents economic opportunities, in cases where competition from integrated, industrial producers and increasingly tight sanitary standards marginalise the livestock-dependent poor and other small-scale farmers by forcing them out of markets [4], [8]–[9], its main social effects on low-income populations may be negative if not carefully managed.

The growing livestock sector also places increased pressure on natural resources and the environment; making a significant contribution to global environmental change [10]–[11]. Also of concern are the public health implications of livestock production and intensification. As pressure on our global resources increases, people and their livestock are pushed into ever-closer proximity with natural areas and the habitats of wild fauna [12]. This increases the chances of emergence and spread, in livestock and people, of infectious zoonotic pathogens originating in wild animals [13]–[15].

To address these and other such issues requires accurate and accessible and information on the distribution and abundance of livestock. Analyses based on reliable livestock sector data can then be used for targeting and impact assessment, epidemiological evaluation, and to inform planning and policy formulation to promote safe, sustainable and equitable livestock sector development.

Until now, the only global estimates mapped at sub-national resolution have been the 2007 Gridded Livestock of the World (GLW) [16], though the distributions of some species have been modelled over more restricted areas. Neumann and colleagues, for example, modelled and validated the distribution of livestock in Europe by downscaling harmonised livestock statistics through a series of explanatory factors [17]. Other studies have modelled the distribution of poultry species in China and of ducks in Monsoon Asia [18]–[19].

The GLW provided modelled livestock densities of the world at a spatial resolution of 3 minutes of arc (about 5×5 km at the equator). The estimated densities were based on statistical relationships between observed densities within administrative units ('training data') derived from survey and census data, and several explanatory variables, including a time-series of remotely-sensed satellite data relating to climate and the environment, and other spatial data relating to demography, land cover and terrain.

Reported sub-national statistics were thus spatially disaggregated and gaps where none were available filled with predictions to provide a complete global distribution map for each species.

The GLW maps have been put to a wide variety of uses. They were the fundamental inputs for mapping current and projected estimates of consumption and production of animal-source foods [7]. They have been applied to the analysis of land use dynamics [20] and of global ecosystem services [21] and have contributed to the estimates of the environmental impacts of livestock [22]–[24] and to the characterization of livestock production systems in Eastern Africa [25]. The GLW data have been used in several epidemiological and ecological studies of livestock disease distributions, including trypanosomosis [26]–[27], highly pathogenic avian influenza [28]–[29] and foot-and-mouth disease [30]. The livestock densities have been combined with information on livestock production systems and on production parameters, to estimate the impacts livestock disease and the economic benefits of removing these constraints to production. Examples from cattle in Africa include brucellosis [31] and trypanosomosis [32]–[33].

Since the GLW datasets were first published, considerable advances have been made to improve a) the sub-national statistics on which livestock distributions are based, b) the predictor variables, masks and stratification layers used in the modelling process, c) the modelling methodology itself, and d) the statistical evaluation of modelled results. Some of initial improvements were described in relation to poultry distributions in Asia [18], [19] and methods have since been further developed. Here, we describe the current revised approach in detail and present new global distribution maps, at 1 km spatial resolution, for cattle, pigs and chickens and a partial duck distribution map.

2. MATERIALS AND METHODS

Producing the global livestock distributions for the GLW was a labour-intensive and time-consuming process. Whilst the majority of time is taken in the laborious and painstaking work of acquiring, cleaning and standardising sub-national statistics and registering these to geospatial files of often-changing sub-national boundaries and international borders, it was clear that to become sustainable the data management and livestock mapping processes had to be automated as far as possible. Improvement in data management has been achieved by the development of GLIMS: the Global Livestock Impact Mapping System [34]. Significant advances in livestock distribution modelling have been made possible through the development of a suite of sequentially-implemented R-scripts that produce

GLW version 2. The overall workflow from the collection of input data to post-processing and dissemination of outputs is illustrated in Fig. 1.

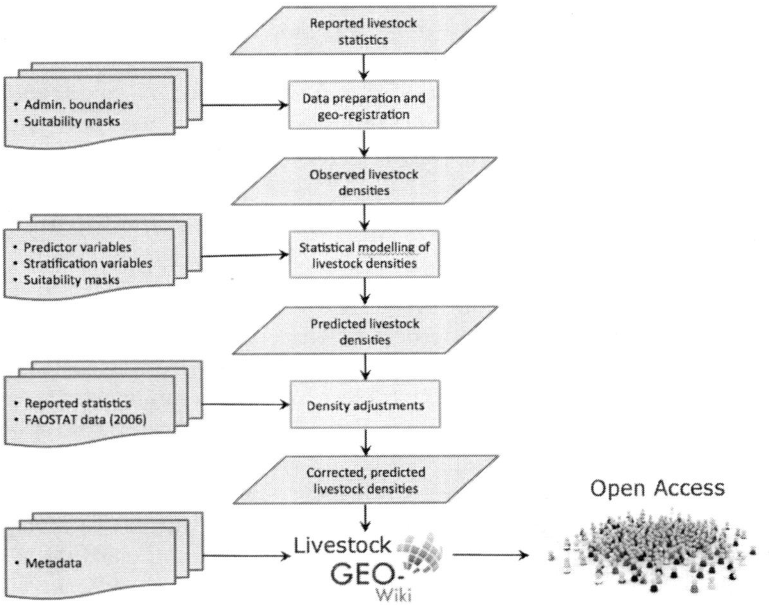

Figure 1. Overall workflow of GLW modelling.

2.1 GLIMS and sub-national livestock data

Wint and Robinson [16] listed and described the variety of sources for sub-national livestock statistics. These data are collected and cross-referenced against other sources (such as earlier statistics and FAOSTAT national totals). They are then linked to Geographical Information System (GIS) files of the administrative areas at the level to which the livestock statistics are reported. The hierarchical structure and nomenclature of these administrative units varies from country to country but typically includes units such as regions, provinces or districts. This involves a careful process of matching place names, often in different languages, and editing the digital boundaries to account for the merging and splitting of administrative areas, as well as the assignation of livestock numbers to newly defined administrative units. Where possible the FAO Global Administrative Unit Layers (GAUL) files are used as source of national boundary files (http://www.fao.org/geonetwork/srv/en/main.home).

The resulting GIS files of administrative areas and attributed sub-national livestock statistics are then entered into GLIMS. The GLIMS data warehouse, hosted at FAO headquarters in Rome, provides a repository and the tools to edit, explore, search and retrieve the collated data. GLIMS is used to manage a variety of data pertaining to the livestock sector, which are then processed and distributed through various channels and in numerous formats [36]. A module has been specifically developed in GLIMS to extract data for a specified livestock species and aggregation of countries, which are then offered as training data, in the form of a geospatial polygon vector file containing livestock numbers, to the modelling procedures.

2.2 GLW 2: a revised livestock distribution modelling system

The procedure builds on the method developed by Wint and Robinson [16] for disaggregating livestock statistics, based on environmental and other spatial data. Spatially stratified, statistical regression models are developed using data from a series of sample points within each training data polygon and these models are then applied to the entire one-kilometre resolution set of predictor variables in order to estimate livestock densities, disaggregated over a defined study area. For GLW 2, this basic methodology has been revised and improved in a number of ways and was first assessed by Van Boeckel et al. [19] and Prosser et al. [18] in a detailed analysis of its performances for modelling domestic ducks in Asia and poultry in China respectively. Van Boeckel et al. (2011) looked at how downscaling performance was influenced by the aggregation level of input domestic duck data in Thailand (no data for the country, only one value for the country, administrative level 1 data, and administrative level 2 data) and comparing the predictions to actual admin level 3 data. The result was that downscaling based on the method outlined below was giving relatively good results provided that the training data were available at administrative level 1, and were degraded with coarser (national-level data) input data. In a separate study, Prosser et al. (2011) compared land-use based downscaling with the GLW 2 methodology to predict chicken ducks and geese in China, and found land-use based methods to give lower performances. In another previous study, though based on a non-stratified implementation of regression methods, Newmann et al. (2009) found comparable results between land-use and regression based downscaling [17]. In human population mapping, the AfriPop and AsiaPop databases are still largely based on land-use weighting of human population density per land use class [35]–[36], whereas new developments of the WorldPop consortium (www.worldpop.org.uk) are moving toward the use of machine learning methods such as random forest or boosted regression trees (Andy Tatem, comm. pers).

The stratified regressions are repeated a specified number of times using random selections of pixels from which to extract dependent and independent variables (bootstraps). This produces multiple models from which the variability as well as the mean values of the model predictions can be calculated for each pixel.

- The procedures for the statistical transformation of variables and their subsequent selection in the regression models has been revised, better to account for non-linearity and to provide a more flexible process of predictor variable selection.
- A method has been developed to combine the results from a number of stratification schemes; choosing for each location the predictions from the best models for that location. This results in an aggregate, mosaic prediction.
- The entire modelling procedure has been streamlined as a suite of scripts that allow new gridded livestock distributions to be produced relatively quickly using the programming language R [37] when updated input data become available. Automation also facilitates the evaluation of the effects of adjusting model parameters.

Because of the sheer volume of input and covariate data, the modelling must be broken down into continental tiles. Whilst bespoke geographic tiles can be created for specific tasks or projects, there are six continental tiles that are processed independently, and the global dataset is updated every time a new continental tile is processed (see file S1).

2.3 Data layers

The GIS data layers used in the modelling process fall into 3 categories: 1) masks and pixel area files; 2) the predictor covariates; and 3) the stratification layers. All variables are un-projected, using latitude and longitude coordinates and the WGS84 datum. The images have a pixel resolution of 30 minutes of arc (0.00833 degrees) – a nominal pixel resolution of approximately 1 km×1 km at the equator. All imagery is maintained and used as scaled integer data in Idrisi format [38].

Suitability masking is an important step of the modelling process; firstly for adjusting the reported livestock numbers into the densities used as dependent variables in the regression models, and secondly for masking out unsuitable areas with densities of zero in the modelled results. If, for example, an administrative area of 100 km^2 is reported to contain 100 cattle, but half of its area is deemed unsuitable for raising cattle, the effective density of cattle on 'usable' land is not 1 but 2 head km^{-2}. Suitability masking in this version is more conservative that it was in the original GLW [16]; areas of forest and desert are no longer explicitly excluded – it is left for the model to determine the suitability of these – so a single mask is used for both ruminant and monogastric species. Land pixels with elevations higher than 4,750 m above sea level or at a slope of gradient higher than 40 percent are deemed unsuitable as well as urban areas and pixels permanently covered with snow or ice. The digital elevation model (DEM) GTOPO30 of the US Geological Survey (USGS) is the source for elevation data and for derived slope information (https://lta.cr.usgs.gov/GTOPO30). The GLC2000 land cover classification developed by the Joint Research Centre (JRC) of the European Commission (EC) is the source for the land cover categories of urban areas and snow and ice [39].

Areas unsuitable for reasons that are not biological or environmental, for example protected areas, are treated differently in the modelling procedure, and are therefore masked separately. A global mask of protected areas is derived from the 2010 version of the World Database on Protected Areas (WDPA)(www.wdpa.org). The International Union for the Conservation of Nature (IUCN) categories Ia and Ib, II, and III were masked as unsuitable as these are characterised by stringent conservation measures and tight regulation of human activity – the encroachment of roaming cattle and other grazing activities is therefore less likely in these than in other areas [40]. Any sample point falling in an area masked for environmental or biological reasons is maintained as a sample point in the model, with a livestock density of zero. If, however a sample point falls in an area deemed unsuitable for legislative reasons, it is dropped from the analysis. The area may, in fact, be quite suitable for livestock, and a zero value for livestock density would mislead the statistical model. A simple land mask is also employed in the modelling process and a file containing the area of each pixel is required for converting livestock numbers into densities, and vice versa. The independent predictor variables offered to the regression analyses are drawn from a spatial data set that allows the modelling process to take advantage of any relationship between livestock densities and climatic, environmental, demographic or topographic factors. The mainstay of these is a Fourier-processed, decadal time series of geo-physical variables derived from Moderate Resolution Imaging Spectroradiometer (MODIS) satellite data (http://modis.gsfc.nasa.gov) from 2001 to 2008. The variables include two vegetation indices, the land surface temperature and the band 3 middle-infra-red, which is particularly suitable for vegetation mapping. Rogers and associates have described the Fourier processing of satellite data in several studies [41]–[44], while a description specific to the MODIS data series is provided by Scharlemann and colleagues [45]. The use of Fourier-processed time-series data is central to the modelling process since the Fourier variables reveal the seasonal characteristics of the environment. Each multi-temporal series is reduced to seven separate, and quite independent, data layers: the mean, and the phases and amplitudes of the annual, biannual and tri-annual cycles of change. Three additional variables combine the amplitude and phase of the three cycles of change and report the contribution of this combination to the total variance in the seasonal phenology. These are further supplemented by the maximum, the minimum and the variance of the satellite-derived geo-physical variables. Two other MODIS-derived products are included as predictor variables: 'green-up' and 'senescence', derived by Boston University from the NASA MOD12Q2 MODIS tiled phenology layers (http://www.bu.edu/lcsc/research/land-cover-dynamics) [46]. These represent the dates when new green vegetation is first detected at the beginning of a growth cycle, and the onset of senescence at the end of that cycle. Detailed information on the MODIS-derived predictor variables is available on file S2. Two climate-related variables are also used: the length of growing period (LGP), which has been modelled to 1 km spatial resolution [47]–[48] and the annual precipitation data (synoptic period 1950 to 2000) from the WorldClim climate dataset [49]. Since human activities are associated with livestock distributions, two demographic variables are also included. The first is the human population density in 2006, for which the source is the Global Rural-Urban Mapping Project (GRUMP) [50] (http://sedac.ciesin.columbia.edu/data/collection/grump-v1), with country totals adjusted to match the United Nations population values in 2006. The second is the travel time to areas populated by 50,000 people or more [51]. Finally, the GTOPO30 dataset is the source for elevation and slope information [52]. Table 1 provides a list of primary variables used in the analysis.

A given predictor variable may be associated quite differently with livestock densities in different ecological zones. For instance, daytime temperature may have a negative association with livestock densities in an ecological zone describing arid environments, whilst it may have a positive association in temperate regions. In GLW 2 the analyses are therefore spatially stratified using three stratification layers selected to represent different types of spatial zonation. The first of these comprises 25 discrete ecological zones (EZ25) and is the result of an unsupervised classification of some of the Fourier-processed MODIS data, achieved by first reducing the data using a principal component analysis (PCA), followed by clustering using the ADDAPIX programme [53]. The zonation is based on the mean values of the temperature and vegetation variables, as well as elevation and a vegetation seasonality index derived from the Fourier processed time-series of the satellite imagery to distinguish zones in each hemisphere. This is very similar to the stratification scheme that was used to produce the 2007 livestock densities. A second stratification scheme, that has proved very useful, is the map of global livestock production systems (GLPS version 5) [5].

Table 1. Summary of the predictor variables.

Type of variable	Predictor variables	Source
Vegetation and climate	14 Fourier-derived variables from MIR*	Scharlemann et al. (2008) [45]
	14 Fourier-derived variables from LSTday and 14 from LSTnight**	
	14 Fourier-derived variables from NDVI*** and 14 from EVI****	
	Green-up (annual cycle 1 and 2)	Zhang et al. (2003) [46]
	Senescence (annual cycle 1 and 2)	
	Length of Growing Period (LGP)	Jones & Thornton (2009) [48]
	Precipitation	Hijmans et al. (2005) [49]
Topography	GTOPO30 Elevation	USGS-EROS (1996) [52]
	GTOPO30 Slope	
Demography	Human population in 2006	CIESIN et al. (2004)**** [50]
	Travel time to places with >50,000 inhabitants	Nelson (2008) [51]

*Middle Infra-Red;
**Land Surface Temperature;
***Normalized Difference Vegetation Index;
****Enhanced Vegetation Index;
****Country totals adjusted to UN values in 2006 (http://www.un.org/esa/population/).

This distinguishes livestock-only, mixed irrigated and mixed rain-fed farming areas, based on land cover, human population densities and data on irrigation. Each of the three categories is then sub-divided, using data on temperature, elevation and LGP, into four sub-categories: hyper-arid; arid and semi-arid; humid and sub-humid; and temperate and tropical highland. A third stratification scheme is provided by a published dataset on the biomes of the world [54]. Olson and associates subdivided the terrestrial world into 867 ecoregions, nested within 14 biomes. The ecoregions were evaluated as a stratification layer but proved too detailed, resulting in many strata failing to contain sufficient sample points to produce models. The biomes, however, have proved to be a useful means of stratification.

2.4 Modelling procedure

The polygon vector data containing the reported livestock numbers derived from GLIMS is combined with the masks and pixel area image and converted to pixel values of animal densities in areas defined as suitable. These observed densities are then converted to $\log_{10}(n+1)$ values in order to normalise the distribution of the dependent variable [55]. A set of sample points is randomly selected at a density of 30 per 10,000 km^2 (though this sometimes needs to be reduced for very large regions), with the constraint that at least one sample point must be selected per input polygon (i.e. if a polygon is smaller than 333 square kilometres). 75% of these points are used to train models and the remaining 25% are used to test their goodness of fit. The number of bootstraps to be used is specified in terms of the number of sample files and the number of point selections within each. Normally, 25 bootstraps are prepared by creating five different files of sample points and, within each, five different selections of the 75% to be used to train the models. Values for the suitability adjusted livestock densities, predictor variables, and stratification layers are extracted for each sample point in each of the five files.

Regressions are then run for each bootstrap, grouping data points according to each stratification scheme and using data points from each stratum of each stratification scheme. A constraint is imposed that a model can only be developed for a stratum if there are at least 30 unique observations for it. Unique, in this context, means from different polygons, each of which represents a single value for the dependent variable (suitability-corrected livestock density). For large polygons, many points may be included in the sample file but all will share the same livestock suitability-corrected density value and are therefore considered a single observation. An un-stratified model is also produced for each continental tile and this is used in areas where there are insufficient data points in any stratum to create a model for it.

Within each stratum a forward stepwise regression is applied. In contrast to standard stepwise regressions though, variables are entered in pairs; each variable being entered simultaneously with its quadratic term. This is intended to account for non-linearity in the relationship between dependent and independent variables. At each step, the pair of variables that best reduces the model Akaike information criterion (AIC) is maintained. In order to avoid over-fitting by including an excessive number of predictor variables, the procedure is halted when the difference in AIC drops below 1 percent in relative terms or when 2 * loglikelihood is below 6 in absolute terms (which corresponds to a p-value of 0.05 of the loglikelihood ratio test for the inclusion of a pair of variables). For the same reason, the maximum number of variables to be entered in a model is estimated by dividing the number of observations by 15. Then, for each bootstrap, the regression models for each stratum (given sufficient points) in each stratification scheme are applied to the predictor variables selected for inclusion in each model. This results in an image of predicted log10 densities for each stratification scheme plus an additional image of predicted log10 densities with no stratification employed.

The next step is to produce a composite of predicted values for each bootstrap by selecting, for each pixel, the prediction resulting from the best performing stratification scheme. Two options are available for choosing the best performing model: the one with the best (highest) R^2 value or the one with the best (lowest) residual standard error (RSE). The RSE provides information on the accuracy of predictions regardless of the slope of the regression line. In a single model, a poor R^2 value may be found concomitantly with a very low (good) RSE around a nearly horizontal line, which would in fact correspond to good prediction. An additional image is produced for each bootstrap, which shows, for each pixel, which of the stratification schemes (if any) contributed the predicted livestock density value used in the composite prediction. The correspondence between the observed and predicted livestock densities for the 25% of sample points reserved for testing is then compared for: a) the un-stratified model; b) the composite models based on best R^2; c) the composite model based on best RSE; and d) the models based on each of the single stratification schemes included. The statistics used for comparison are the simple Pearson correlation coefficient and the root mean square error (RMSE) between the predicted and observed densities. It is normally the case that the best R^2or best RSE composite models out-perform all other models (and of these the best RSE model is preferred for the reasons given above). Sometimes, however, a single stratification scheme consistently performs comparably. In such cases, where there is no significant improvement to be had by using a composite model, the predictions from the single stratification are taken; following the principles of parsimony.

Based on either an individual stratification or an aggregate prediction for each bootstrap the predicted (log10) densities are then averaged across the 25 bootstraps to give the mean and standard deviation of modelled predictions. It is this mean value on which the final predicted densities are based, and the standard deviation of this mean shows the consistency in predictions of the bootstraps.

The first step in post-processing is to adjust the predicted densities within polygons for which observations were available such that the predicted totals (of livestock numbers) match those of the reported statistics in the input polygon data (where no observed data are indicated for a polygon by a value of −9,999 the predicted density is retained). In a second post-processing stage the polygon-corrected density maps are further adjusted to match FAO's official national statistics (from FAOSTAT, accessed in 2013), providing a time-standardised datasets for the year 2006. The year 2006 was selected for these livestock distribution maps since it is the base year from which FAO's revised agricultural projections to 2030 and 2050 have been made [3].

Further corrections are made to the predicted densities for pigs, for which a value of zero is applied to all pixels in countries where Islam is the majority religion. Masking only occurs in countries where all of the following conditions are met: 1) more than 50 percent of the population is Muslim; 2) sub-national GLIMS data indicate zero or no data; and 3) FAOSTAT indicates zero or no data for pigs but does report data for other livestock species (see file S3).

2.5 Accuracy of predictions

Beyond the intrinsic ability of the available variables to explain the variance in livestock densities, the goodness of fit of the regression models depends on the level of detail in the input data. Better model performances are expected with more observed data obtained at a lower administrative level [19]. To illustrate this, two tests were undertaken. The first example analysed the predictive accuracy of separate models for cattle in South America by degrading the input data from Brazil to increasingly coarse levels. Recent statistics (2009) are available for Brazil at municipality level (administrative level 2). Brazil accounts for 5,510 out of 7,217 total input polygons in South America so it is expected strongly to influence the continental model for cattle. A similar approach was used in the second test by evaluating the impact of using increasingly coarse training data from Thailand to predict chicken distributions in the country. Sub-district level (locally called *tambon*, corresponding to administrative level 3) statistics on chicken distributions are available for Thailand and, at this level, the country accounts for nearly one third of all the input polygons of the Asia tile. In both examples, the goodness of fit of predicted models was calculated against the input data at the highest administrative level.

3. RESULTS

Figure 2 shows the global distributions of cattle, pigs and chickens and the partial distribution of ducks respectively; created by merging the results from the continental tiles for each species. These global maps represent the predicted data, first corrected to match the polygon values of the observed data and then to match the FAOSTAT country values in 2006. The highest cattle densities (Figure 2a) are found in India, in the East African highlands (particularly in Ethiopia), in Northern Europe and in South America. Desert areas and the tropical rain forests of Amazonia and of the Congo Basin have practically no cattle. The highest concentrations of pigs are found in China and in other Eastern Pacific countries (Figure 2b). Pigs are also densely distributed in European countries while only a few countries in Africa (e.g. Uganda, Burkina Faso, Ghana, Nigeria and Togo) have significant densities. Relatively high concentrations are also found in Central America and in Brazil. The distribution of chickens (Figure 2c) closely follows that of the human population. The highest concentrations are found in eastern China, in Pakistan and India, and in western Europe. In Africa, the countries facing the Gulf of Guinea and Madagascar also have high chicken densities. The densely populated east coast of the United States also shows high numbers whilst chickens are only sparsely distributed in the central and western states. The heavily populated areas of southern Brazil also show high concentrations of chickens. The distribution of ducks (Figure 2d), for those regions for which sub-national statistics were available, adds information to previous national and regional duck mapping efforts [18], [19]. Ducks are far less common than chickens worldwide although high densities are to be found in South-East Asia and China where duck production is often integrated with rice cropping and fish farming [56]–[57]. Though to a lesser extent, duck densities are also quite high in a few European countries (e.g. France). file S4 provides a summary of the sub-national statistics used for the modelling and file S5 provides detailed metadata for the sub-national statistics used to develop the livestock distribution models. file S6 provides two graphic summaries of data availability for the modelled species (a) the average spatial resolution of the training data and (b) administrative level of the training data.

The thematic detail in GLW 2 has been significantly improved compared to the previous version; GLW 2007. Figure 3 zooms in on Thailand and its neighbouring countries and compares the original 2005-corrected distribution of poultry (a) with the new 2006-corrected maps of chickens

(b) and ducks (c). Such thematic disaggregation is particularly important in this region because of the abundance of ducks and their important role in farming systems in this part of the world. The figure also illustrates that whilst GLW 2007 explicitly excluded the unsuitable areas based on expert opinion, GLW 2 applies a more conservative approach and leaves to the model to predict the livestock distribution. The two approaches however agree substantially in their mapping of areas with low or zero densities.

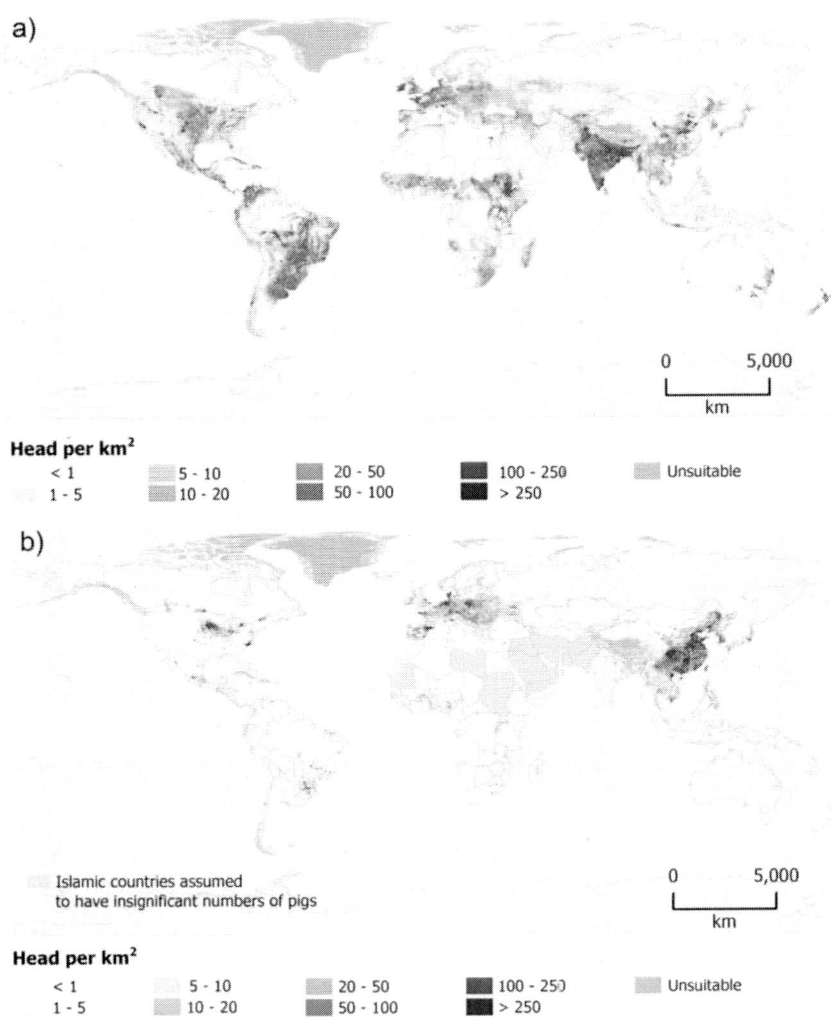

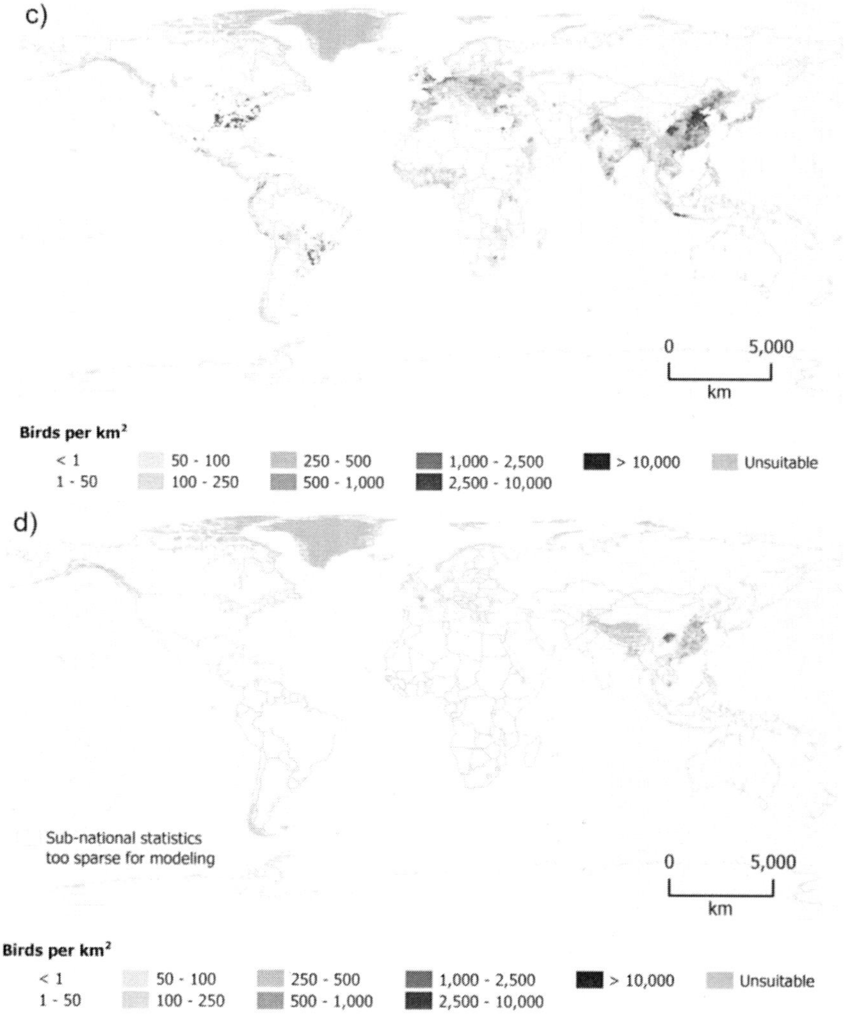

Figure 2. GLW 2 global distributions of a) cattle; b) pigs; c) chickens; and d) distribution of ducks, excluding South America and Africa.

The finer spatial resolution of the GLW 2 maps (approximate cell size of 1 km^2 at the equator) also significantly increases the detail of predictions compared to the original version (approximate cell size of 5 km^2 at the equator). Figure 4 helps to illustrate this with an example from central Uganda. In this case, the finer pixel size combines with a higher disaggregation of the training data; input data were at administrative levels 1 and 4 respectively in the GLW 2007 and GLW 2.

As so very often the case, it is the gathering and processing of primary data – the sub-national statistics on livestock numbers – that is the most arduous and time-consuming part of the process. This operation is necessarily very labour-intensive and usually requires cases-by-case judgements to be made, in particular regarding the linking of reported statistics to geographical units. Few advances could be made, therefore, with respect to earlier approaches. The development of GLIMS, however, has greatly reduced the time and effort needed to store, query and prepare the livestock statistics for modelling. More importantly, the modelling procedure is easily repeated; the suite of R scripts is readily adapted to modelling recurrent updates, different species, different geographical tiles and different technical specifications for statistical modelling and aggregation of results.

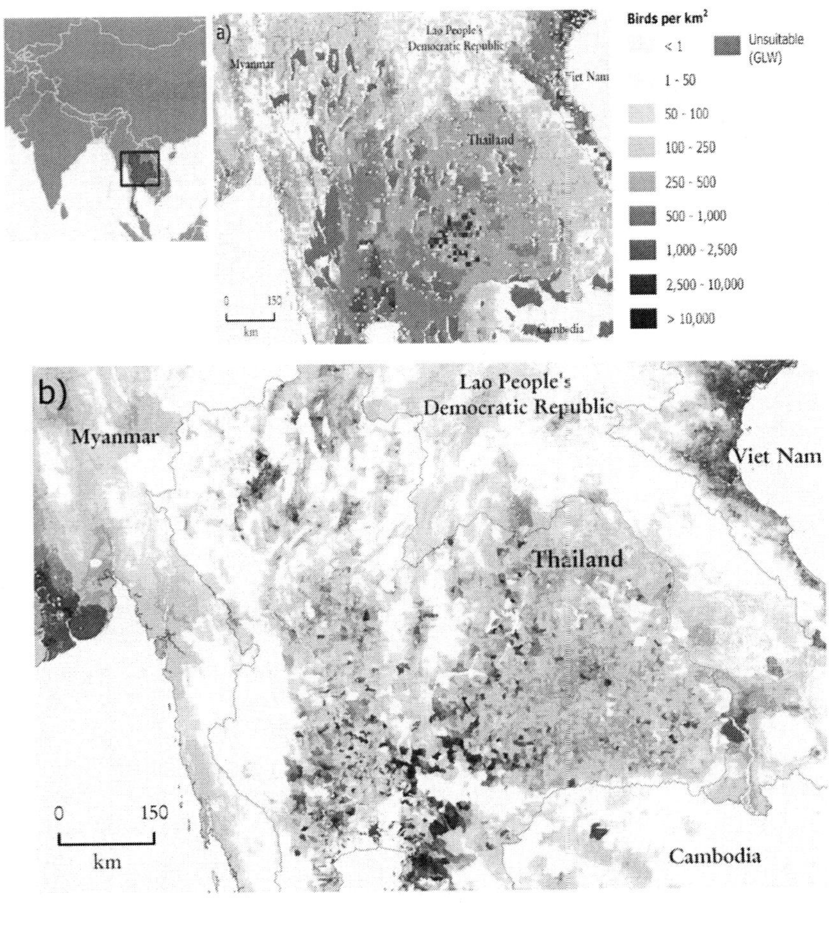

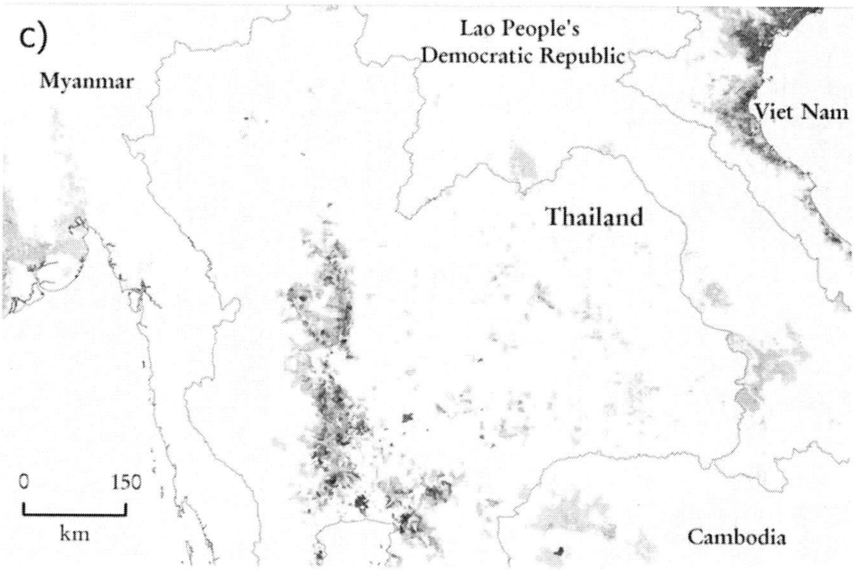

Figure 3. Thailand, visual comparison of a) poultry distribution as mapped in GLW at 5 km; against b) chicken and c) duck distributions mapped separately in GLW 2 at 1 km spatial resolution.

Automated post-processing activities further allow the rapid production of country-corrected maps for any specified year, global merging of continental tiles, and aggregation to coarser spatial resolutions, as needed.

Computer processing time was primarily a function of the extent of the continental window (i.e. the number of pixels), of the number of polygons contained in the input shape file and of the technical characteristics of the machines used for modelling. Processing time varied from a minimum of 17 hours for the duck model in Europe to a maximum of 210 hours (or 8.7 days) for the cattle model in Asia. file S7 reports the running times for each species and tile, the number of polygons processed in the input training data and the technical specifications of the computers that were used.

Table 2 summarises the fitting metrics for each tile, species and stratification scheme. It reports both the correlation coefficients and the RMSE between observed and predicted values in the validation data sets. The correlation coefficient is an indication of the precision of the predictions, i.e. the extent to which the observed and predicted values are proportional to each other. However, even with a nearly perfect correlation, the predicted values can be wrong in absolute terms if, for example, they systematically overestimate the population.

The RMSE, in contrast, is an indicator of the accuracy of the predictions, i.e. how far they are, on average, from the observed values. The highest correlation coefficients between observed and predicted values are typically found for the species and tiles for which observed data are at a higher spatial resolution and are evenly distributed within the modelling window. The models for the Asian tile had the best correlation coefficients compared to models for other continents: pigs and ducks (0.81), chickens (0.74), and cattle (0.63).

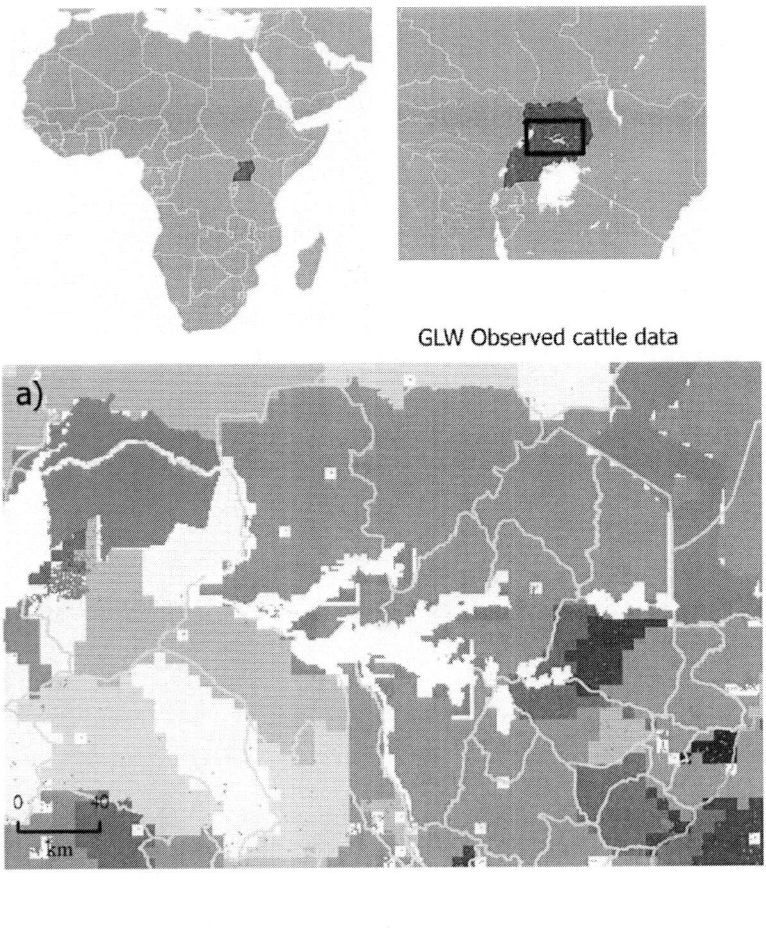

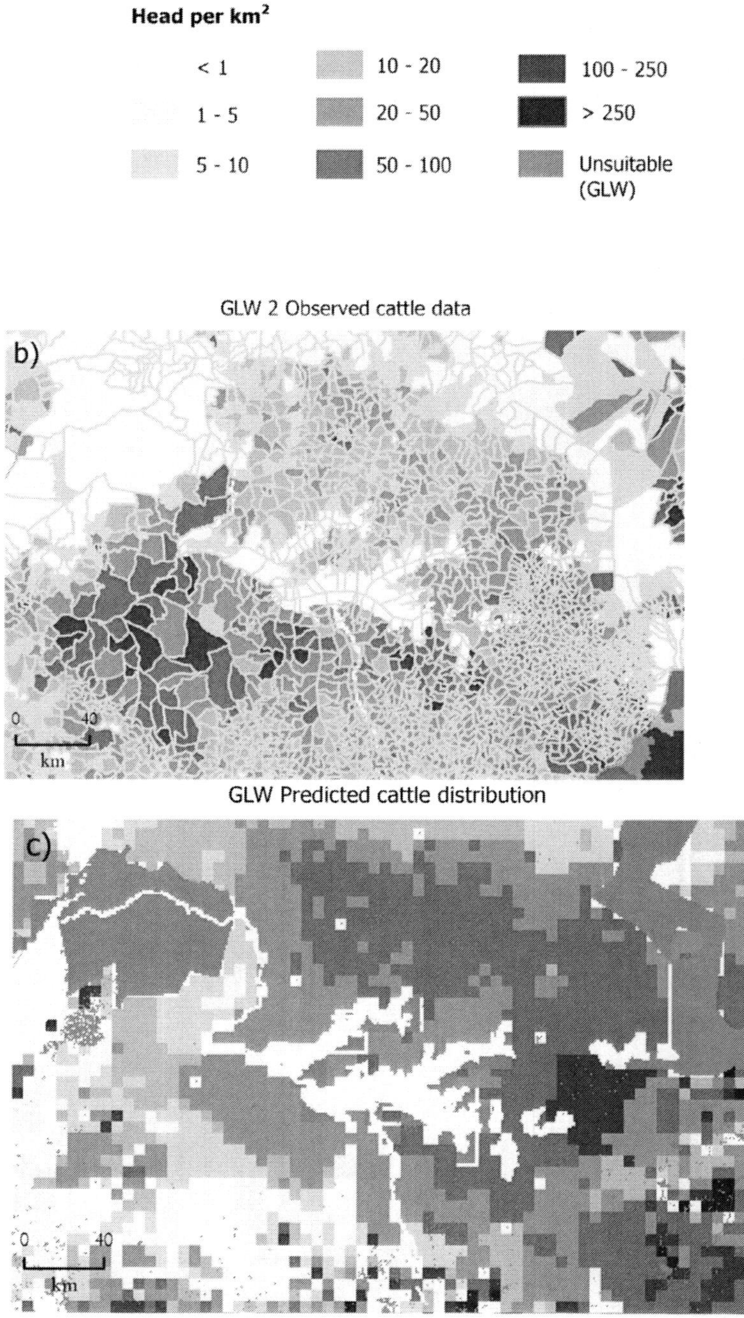
Head per km²
< 1
1 - 5
5 - 10
10 - 20
20 - 50
50 - 100
100 - 250
> 250
Unsuitable (GLW)
GLW 2 Observed cattle data
b)
0
40
km
GLW Predicted cattle distribution
c)
0
40
km

GLW 2 Predicted cattle distribution

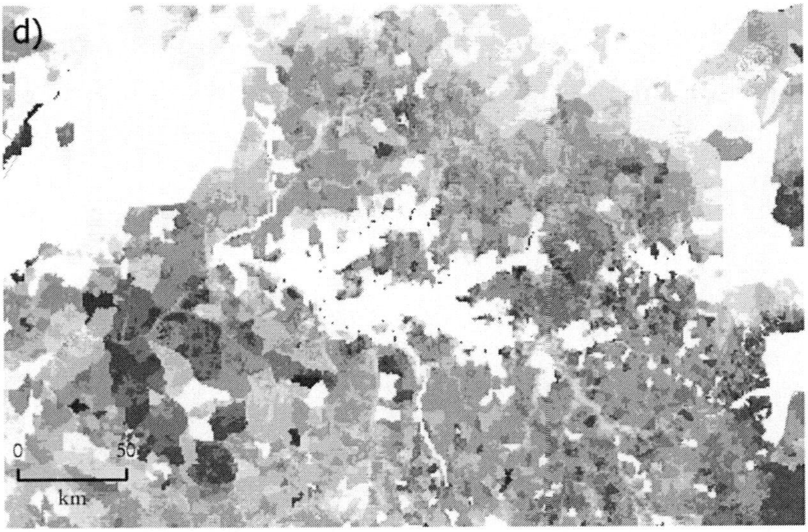

Figure 4. Uganda, visual comparison of observed cattle data a) in GLW 2007 (level 1) and b) in GLW 2 (level 4), and the resulting predicted distribution c) for GLW 2007 and d) for GLW 2.

Better accuracy of predictions was generally found for cattle compared to other species with RMSE values as low as 0.33 and 0.35 in South and North America respectively. RMSE values for cattle and pigs were consistently lower than they were for chickens. Results in Table 2 also indicated, for a given species, that the best stratification differed across the six continental tiles. For the duck models, the biomes stratification scheme gave the lowest RMSE values in the European and North American tiles and performed similarly to the GLPS stratification in Oceania. Composite stratifications had the lowest RMSE values for all species in Asia, but this was the only tile for which the composite prediction performed consistently better than one of the individual stratification schemes. The biomes stratification performed the best in the North American tile, regardless of the species being modelled, and the EZ25 stratification consistently best in South America.

The results of the tests carried out to evaluate the influence of the administrative level of input data on the accuracy of the predictions are presented in Figure 5. In both countries and species tested, the accuracy of the predictions decreased (increasing RMSE values) as the administrative level of the input data became coarser. It is notable also that the RMSE values for the cattle predictions in Brazil were lower at all administrative levels than those for the chicken predictions in Thailand.

Table 2. Summary of the fitting metrics for each continental tile and species modelled.

Tile	Species	Coefficient of correlation	Stratification	RMSE
Africa	Cattle	0.63	GLPS	0.42
	Chickens	0.66	GLPS	0.45
	Pigs	0.60	EZ25	0.40
	Ducks	n.a.	n.a.	n.a.
Asia	Cattle	0.63	Best RSE	0.46
	Chickens	0.74	Best RSE	0.51
	Pigs	0.81	Best RSE	0.45
	Ducks	0.81	Best RSE	0.57
Europe	Cattle	0.77	EZ25	0.42
	Chickens	0.42	EZ25	1.00
	Pigs	0.57	GLPS	0.67
	Ducks	0.55	Biomes	0.54
North America	Cattle	0.63	Biomes	0.35
	Chickens	0.59	Biomes	0.88
	Pigs	0.66	Biomes	0.45
	Ducks	0.57	Biomes	0.22
Oceania	Cattle	0.50	Biomes	0.48
	Chickens	0.63	Best RSE	0.59
	Pigs	0.72	Biomes	0.54
	Ducks	0.54	GLPS	0.59
South America	Cattle	0.73	EZ25	0.33
	Chickens	0.56	EZ25	0.77
	Pigs	0.64	EZ25	0.41
	Ducks	n.a.	n.a.	n.a.

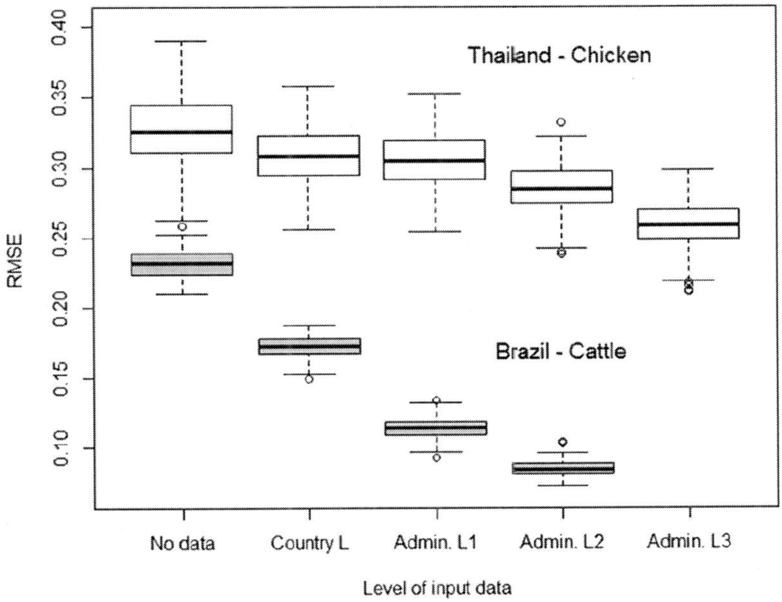

Figure 5. Residual Mean Square Error (RMSE) for predicted versus observed cattle distributions in Brazil, and chicken distributions in Thailand, by administrative level of training data.

4. DISCUSSION

The new GLW 2 livestock density maps described above provide a timely update of the GLW 2007 livestock distributions [16] and the enhnaced methods and automated procedures mean that updates will, in future, be more frequent than they have been to date. The differences in the modelling procedures, the type and resolution of predictor variables and the different unsuitability masks prevent an explicit, quantitative assessment of change between the livestock densities as mapped by the old and new GLW version. However, we can discuss the improvements that have been made over the original GLW, the accuracy of the predictions and areas where further advances could be made.

4.1 Improvements

The first and most obvious improvement is that GLW 2 provides more contemporary data on the distribution of livestock. The original GLW was published in 2007, but many of the reported statistics on which the predictions were based dated back to the 1990's. Whilst some data in the current set of reported statistics remain relatively old, especially in poor countries where censuses have not been carried out for many years, the median year for which livestock statistics have been obtained is much more recent, with most input data dating being more recent than 2000 (see file S4).

The second important improvement is in the spatial resolution of the training data, allowing the full potential of the higher spatial resolution (1 km) in the predictor variables to be availed. This is clearly shown in Figures 5, which illustrates how the use of higher resolution training data and predictor variables have allowed the spatial disaggregation of livestock census data at an unprecedented spatial resolution. Whilst the 1 km resolution may not be required for many applications, such as for example risk or impact assessments at global or continental scales, the higher resolution provides more opportunities for analysis and modelling to be carried out at the scale of individual countries; greatly extending the use of GLW 2. In some countries, the level of detail in the input data is so fine as to bring into question the usefulness of further disaggregating livestock densities to pixel-level. The reasons for continuing with the modelling, even for small polygons, are four in number. Firstly, even small administrative units are rarely homogeneous in terms of land use and farming so, provided that the covariates are accurate at the finer resolution, the data can still be spatially disaggregated in a meaningful way. Secondly, the training data come in heterogeneous administrative units, which would be otherwise difficult to compare between countries. Spatial disaggregation to pixel level harmonises the spatial scale of the estimates and thus facilitates comparison. Thirdly, even fine-scale input data often have gaps, caused for example by administrative units where no census data are available or are of restricted access. The modelled distributions allow prediction of livestock densities to be made in those areas. The fourth reason is possibly the most important. Statistical models based on smaller administrative units will capture more of the underlying environmental variability, and are statistically more robust.

A third important improvement in GLW 2 is the estimation of uncertainty around the predicted livestock densities. Whilst the modelling procedure does not provide a full and explicit integration of uncertainty based on a Bayesian framework, the bootstrapping procedure allows coefficients of variation to be calculated around the mean estimates; provides an indicative estimate of the degree of consensus among the different models around the mean estimate.

Finally, the breakdown of species and species groups has also been changed. For example, the previous estimates of poultry densities have been replaced by separate estimates for chickens and ducks. Whilst only cattle, pigs, chicken and ducks have been presented here, GLW 2 is currently being applied to produce distribution maps for other species including goats, sheep and buffaloes, as well as regional maps of camels and equines. This level of species detail is essential for some epidemiological investigations, for example in understanding the geographical distribution of highly pathogenic avian influenza (HPAI) H5N1, where treating chickens and ducks separately has greatly enhanced our ability to predict risk [29].

These new GLW 2 outputs are already finding applications in diverse fields. The new pig distribution maps are currently being applied to a global analysis of current and future pig production, in which the pig distributions are further disaggregated by production system. The latest version (5) of the GLPS [5], which is based on land cover data and agro-ecological conditions, maps potential systems rather than actual livestock systems. Integrating these livestock distribution maps with the GLPS will help bring these estimates closer to reality by showing where livestock actually are, rather than where we think conditions are suitable for them to occur. In combination with the revised agricultural projections to 2030 and 2050, recently released by the Agricultural Development and Economic Division (ESA) of FAO [3], these new maps are proving important inputs for further efforts to map the demand and supply of animal-source foods [7]. Likewise, updated global maps of livestock distributions will continue to support analyses that quantify greenhouse gas (GHG) emissions from both the ruminant and monogastric livestock sectors [58]. Finally, the revised maps will find many important epidemiological applications to study disease risk and estimate the impacts of diseases not only in livestock but also for those zoonotic diseases that spill over into the human population. The poultry distributions have already been incorporated into a number of HPAI H5N1 risk assessments [35], [59], and their importance has been highlighted in assessing the risk of spread of the recently emerged H7N9 virus in China and beyond [60]–[61].

Future updates of GLW 2 to include new data as they become available will be greatly facilitated by the integration of the modelling procedure into a fully scripted workflow. The post-processing is part of this automated workflow and allows a variety of outputs to be derived with relative ease and speed. The reference output presented here is a global map of animal densities at 1 km resolution adjusted at the country level to match FAOSTAT 2006 totals. However, outputs expressed in absolute numbers (rather than densities), aggregated to different spatial resolutions (5 km, 10 km or 20 km), or matching different FAOSTAT country totals can also be derived.

4.2 Prediction accuracy

The accuracy of the predictions varied significantly according to species and the level of detail in the input training data, and were comparable to those found by Van Boeckel et al. (2011) in Asia, Van Boeckel et al. (2012) in Thailand, and Prosser et al. (2011) in Asia. The administrative level of the training data had a strong influence on model accuracy; with greater accuracy observed with higher administrative levels of training data for both cattle in Brazil and chickens in Thailand (Figure 5). The modifiable areal unit problem (MAUP), well known to quantitative geographers [62], has been demonstrated to have varying effects with different levels of aggregation of input data. Even though the absolute levels of accuracy in prediction for chickens in Thailand and cattle in Brazil were quite different, the fact that we found a positive association between model accuracy and the administrative level at which the training data were used suggests that, wherever possible, data should be sought and collected at the highest possible resolution to train models, even if this results in heterogeneity in administrative levels used to train models, a result comparable to the one found by Van Boeckel et al. (2011) for domestic ducks in Thailand.

Another observation is that the predictions in Europe and North America for cattle and ducks, whose distributions are conditional on presence of suitable pasture and access to water, were consistently better (i.e. had lower RMSE values) than those for chickens and pigs, whose production is highly intensive in these parts of the world (Table 2). In contrast, in a continent like Africa, where the majority of chicken and pig production is extensive, the accuracies of the models for cattle, pigs and chicken were comparable. These observations relate to the influence of the intensity of livestock production on the accuracy of the predictions. As production intensifies it becomes increasingly detached from the land resource base (for example as feeds are brought in that are grown in completely different places) and thus more difficult to predict based on spatial, agro-ecological variables. This effect is particularly marked for chickens and pigs, where the locations of intensive farming units often have more to do with accessibility to markets or to inputs of one sort or another, than to the agro-ecological characteristics of the land that can be quantified through remotely sensed variables. In support of this, Van Boeckel and colleagues [63] found significantly lower predictability for models of intensive chicken production than for models of extensive chicken production in Thailand. As chicken production becomes intensified in many countries, chicken distributions will become increasingly difficult to predict geographically using largely agro-environmental predictors. This inverse relationship between predictability and the level of intensification is well illustrated by the case of chickens in Europe (the model which performed least well, with an RMSE value of 1.0). The regression statistics suggest that the environmental predictors can only partially explain the distribution of chickens and that other factors that are not currently included in the regression models, relating to policy and economics, for example, are likely to be relatively more important in determining the distributions of chickens in such settings. With general trends toward intensified production, the modelling could be improved in the future by incorporating a wider set of anthropogenic, socio-economic and perhaps demand- or trade-related predictor variables. For intensive production systems, the intrinsic variability is also expected to be higher. A chicken production unit, falling into a single 1 km^2 pixel, may contain up to a million birds. The characteristics of that pixel, in all covariates, may be identical to those of a neighbouring pixel where an equivalent production plant was not installed. However, the difference in numbers will be so high that the stochastic decision to place the plant in one pixel rather than the other equivalent one will increase the variability that the model cannot capture, even with the most pertinent covariates.

4.3 Validation

A true validation of the predictive accuracy of these models would involve field observation of livestock densities in different pixels and testing those observations against predicted values. However, livestock census data are generally collected and distributed by area (administrative units) and so validation would have to be done on re-aggregated model predictions. Moreover, such validation would be extremely costly and time-consuming. Aside from artificially degrading the training data contributing to models, as we have done here (Figure 5), to make an internal validation of the disaggregation efficiency, there are few options for validation. One possibility which is being explored is the Livestock Geo-Wiki, currently under development along similar lines to the Geo-Wiki produced by the International Institute for Applied Systems Analysis (IIASA) [64]–[65]. Geo-Wiki uses Internet crowd-sourcing to validate and modify land cover information. It is a web-based application with Google Earth as a backdrop, over which various global land cover datasets are draped. Discrepancies in land cover assignation among the datasets can be highlighted and values may be either corroborated or changed by users logged onto the wiki. The development of the Livestock Geo-Wiki was proposed by Robinson and colleagues [5] as a way to validate livestock production systems information and it is hoped that some form of validation will also be possible for livestock distributions. The roles of the Livestock Geo-Wiki for the livestock distribution data described here are, however, in viewing and providing open access to the data.

4.4 Dissemination and future developments

An important objective of the Livestock Geo-Wiki is to disseminate the data. Currently, data on livestock production systems and livestock densities are disseminated via the Gridded Livestock of the World (GLW) website, hosted by FAO (http://www.fao.org/ag/againfo/resources/en/glw/home.html). Whilst this has provided a valuable information resource it is not interactive so the opportunities for users to provide feedback are limited. The Livestock Geo-Wiki has several advantages over more static ways of disseminating data, including, for example: (a) providing a central repository where many different aspects of livestock information may be explored in a highly interactive way; (b) publicising and disseminating open-access livestock sector data at a range of spatial scales; (c) using crowd-sourcing approaches to validate and improve livestock data; and (d) providing innovative data visualisation and analysis tools that will facilitate the investigation of dependencies among the data sets and address specific requirements of diverse groups of users. Instructions on the procedure to download the livestock data from the Livestock Geo-Wiki (http://www.livestock.geo-wiki.org) are provided in file S8.

The improvements to global livestock distribution data, presented here, have been motivated by the pressing need for higher resolution and more contemporary outputs than provided by the original GLW 2007 livestock distributions. Although the methodology has been considerably revised in many aspects, it remains similar to that developed by Wint and Robinson for GLW 2007 [16], involving the use of several stratified multiple regressions linking observed livestock densities to environmental data. New machine-learning methods such as Boosted Regression Trees, or Random Forests have recently demonstrated their increased predictive capacities over more standard statistical methods for species distribution modelling [66]. These methods can better account for non-monotonic relationships between dependent and predictor variables, and can better incorporate the effects of interactions among predictors. Currently, the comparative strength of the stratified regressions lies in the relatively fast fitting of models, and their rapid application to large data sets. However, with the advent of parallelised computing, the use of such machine-learning techniques for modelling livestock distributions is becoming a realistic prospect. These approaches should be investigated in relation to their potential superiority in terms of predictive capacity. In addition, alongside the current set of predictor variables, which are principally environmental in nature, incorporating more socio-economic and anthropogenic predictor variables may confer considerable improvements to models for which the livestock distributions are dominated by more intensified modes of production.

Considerable progress has been made in advancing the spatial modelling of livestock distributions, and there are clearly areas where further improvements can be made. No matter how good the predictive model though, it will always be limited by the quality of the training data on which is based. With this in mind, efforts must be made to identify areas where reported statistics on livestock distributions are out-dated, of low spatial resolution or of doubtful quality, and steps must be taken to address these shortfalls.

ACKNOWLEDGMENT

The work presented here is the culmination of many years of collaboration between FAO, ILRI, the University of Oxford and the Université Libre de Bruxelles. The work has been facilitated in different ways by numerous colleagues and we would like in particular to acknowledge contributions from Samuel Jutzi, Joachim Otte, Henning Steinfeld, Jan Slingenbergh, Pierre Gerber, Jeroen Dijkman, Giuliano Cecchi, Russ Kruska, Philip Thornton, David Bourne, David Rogers, Steffen Fritz, Martina Duerauer and Christoph Perger.

REFERENCES

1. FAO (2009) The State of Food and Agriculture. Livestock in the Balance. Rome, Italy: Food and Agriculture Organization of the United Nations (FAO).
2. FAO (2011) World Livestock 2011 - Livestock in food security. Rome, Italy: Food and Agriculture Organization of the United Nations (FAO).
3. Alexandratos N, Bruinsma J (2012) World Agriculture Towards 2030/2050. The 2012 Revision. Global Perspective Studies Team. ESA Working Paper No. 12-03. Rome, Italy FAO, Food and Agriculture Organization of the United Nations.
4. de Haan C, Gerber P, Opio C (2010) Structural Change in the Livestock Sector. In: Steinfeld H, Mooney HA, Schneider F and Neville LE, editors. Livestock in a Changing Landscape. Drivers, Consequences, and Responses. Volume 1. Washington D.C., USA: Island Press. pp. 35–51.
5. Robinson TP, Thornton PK, Franceschini G, Kruska RL, Chiozza F, et al.. (2011) Global livestock production systems. Rome: Food and Agriculture Organization of the United Nations (FAO) and International Livestock Research Institute (ILRI). pp. 152.
6. Otte J, Costales A, Dijkman J, Pica-Ciamarra U, Robinson T, et al.. (2012) Livestock sector development for poverty reduction: an economic and policy perspective. Livestock's many virtues. Rome: Food and Agriculture Organization of the United Nations (FAO). 161 p.
7. Robinson TP, Pozzi F (2011) Mapping supply and demand for animal-source foods to 2030. Animal Production and Health Working Paper No. 2. Rome: Food and Agriculture Organisation (FAO) of the United Nations. pp. 164.
8. Narrod C, Tiongco M, Delgado C (2010) Socioeconomic Implications of the Livestock Industrialization Process. In: Steinfeld H, Mooney HA, Schneider F and Neville LE, editors.

Livestock in a Changing Landscape. Drivers, Consequences, and Responses. Volume 1. Washington D.C., USA: Island Press. pp. 269–285.
9. World Bank (2005) Managing the livestock revolution. Report no. 32725-GLB. Washington D.C. USA: World Bank. pp. 63.
10. Solomon S, Qin D, Manning M, Chen Z, Marquis M, et al.. (2007) Climate Change 2007: The physical science basis: Contribution of Working Group I to the Fourth Assessment Report of the Intergovernmental Panel on Climate Change. In: Solomon S, Qin D, Manning M, Chen Z, Marquis M, Averyt KB, Tignor M and Miller HL, editors. Cambridge, United Kingdom and New York, NY, USA: Cambridge University Press pp. 996.
11. .Steinfeld H, Mooney HA, Schneider F, Neville LE (2010) Livestock in a Changing Landscape. Drivers, Consequences, and Responses. Volume 1. Washington D.C., USA: Island Press. pp. 396.
12. Bonfoh B, Schwabenbauer K, Wallinga D, Hartung J, Schelling E, et al.. (2010) Human Health Hazards Associated with Livestock Production. In: Steinfeld H, Mooney HA, Schneider F and Neville LE, editors. Livestock in a Changing Landscape. Drivers, Consequences, and Responses. Volume 1. Washington D.C., USA: Island Press. pp. 197–221.
13. Herrero M, Thornton PK, Notenbaert AM, Wood S, Msangi S, et al. (2010) Smart Investments in Sustainable Food Production: Revisiting Mixed Crop-Livestock Systems. Science 327: 822–825.
14. Ito T, Couceiro JN, Kelm S, Baum LG, Krauss S, et al. (1998) Molecular basis for the generation in pigs of influenza A viruses with pandemic potential. J Virol 7367–7373.
15. Ma W, Kahn RE, Richt JA (2009) The pig as a mixing vessel for influenza viruses: Human and veterinary implications. J Mol Genet Med 158–166.
16. FAO (2007) Gridded Livestock of the World, 2007, by G.R.W. Wint and T.P. Robinson. Food and Agriculture Organization of the United Nations, Rome. 131 p.
17. Neumann K, Elbersen BS, Verburg PH, Staritsky I, Pérez-Soba M, et al. (2009) Modelling the spatial distribution of livestock in Europe. Landsc Ecol 24: 1207–1222.
18. Prosser DJ, Wu J, Ellis EC, Gale F, Van Boeckel TP, et al. (2011) Modelling the distribution of chickens, ducks, and geese in China. Agric Ecosyst Environ 141: 381–389.
19. Van Boeckel TP, Prosser D, Franceschini G, Biradar C, Wint W, et al. (2011) Modelling the distribution of domestic ducks in Monsoon Asia. Agric Ecosyst Environ 141: 373–380.
20. Letourneau A, Verburg PH, Stehfest E (2012) A land-use systems approach to represent land-use dynamics at continental and global scales. Environ Model Softw 33: 61–79.
21. Naidoo R, Balmford A, Costanza R, Fisher B, Green RE, et al. (2008) Global mapping of ecosystem services and conservation priorities. Proc Natl Acad Sci U S A 105: 9495–9500.
22. Herrero M, Thornton PK, Kruska R, Reid RS (2008) Systems dynamics and the spatial distribution of methane emissions from African domestic ruminants to 2030. Agric Ecosyst Environ 126: 122–137.
23. Steinfeld H, Gerber P, Wassenar T, Castel V, Rosales M, et al.. (2006) Livestock's long shadow: Environmental issues and options. Rome, Italy: Food and Agriculture Organization (FAO).
24. Thornton PK, van de Steeg J, Notenbaert A, Herrero M (2009) The impacts of climate change on livestock and livestock systems in developing countries: A review of what we know and what we need to know. Agric Syst 101: 113–127.
25. Cecchi G, Wint W, Shaw A, Marletta A, Mattioli R, et al. (2010) Geographic distribution and environmental characterization of livestock production systems in Eastern Africa. Agric Ecosyst Environ 135: 98–110.
26. Cecchi G, Mattioli R (2009) Geospatial datasets and analyses for an environmental approach to African trypanosomiasis. PAAT (Programme Against African Trypanosomiasis). Technical and Scientific Series. No. 9. Rome: Food and Agriculture Organization of the United Nations. pp. 80.
27. Cecchi G, Mattioli RC, Slingenbergh J, de la Rocque S (2008) Land cover and tsetse fly distributions in sub-Saharan Africa. Med Vet Entomol 22: 364–373.

28. Gilbert M, Xiao X, Pfeiffer DU, Epprecht M, Boles S, et al. (2008) Mapping H5N1 highly pathogenic avian influenza risk in Southeast Asia. Proc Natl Acad Sci U S A 105: 4769–4774.
29. Muzaffar SB, Takekawa JY, Prosser DJ, Newman SH, Xiao X (2010) Rice Production Systems and Avian Influenza: Interactions between Mixed-Farming Systems, Poultry and Wild Birds. Waterbirds 33: 219–230.
30. Sumption K, Rweyemamu M, Wint W (2008) Incidence and Distribution of Foot-and-Mouth Disease in Asia, Africa and South America; Combining Expert Opinion, Official Disease Information and Livestock Populations to Assist Risk Assessment. Transbound Emerg Dis 55: 5–13.
31. Mangen MJ, Otte J, Pfeiffer D, Chilonda P (2002) Bovine brucellosis in sub-Saharan Africa: estimation of sero-prevalence and impact on meat and milk offtake potential. Livestock Policy Discussion Paper No. 8 Rome: Food and Agriculture of the United Nations - Livestock Information and Policy Branch, AGAL. pp. 58.
32. Hendrickx G, de La Rocque S, Reid R, Wint W (2001) Spatial trypanosomosis management: from data-layers to decision making. Trends Parasitol 17: 35–41.
33. Shaw A, Cecchi G, Wint G, Mattioli R, Robinson T (2014) Mapping the economic benefits to livestock keepers from intervening against bovine trypanosomosis in Eastern Africa. Prev Vet Med 113: 197–210.
34. Franceschini G, Robinson TP, Morteo K, Dentale D, Wint GRW, et al. (2009) The Global Livestock Impact Mapping System (GLIMS) as a tool for animal health applications. Vet Ital 491–499.
35. Gaughan AE, Stevens FR, Linard C, Jia P, Tatem AJ (2013) High Resolution Population Distribution Maps for Southeast Asia in 2010 and 2015. PLoS ONE 8: e55882.
36. Linard C, Gilbert M, Snow RW, Noor AM, Tatem AJ (2012) Population Distribution, Settlement Patterns and Accessibility across Africa in 2010. PLoS ONE 7: e31743.
37. R Core Team (2012) R: A language and environment for statistical computing. R Foundation for Statistical Computing, Vienna, Austria. ISBN 3-900051-07-0, URL http://www.R-project.org/.
38. Eastman JR (2009) IDRISI Taiga (Worcester, MA: Clark University).
39. European Commission Joint Research Centre (2003) Global Land Cover 2000 database. Available: http://bioval.jrc.ec.europa.eu/products/glc2000/glc2000.php. Accessed 2014 April 16.
40. Dudley N (2008) Guidelines for Applying Protected Area Management Categories. International Union for the Conservation of Nature (IUCN). Gland, Switzerland. pp. 106.
41. Rogers DJ (1997) Satellite Imagery and the Prediction of Tsetse Distributions in East Africa. In Diagnosis and Control of Livestock Diseases using Nuclear and related techniques Vienna, Austria: International Atomic Energy Agency. pp. 397–420.
42. Rogers DJ (2000) Satellites, space, time and the African trypanosomiases. Adv Parasitol 47: 129–171.
43. 43.Rogers DJ, Hay SI, Packer MJ (1996) Predicting the distribution of tsetse flies in West Africa using temporal Fourier processed meteorological satellite data. Ann Trop Med Parasitol 90: 225–241.
44. Rogers DJ, Williams BG (1994) Tsetse distribution in Africa: seeing the wood and the trees. In: Edwards PJ, May RM, Webb NR, editors. Large-scale ecology and conservation biology. 35th symposium of the Britush Ecological Society with the Society for Conservation Biology. Chapter 11. pp. 249–273. Oxford, UK: Blackwell Scientific Publications.
45. Scharlemann JPW, Benz D, Hay SI, Purse BV, Tatem AJ, et al. (2008) Global data for ecology and epidemiology: A novel algorithm for temporal fourier processing MODIS data. PLoS ONE 3.
46. Zhang X, Friedl MA, Schaaf CB, Strahler AH, Hodges JCF, et al. (2003) Monitoring vegetation phenology using MODIS. Remote Sens Environ 84: 471–475.
47. Jones P (1987) Current availability and deficiencies in data relevant to agro-ecological studies in the geographic area covered by the IARCS. In: A.HBunting (Ed), Agricultural Environments. CAB International, Wallingford, UK, pp. 69–83.

48. Jones PG, Thornton PK (2009) Croppers to livestock keepers: livelihood transitions to 2050 in Africa due to climate change. Environ Sci Policy 12: 427–437.
49. Hijmans RJ, Cameron SE, Parra JL, Jones PG, Jarvis A (2005) Very high resolution interpolated climate surfaces for global land areas. International Journal of Climatology 25: 1965–1978.
50. CIESIN, IFPRI, World Bank and CIAT(2004) Global Rural Urban Mapping Project (GRUMP), Beta Version. New York, Center for Internation Earth Science Information Network (CIESIN), Columbia University; International Food Policy Research Institute (IFPRI); The World Bank & Centro Internacional de Agricultura Tropical (CIAT). Palisades, NY: Socioeconomic Data and Applications Center (SEDAC), Columbia University. Available: http://sedac.ciesin.columbia.edu/data/sets/browse. Accessed 2014 April 16.
51. Nelson A (2008) Travel time to major cities: A global map of Accessibility. Global Environment Monitoring Unit - Joint Research Centre of the European Commission, Ispra Italy. Available: http://gem.jrc.ec.europa.eu/. Accessed 2014 April 16.
52. USGS-EROS (1996) Global 30 Arc-Second Elevation Data Set. 1996. US Geological Survey (USGS) - Earth Resources Observation Systems (EROS). EROS Data Center (EDC) Distributed Active Archive Center (DAAC), Sioux Falls, South Dakota, USA. Available: http://eros.usgs.gov/#/Find_Data/Products_and_Data_Available/GTOPO30. Accessed 2014 April 16.
53. Griguolo S, Mazzanti M (1996) ADDAPIX: pixel-by-pixel classification for zoning and monitoring. Rome: FAO Technical Report SD:GCP/INt/578/NET. Revised May 2000. Rome, Italy. pp. 76.
54. .Olson DM, Dinerstein E, Wikramanayake ED, Burgess ND, Powell GVN, et al. (2001) Terrestrial ecoregions of the world: A new map of life on Earth. BioScience 51: 933–938.
55. Zuur A, Ieno EN, Smith GM (2007) Analysing Ecological Data. XXVI, 672 p.
56. Gilbert M, Xiao X, Chaitaweesub P, Kalpravidh W, Premashthira S, et al. (2007) Avian influenza, domestic ducks and rice agriculture in Thailand. Agric Ecosyst Environ 119: 409–415.
57. Weimin M (2010) Distribution and characteristics of duck-fish farming systems in Eastern China. Smallholder Poultry Production. Rome: Food and Agriculture of the United Nations (FAO). pp. 28.
58. MacLeod M, Gerber P, Opio C, Henderson B, Dietze K, et al.. (2013) Greenhouse gas emissions from pig and chicken supply chains – A global life cycle assessment Rome: Food and Agriculture Organization (FAO) of the United Nations. pp. 196.
59. Martin V, Pfeiffer DU, Zhou X, Xiao X, Prosser DJ, et al. (2011) Spatial distribution and risk factors of highly pathogenic avian influenza (HPAI) H5N1 in China. PLoS Pathogens 7.
60. Butler D (2013) Mapping the H7N9 avian flu outbreaks. Nature
61. Horby P, Tatem AJ, Huang Z, Gilbert M, Robinson TP, et al. (2013) H7N9 is a virus worth worrying about. Nature
62. Openshaw S (1983) The modifiable areal unit problem. Concepts and techniques in modern geography. Norwich, UK: Geo Books Vol. 38
63. Van Boeckel TP, Thanapongtharm W, Robinson T, D'Aietti L, Gilbert M (2012) Predicting the distribution of intensive poultry farming in Thailand. Agric Ecosyst Environ 149: 144–153.
64. Fritz S, McCallum I, Schill C, Perger C, Grillmayer R, et al. (2009) Geo-wiki.org: The use of crowdsourcing to improve global land cover. Remote Sens 1: 345–354.
65. Fritz S, McCallum I, Schill C, Perger C, See L, et al. (2012) Geo-Wiki: An online platform for improving global land cover. Environ Model Softw 31: 110–123.
66. Elith JH, Graham CP, Anderson R, Dudík M, Ferrier S, et al. (2006) Novel methods improve prediction of species' distributions from occurrence data. Ecography 29: 129–151.

CHAPTER 2

Securing Sustainable Livestock Production Systems in an Uncertain Economic Climate: Nurturing Flexibility and Resilience

Stéphane Ingrand[1, 2, 3, 4*], Laura Astigarraga[5], Eduardo Chia[6], Xavier Coquil[7], Christophe David[8] and Jean-Louis Fiorelli[7]

1 Inra, UMR, Métafort, Méthamis, France

2 Agro Paris Tech, UMR, Aubière, France

3 Irstea, UMR, Aubière, France

4 Clermont Université, VetAgro Sup, UMR, Clermont-Ferrand, France

5 Universidad de la Republica, Facultad de Agronomia, Montevideo, Uruguay

6 UMR Innovation, Montpellier Cedex, France

7 Inra UR Aster, Mirecourt, France

8 Isara, Lyon Cedex, France

ABSTRACT

Resilience is one of the three core properties of social-ecological systems, mixing adaptability and transformability. Flexibility can be defined in terms of diversity of procedures and the speed at which they can be mobilized by one organization. The analyses performed are presented in terms of levers that farmers can deploy to protect their management systems against market uncertainty. These levers differ depending on farmer standpoints, objectives, lessons learned, the collective organizations they work with, the standards and specifications they work to, etc. It is equally important to identify the interplays between overarching and underlying scale levels for the system studied and to hone in on the dynamics at work during periods of transition. Adaptive capacities of farm systems are closely linked to how the farmer perceives the situations to manage, according to his aims, to his behaviour face to risk and to his idea of what is his job. We propose to use different words to describe the properties of farming systems to cope with changes, according to the level within the system: "adaptive capacity" or "plasticity" for the animal level, "resilience" for the biotechnical level and "flexibility" for the whole system, including the manager. We think there is a real challenge working at each level on transition periods and processes, as farming systems will have more and more to adapt face to unpredic events.

Keywords: resilience, flexibility, adaptive capacities, uncertainty, crops systems, animal system

1. INTRODUCTION

Farm businesses, just like any other business enterprise, develop response strategies in order to cope with the many demands imposed on them and the uncertainties they face. The challenge for farmers lies in securing sustainability for their business, in a context where farming is subject to wide-reaching change and where farms are increasingly exposed to agronomic trends and climatic risks that the agricultural productivity model generally seeks to overcome by controlling processes and disengaging the effects of environmental disturbance.

Incorporating the precepts of sustainable development in order to build and assess new technical agricultural systems hinges on breaking away from the rationales underpinning these systems and moving towards more holistic objectives encompassing far more than the simple production output function [1]. There are two key drivers to this breakaway: (i) reinventing how researchers interact with the other actors involved in the process of developing new systems and their multiple outcomes [2, 3], and (ii) producing tools capable of quickly rendering a priori system assessments [4, 5] as a first step towards subsequently deploying the systems in compliance with complex multicriteria specifications [6]. This means that agronomists face the challenge of translating the impacts of integrating these dimensions into terms that farmers can understand and use to reshape their farm systems, taking into account new social and environmental factors [7, 8].

This reshaping redefines the farm business as a complex system that needs to be analysed not just in terms of its type but also the rationales driving how it operates [9, 10]. A few years ago, farming system researchers started using the notion of flexibility to define the capacity of a business to weather and adapt to economic uncertainty. The concept of resilience, as pioneered by Holling [11], has also been analysed in this setting, particularly when applied in more recent social-ecological systems [12]. "Flexibility" has been researched extensively in management science and industrial economics, whereas "resilience" has mainly been used in ecology (but also in social psychology; [13]). Our study will draw on illustrative examples to highlight how the notion of flexibility can prove useful for designing and assessing innovative technical systems.

2. FLEXIBILITY IN MANAGEMENT SCIENCES

Industrial economics and management sciences understand the concept of flexibility [14, 15] as the capacity of a business or organization to re-adapt its structure and projects in response to environmental challenges (strategic flexibility) and to re-adapt its skillsbase, reorganize its workflows (workflow flexibility) and/or adjust its production methods as a response to unforeseen variations in *inputs* from outside (operational flexibility). The concept therefore appears relevant when analysing farmers' response strategies in the current climate governing agricultural production (characterized by regulatory developments, volatile agricultural prices, climatic variability, etc.). Tarondeau [15] (*ibid.*) stratified different sources of flexibility: product flexibility (product range), process flexibility and inputs flexibility. The basic idea is that the capacity to cope with unknowns and carry the business forward is dependent on several factors, both material and non-material: the configurations of their technical production systems, their structures, their projects and their objectives [16]. Reix et al. [17] suggest that the drive for flexibility can be seen as the drive to maintain consistency in how the business is managed in response to a changing environment. Flexibility, as a system property, is not "given": it is built, shaped and "nurtured," and it has a cost [18]. Flexibility can be considered a competitive advantage insofar as it enables performance levels to be sustained in situations of uncertainty [19].

Different commentators use different terms as synonymous with or acceptances of the concept of flexibility, but there is a body of ideas that remain recurrent. Flexibility refers to organizational capacity [14, 15, 20–23, 28]. This means that the systems described are always management-led and that the organizational procedures governing their management constitute a source of flexibility for the system. In each case, flexibility is defined as an attribute that is inherent to humans, dependent on how they perceive situations to be addressed, their objectives, their level of risk aversion and the perception they hold of their business. Flexibility is a property that has to associate both change and stability, forming a paradox between permanence (continuity, mainstay) and change [16]. The authors see management flexibility as the result of constructive tension between what needs to be held onto and what needs changing. This same idea has been explored through analyses of how livestock farming systems work, with the notion of invariants [24]. The invariant acts as a backbone, a basis, a bottom line and the frame of reference for handling change (not everything has to change at the same time, otherwise the system risks getting disorganized or even collapsing into chaos). Flexibility is intrinsically dynamic. It can only be meaningfully studied in the long term, at multiperiod scale. Integrating flexibility into the analysis of a system or an organization presupposes that the decision-maker is looking to achieve short-term objectives while also securing a range of opportunities for the longer term [25]. In other words, a given decision may appear non-rational (or non-optimal) when analysed at timepoint t, but become entirely rational once events liable to arise at some point in the future are factored in (uncertainty preparedness). Indeed, the speed of response to these events is one of the key components of flexibility [15].

Furthermore, in every scenario, the concept of flexibility is also linked to the notion of interaction between the system/organization and its environment. It can therefore be measured and thus assessed, by quantifying the degree of control (according to the dual flexibility concept proposed by De Leeuw and Volberda [20]: controlled systems vs. independent systems) over environmental inputs (**Figure 1**).

The two paradigms coexist within a single system (controlling-controlled) and must therefore be analysed in tandem. However, the extent to which one paradigm dominates the other reveals specific system behaviours.

The organization as an environment-controlled system: in this configuration, the organization "copes with" environmental factors [16]. Flexibility hinges on accommodative processes [26], which hallmarks defensive behaviour in response to external perturbation [27]. The target objective for the system will be adaptation, stability, resilience to environmental forces and robustness. Systems unable to achieve this objective would be defined as vulnerable.

The organization as an environment-independent system: in this configuration, the organization seeks to subordinate all changes in its environment to the task of maintaining its objectives and its identity. Interactions with the environment are specified internally, and on a certain level, the environment is integrated into the organization. The processes deployed in the search for flexibility are assimilative processes, which hallmarks a pro-active pattern of behaviour that will respond to each perturbation by generating new behaviours, thereby expanding the range of adaptation options possible. These configurations define self-learning organizations with self-directed learning capacities.

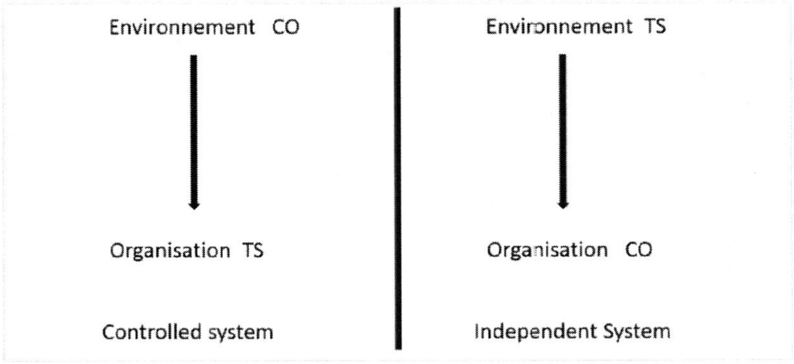

Figure 1. Organization of an environment-controlled system (left) and an environment-independent system (right) (concepts taken from Ref. [20]). TS = target system; CO = controlling organ. The arrows illustrate the direction of control exerted by the CO over the TS.

De Leeuw and Volberda [20] encapsulated these two configurations by defining flexibility in terms of diversity of procedures and the speed at which they can be mobilized: (i) to increase the organization's environmental control capacities and (ii) to decrease the organization's environmental vulnerability. The authors define different types of flexibility according to the number of planned procedures and the speed at which they can be implemented (**Figure 2**).

3. SOCIAL-ECOLOGICAL RESILIENCE: A KIND OF FLEXIBILITY?

The concept of resilience is borrowed from material physics as well as ecology as a means of describing the transformation and/or adaptive capacity of a material or ecosystem in response to stressors. In ecology, Holling [11] described resilience as the capacity of an ecological system or species to absorb challenges and then recover its initial configuration. The concept was then broadened to encompass shifts, learning and human-nature interactions [29]. Resilience was then extended to describe the mechanics of "anthropized" systems [30]. More recently, the concept of resilience has been applied to social-ecological systems, where humans are a governing actor [2, 12, 31–33]. The system is thus considered as a "learner," with a shift in the underlying idea from a return to the initial state following the perturbation towards a capacity to reconfigure itself while maintaining the core objectives and projects, where stakeholders can continue to plan for the future [2]. According to Ref. [34], there are three potential strategies capable of increasing the resilience of actively governed systems: increasing the system's buffer capacity (room for manoeuvre), scale-based governance (spatial and temporal scales) and creating opportunity for innovation (sources of change to system properties, learning capacity). These systems therefore have the ability to respond to perturbation by shifting into different stability domains rather than a single, "initial" steady state.

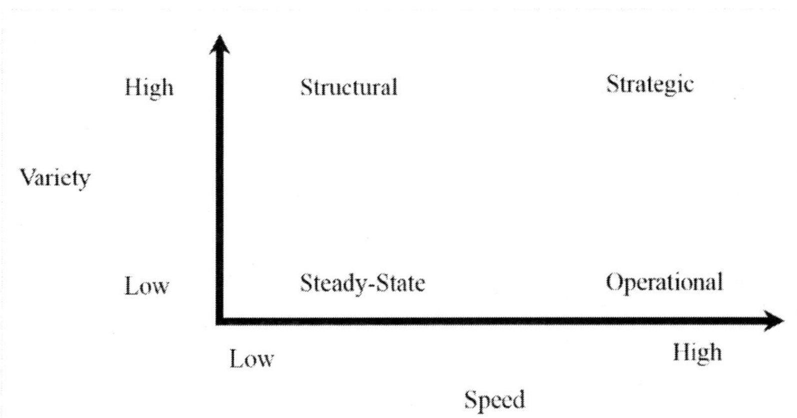

Figure 2. Different types of flexibility according to the number of planned procedures (vertical axis) and the speed at which they can be implemented (horizontal axis); adapted from Ref. [28].

Walker et al. [35] outlined four main features of system resilience connected to the notions of steady state and initial state: (i) the amount of change that the system can tolerate without collapsing into an essentially different state, this idea works on the assumption that there is a threshold beyond which the system can no longer recover its initial configuration; (ii) the capacity to resist change, which is connected to properties like rigidity and robustness; (iii) vulnerability (precariousness), which is how close the system state is to the threshold cited under point 1; (iv) panarchy, which describes a system integrating a great many elements undergoing cross-scale interactions, and that the level of resilience depends on the different states and dynamics interplaying at the scales above and below.

Resilience can also be described in terms of successive system states over time. Holling [36] and Walker et al. [2] consider that ecological systems follow adaptive cycles comprising four successive phases. They posit that actively governed systems reproduce cyclic patterns of behaviour aligned to these four phases: a phase of accelerated growth (annotated r), followed by a longer phase of steady accumulation towards stability, associated with a progressive decline in resilience (K), then a sharp structural collapse (Ω) before another short phase of rebuilding and reorganization (α). Depending on the current phase of the system, a given disturbance (which can in fact be seen positively as the introduction of an accommodative stance) will not have the same effect.

4. LEVERAGES TO ENHANCE FLEXIBILITY IN LIVESTOCK SYSTEMS

4.1. Different levers according to scale

Aaker and Mascarenhas [37] focusing on the means to enhance organizational flexibility outlined the following four levers centred on products, resources and management: (i) diversification of processes, business activities and products, running from broadening the range but also including activity in different marketplaces and extended use of different process technologies. In Ref. [38], the authors assert their notion of "relational flexibility" to account for the sources of adaptive capacity employed by livestock farmers through their marketing networks and the circuits they build or exploit to sell livestock; (ii) increasing inter-independence between production units; (iii) developing a base of potentially useful resources that are deployed not continually but on a case-by-case basis "should the need arise": functional redundancies, latent competencies, room for manoeuvre; (iv) minimizing workflow specialization, steering away from situations where tasks are accomplished by staff who have competencies deemed "necessary and sufficient" to complete the task. For example, Madelrieux et al. [39] clearly illustrate the flexibility achievable by a more collective workplace organization and workload breakdown in livestock farming systems.

Using two examples of farm systems (crop and livestock), we illustrate how these flexibility leverages can be deployed to minimize vulnerability to changes in the systems' environments. These two examples were chosen to demonstrate how the internal organization of the system (the sequencing of the system's structural components) and the system manager's perception of the environment act as complementary leverage for lending flexibility to farm production systems.

4.2. Animal contribution (plasticity) to system flexibility in an organic dairy system

The Mirecourt (INRA) research team prototypes sustainable dairy systems focused on agro-environmental sustainability. One system, tested since 2004, is a low-input grass-only system, in accordance with the specifications governing organic farming and based on the hypothesis that pasture-based systems are more sustainable [40].

This system is designed to introduce rulesets and animal and farmland management modes for achieving the objectives assigned to the system at the outset. In other words, the system aims to define how to achieve a result targeted at the outset without having to run through the conventional pattern of conducting experimental trials to measure results from different management condition sets established at the outset. Systems employing this strategy are designed to be sustainable in agro-environmental terms. More operationally, we posit that in order to cope with these objectives, the systems have to be self-sufficient (no importation of fertilizers or pesticides) and able to cope with unanticipated events, especially climatic events, since self-sufficiency can render systems more sensitive to natural variations in farmland properties.

The herd breed is split equally between two breeds (Holstein and Montbeliarde) in order to test the capacities of each breed to enable the system to achieve the objectives set. Maximizing grazed grass in the cow diet led to grouped calvings in late winter (February to April) in order to match the animals' energy requirements with grass availability. Under this management policy, cows produced 5132 kg milk/cow/year on average in 2005 and 2006: Holstein cows milked on average 400 kg milk/cow/year higher than Montbéliarde cows (respectively 5347 and 4947 kg milk/cow/year). However, at the end of the breeding period, 65% of dairy cows were pregnant in 2005 but only 27% at the corresponding timepoint in 2006. These very poor ratios affected herd sustainability, even though performance levels for replacement heifers were better (**Table 1**).

An analysis of individual animal management within the cow herd highlighted different groups. Each group corresponds to a specific calving date, which, in relation to turnout date, determines the feed diet at the beginning of lactation: a switch from winter feed to pasture grass.

Table 1. Reproductive performances of dairy cows in 2005 and 2006, according to breed.

Year	**2005**			**2006**		
	Herd	**Hn**	**Mo**	***Herd***	**Hn**	**Mo**
Cows						
Success AI1 and AI2 (%) [1]	*54*	38	75	*23*	8	33
Fertility (%) [2]	*65*	52	81	*27*	17	33
Heifers						
Success AI1 and AI2 (%)	*79*	80	78	*71*	75	67
Fertility (%)	*86*	80	89	*88*	75	100

[i] - Hn: Holstein; Mo: Montbeliarde; AI: artificial insemination.

[ii] - [1]Percentage of pregnant cows served once or twice.

[iii] - [2]Percentage of animals calving after being served during the breeding period.

The milk production of dairy cows calved after turnout increased very quickly (2–4 weeks) to maximum daily production, generating high energy requirements, which is detrimental to reproduction. The milk production of dairy cows calving at least one month before turnout showed a slower increase to maximum daily production (taking 8–12 weeks), with a smoother effect on energy balance and reproduction. Within these two configurations, Montbeliarde cows gave smoother lactation curves than Holstein cows (**Figure 3**). They were able to limit milk production, even when stimulated by turnout to grass, and thus gave better reproduction performances than Holstein cows.

In the grass-based systems, Montbeliarde cows offer more plasticity than Holstein cows. Secondly, shifting the calving period (January to March instead of February to April) should maximize the number of calvings before turnout to grass, thus lending the system more flexibility by enhancing reproductive performance.

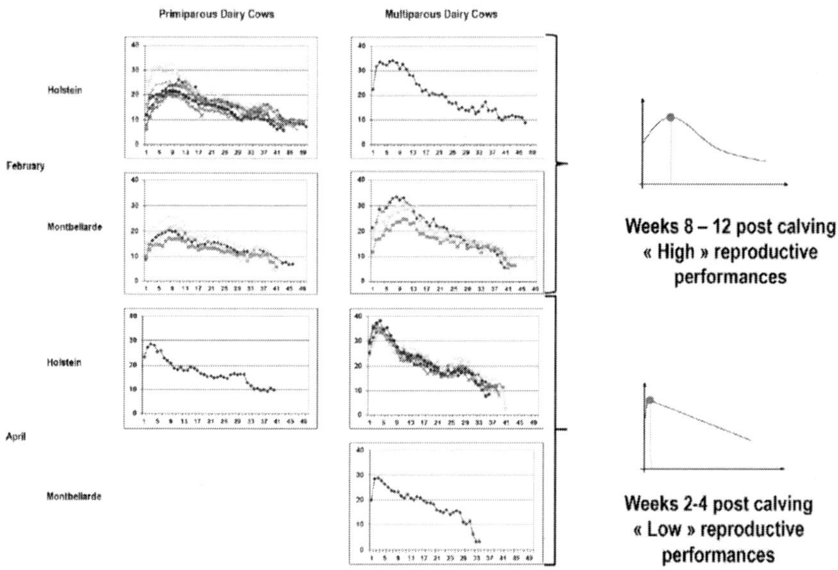

Figure 3. Individual lactation curves (milk yield in kg/cow/day throughout time after calving, in weeks) of Montbeliarde (Mo) and Holstein (Ho) dairy cows in 2005, according to the parity and to the calving period (February = at least 1 month before turnout vs. April = after turnout). On the right side, the average shape of curves for each period.

4.3. The collective workflows lever: flexibility in response to market uncertainty

The flexibility of suckler cattle farms is induced by commercial circuits: one of the features of suckler cattle farms is that they offer the possibility of selling livestock, and particularly females, at virtually any age. There are potentially over 15 different categories, with some breeders selling a minimal number of animal categories (n = 3: male calves, female calves and cows), whereas other systems offer a broader range comprising four or more different categories. Some systems always produce the same types of animal, whereas others gear themselves with options to change in response to climate events or market openings. There is also a heavy and practically range-independent variability in the number of buyers for the animals produced (**Figure 4**): a 2005 survey sampling livestock farmers ranged from one buyer for all animals up to seven different partners. Over and above buyer numbers, buyer status is also a critical criterion for livestock breeders. We have identified two different sets of strategic choices:

- Cooperatives vs. private buyers: some livestock farmers are convinced that cooperatives rob them of their freedom to market their products and thus refuse to help finance the running costs (premiums), in contrast to other farmers who strongly believe the cooperative represents their best interests, offering them a voice and a channel through which they can take action if problems arise. Finally, there is another category of livestock farmers who attach little importance to buyer status and who choose to sell their animals based on the prices they can get and how well they know and trust the buyer;
- Single buyer vs. several buyers: for farmers who work with a single buyer, the driving factor is the relationship of trust: the buyer understands how the farmer works and knows what animals are produced: negotiations are relatively straightforward, and sometimes a phone call is all that is needed. While the cattle farmer does need to make efforts to protect this special relationship (trust-system payments, sales spread across the year, etc.), in return they can expect the buyer to step in and make priority purchases when business is bad (security factor). In contrast, other farmers see the option of juggling between buyers as a way to take advantage of competition. If the market goes through a crisis, the farmer hopes to weather the storm by having a number of available buyers in order to sell their total livestock.

The components of biophysical systems (plants, animals and soils; **Figure 5**) confer a relatively greater level of system-wide flexibility through their own, intrinsic properties: (i) delayed differentiation process: unicity, particularly for females from suckler cattle breeds, regardless of their end purpose and their age at sale [41]; (ii) plasticity, breed diversity and ability to adapt to different management strategies [42, 43]. Gaillard et al. [44] showed how Simmental breed diversity offered dairy farmers options to take up a more or less marked position on the intensified fodder system gradient, ranging from extensive 100% grassland systems to intensive corn silage-based systems.

Depending on the flexibility leverage deployed by the farmer [7], both the system components (structural dimensions) and their interplays (functional dimensions) will take on a certain measure of specificity.

Furthermore, this distinction picks up on the distinction made by Alcaras and Lacroux [16] between the stability of an organization's structure and the stability of an organization's target objectives: (i) the "size" lever: reproductive capacities, useful lifespan and carcass yield, for animals that farmers can no longer select to work with once they opt to increase the size of their holding through internal growth (zero buy-in); (ii) the "responsiveness" lever (short-range opportunity-taking): versatility, ability to handle change (feed type and volume), malleability, breed mix, capacities for out-of-season production; (iii) the "collective workflows/technicity" lever: quantitative performance, standardized high-tech information system, records; (iv) the "room for manoeuvre" lever: versatility, simplicity, hardiness.

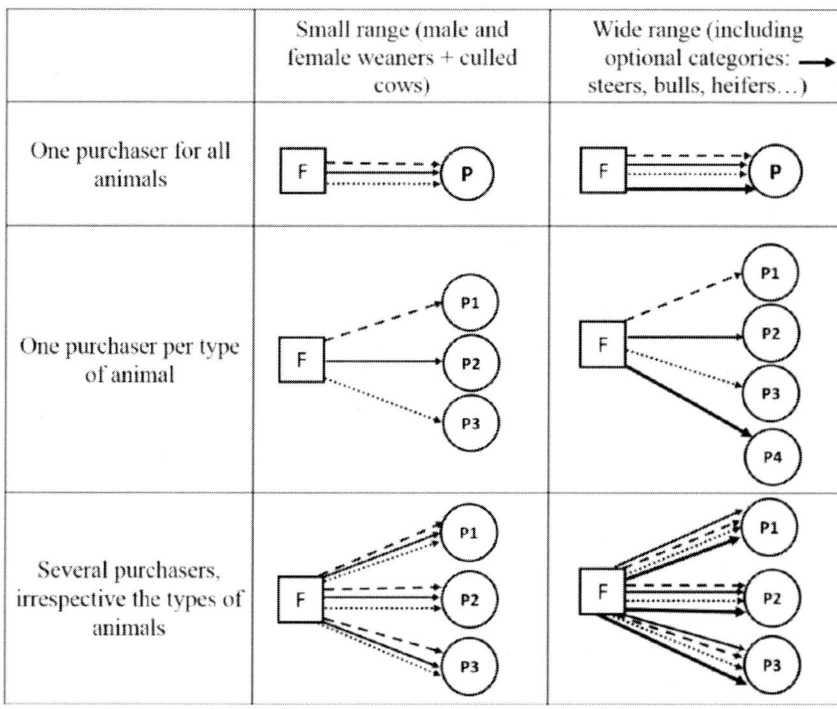

Figure 4. Different farmer (F) strategies for animal sales in livestock farming systems, combining range and number of purchasers (P); (one arrow corresponds to one specific category of animals sold, i.e. culled cows, weaned calves, heifers, bulls and steers).

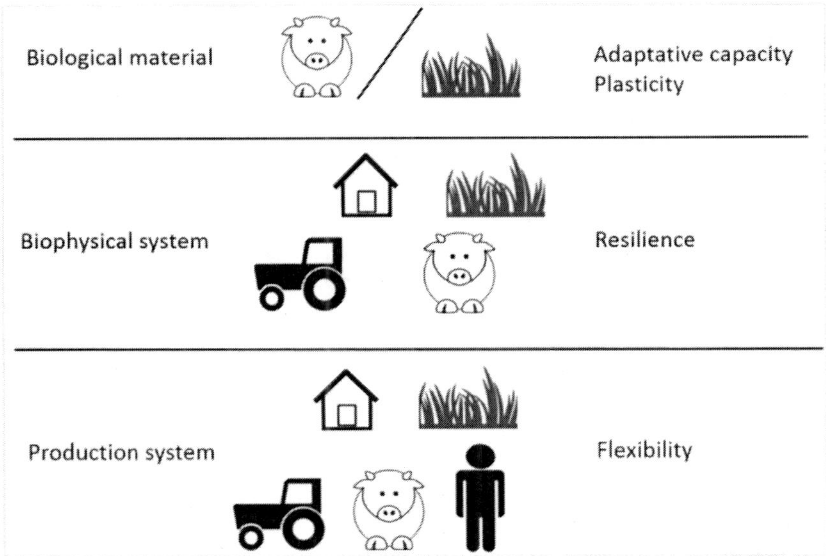

Figure 5. Descriptors assigned to adaptive capacities according to level of organization in the functional analysis of production systems.

5. DISCUSSION AND CONCLUSION

The foundations of resilience analysis have progressively shifted towards the foundations of flexibility analysis. Our assertion is based on qualifying the set of properties that will enable a system to secure sustainability by restricting the use of the two terms to different levels of organization (**Figure 5**): "flexibility" to cover the level overarching the entire production system and "resilience" to cover the underlying level of the biophysical (or operant learning) system. The terms used at the next level down, comprising the organic system entities such as plants and animals, would be "plasticity" and "adaptive capacity" as employed in Ref. [42]. The three examples of production systems highlighted earlier share a common denominator in that they are all "extensive" systems, that is, where productivity per surface unit of land is not maximized compared to intensive systems. A clear pattern emerged, wherein the adaptive capacities of these systems are perceived differently under the two scenarios. The design and development of intensive systems (high production per surface unit of land) consisted then, as now, in targeting measures capable of absorbing the negative effects of increasing performance. This means that for the animals, the primary property needed is "robustness," that is, the ability to produce a lot and regularly, regardless of the environmental disturbances.

The levers that farmers can deploy to protect their management systems against market uncertainty will differ depending on farmer standpoints, objectives, lessons learned, the collective organizations they work with, the standards and specifications they work to, etc. Therefore, in order to properly analyse the attributes of systems that make them less vulnerable to unknowns, the focus should be directed towards the information systems employed by farm system managers [45]. It is equally important to identify the interplays between overarching and underlying scale levels for the system studied (panarchy) and to hone in on the dynamics at work during periods of transition.

Literature review combined with the examples compiled reveals that studies directed at developments and changes in farm systems harnessing ecological-biological (animals, plants, etc.) and human-social (farmers' strategies and objectives) dimensions can use the notion of flexibility to gain a sharper and more explicit analysis of the interactions between these dimensions.

The move to revitalize the analytical framework governing livestock farming systems has to explicitly factor in dimensions stemming from interactions between animal production science and social sciences (formalization of livestock farmer strategies, workflow organization; [46]) as well as between ecology (resilience) and management science (flexibility). The target is to combine the analytical perspectives on (i) the regulatory properties of management-led biological systems (such as the herd, whose dynamics are shaped by interactions between human decisions and the biological functions of the animals; [43, 47]) and the leverages capable of parrying the effects of climatic risks and economic unknowns (types of product, relations with downstream factors, socio-technical networks).

There has been a key turning point in the way agronomics researchers have addressed the issue of performance in farm production systems. There has been a move away from focusing on ways to control or increase quantitative performance metrics (although there are shades of ecological intensification policy that still encourage this kind of outlook; [48]) and towards other rationales, such as "multicriteria" system design and assessment frameworks. Looking at the issues left unresolved and the various standpoints on offer, we have identified at least two courses of action:

- The interplay, or rather the fitting of abilities between production system components (system entities) and the type of system environment. This standpoint leads to a subsequent issue of whether there are advantages to be drawn from preserving certain specific animal or plant genotype characteristics that are underrepresented or tend to pale in comparison when balanced against the yield capacities of different breeds and the so-called improved crop varieties.

- The advantages of mixed farm systems combining different animal breeds/plant species, where the farmer is hedging on complementarity between the properties of each breed/species to cope with climatic unknowns (species offering different hardiness or which develop at different periods of the year) or variations in market prices (which have different effects on different farm outputs).

Approaches based on concepts and theories borrowed from disciplines such as ecology and management science are particularly fruitful for fuelling reflective thinking and reframing analyses in agronomics science when the aim is to investigate the dynamics of change and the adaptability of farm in response to situations of uncertainty.

For farmers, the art of farm management resides in tackling head-on how they define and readjust the production objectives set, how they lead negotiations with other farm stakeholders in order to achieve these objectives given the resources available, how they tackle uncertainty and how they tackle opportunity. These are all complex adaptation processes occurring at the interface between the farm and its environment, which emerge not only in the decisions taken but also in the short-term and long-term practices that we have termed "flexibility." Our analysis of these processes applied to three real-world systems enabled us to highlight a handful of principles governing farm business flexibility. First, the situational contextualization: flexibility is dependent not only on the technical features of the production system components (plasticity) but also on the socio-economic environment in which the businesses evolve; second comes the collectiveness component: flexibility becomes greater as the business integrates the collective dimension of farm activity, even if the overriding aim is to maintain decision-making autonomy over the production system. Finally, from the methodology standpoint, trials led at our experimental farm station have prompted us to continue investigations into methods for qualifying and if possible even quantifying the sustainability of farm structures in interaction with their environment, factoring in the different farm-structuring organizational levels. This research will ultimately be used for inter-farm comparisons integrating on-farm production system adaptability over time.

ACKNOWLEDGEMENTS

This work was carried out with the financial support of the "ANR—Agence Nationale de la Recherche—The French National Research Agency" under the "Programme Agriculture et Développement Durable," project "ANR-05-PADD-04-01, Discotech."

REFERENCES

1. Sabatier R, Durant D, Hazard L, Lauvie A, Lecrivain E, Magda D, Martel G, Roche B, de Sainte Marie C, Teillard F, Tichit M. Towards biodiversity-based livestock systems: review of evidence and options for improvement. CAB Reviews. 2015;10(20):1–13. doi:10.1079/PAVSNNR201510025
2. Walker B, Carpenter S, Anderies J, Abel N, Cumming G, Jansen M, Lebel L, Norberg J, Peterson GD, Pritchard R. Resilience management in socio-ecological systems: a working hypothesis for a participatory approach. Conservation Ecology and Society. 2002;6(1):14. Available from:http://www.consecol.org/vol6/iss1/art14
3. Godart O, Hubert B. Le développement durable et la recherche scientifique à l'Inra: Inra Paris; 2002. 58 p. Available from: http://www.youscribe.com/catalogue/ presentations/ savoirs/techniques/ le-developpement-durable- et-la-recherche- scientifique-a- l-inra-395481
4. Sadok W, Angevin F, Bergez JE, Bockstaller C, Colomb B, Guichard L, Reau R, Doré T. Ex ante assessment of the sustainability of alternative cropping systems: implications for using multicriteria decision-aid methods. A review. Agronomic Sustainable Development. 2008;28:163–174. doi:10.1051/agro:2007043
5. Gouttenoire L, Cournut S, Ingrand S. Farm-scale models built with farmers converting and converted to organic farming. Agronomy for Sustainable Development. 2012;33(3):609–619. doi:10.1007/s13593-012-0125-6
6. Coquil X, Fiorelli JL, Mignolet C, Blouet A, Foissy D, Trommenschlager JM, Bazard C, Gaujour E, Gouttenoire L, Schrack D. Assessing the agro-environmental sustainability of organic mixed-crop dairy systems on the basis of a multivariate approach. Innovations Agronomiques. 2009;4:239–247. Available from: https://www.researchgate.net/ publication/277800762_ Assessing_the_agro-environmental_sustainability_of_ organic_ mixed-crop_dairy_systems_on_the_basis_of_a_multivariate_approach
7. Lémery B, Ingrand S, Dedieu, B, Degrange B. Agir en situation d'incertitude: le cas des éleveurs de bovins allaitants. Economie Rurale. 2005;288:57–69. doi:10.4000/ economierurale.2718
8. Rigolot C, Roturier S, Dedieu B, Ingrand S. Climate variability drives livestock farmers to modify their use of collective summer mountain pastures. Agronomy for Sustainable Development. 2014;34:899–907. doi:10.1007/s13593-014-0224-7
9. Gafsi M. Exploitation agricole et agriculture durable. Cahiers Agricultures. 2006;15(6): 491–497. doi:10.1684/agr.2006.0035
10. Chia E, Dugué P, Sakho-Jimbira S. Can family farms be considered as institutions? Cahiers Agricultures. 2006;15(6):498–505. doi:10.1684/agr.2006.0027
11. Holling CS. Resilience and stability of ecological systems. Annual Review of Ecological Systems.1973;4:1–23. doi:10.1146/annurev.es.04.110173.000245
12. Anderies JM, Jansen, M.A, Ostrom E. A framework to analyze the robustness of social-ecological systems from an institutional perspective. Ecology and Society. 2004;9(1):18. Available from: http://www.ecologyandsociety.org/vol9/iss1/art18/

13. Tisseron S. La résilience. Coll. Que sais-je? Edition PUF; 2007. 128 p. Available from: http://www.cairn.info/la-resilience--9782130577959.html
14. Cohendet P, Llerena P, editors. Flexibilités, risque et incertitude dans la théorie de la firme. In Flexibilité, information et décision, Paris, Éditions Économica;1989
15. Tarondeau JC. Approches et formes de flexibilité. Revue française de gestion, mars-avril-mai.1999;66–71.
16. Alcaras JR, Lacroux F. Planifier, c'est s'adapter. Economies et sociétés Sciences de gestion, Série S.G. 1999;26–27:7–37.
17. Reix R. « Flexibilité », Simon, Y, Joffre P. (sous la direction de), Encyclopédie de Gestion, Tome II, Paris, Economica; 1997. p. 1407–1420.
18. Chia E, Marchesnay M. Un regard des sciences de gestion sur la flexibilité: enjeux et perspectives. In: Dedieu B, Chia E, Leclerc B, Moulin C.H, Tichit M, (Eds), "L'élevage en mouvement: flexibilité et adaptations des exploitations d'herbivores". Editions Quae; 2008. p. 23–36. Available from: http://www.quae.com/fr/r1308-l- elevage-en-mouvement.html
19. Dreyer B, Gronhaug K. Uncertainty, flexibility and sustained competitive advantage. Journal of Business Research. 2004;57(5): 484–494. doi:10.1016/s0148- 2963 (02)00315-6
20. De Leeuw ACJ, Volberda HM. On the concept of flexibility: a dual control perspective. Omega.1996;24(2):121–139.doi:10.1016/0305-0483(95)00054-2
21. Evans JS. Strategic flexibility for high technology manoeuvres: a conceptual framework. Journal of Management Studies. 1991;28(1): 69–89.doi:10.1111/j.1467- 6486. 1991.tb00271.x
22. Sanchez R. Strategic flexibility in product competition. Strategic Management Journal. 1995;16(S1):135–159. doi:10.1002/smj.4250160921
23. Weick KE. Psicología social del proceso de organización. Colombia: Fondo Educativo Interamericano. 1982
24. Moulin CH, Ingrand S, Lasseur J, Madelrieux S, Napoléone M, Pluvinage J, Thénard V. Comprendre et analyser les changements d'organisation et de conduite de l'élevage dans un ensemble d'exploitations: propositions méthodologiques. In: Dedieu B, Chia E, Leclerc B, Moulin CH, Tichit M, editors. L'élevage en mouvement: flexibilité et adaptations des exploitations d'herbivores. Editions Quae; 2008. p. 181–196.
25. Lev L, Campbell DJ. The temporal dimension in farming systems research: the importance of maintaining flexibility under conditions of uncertainty. Journal of Rural Studies. 1987;3(2): 123–132. doi:10.1016/0743-0167(87)90028-3
26. Piaget J. La naissance de l'intelligence chez l'enfant, Delachaux et Niestlé. 1977.
27. Le Moigne JL. La modélisation des systèmes complexes, Dunod; 1990.
28. Volberda H. Towards the flexible form: how to remain vital in hypercompetitive environments. Organisation Science. 1996;7(4):359–374. doi:10.1287/orsc.7.4.359
29. Gunderson LH, Holling CS, Light SS editors. Barriers and bridges to the renewal of ecosystems and institutions. Columbia University Press. New York; 1995.
30. Ludwig D, Walker B, Holling CS. Sustainability, stability and resilience. Conservation Ecology. 1997;1(1):7. Available from: http://www.ecologyandsociety.org/vol1/ iss1/ art7/
31. Carpenter S, Walker B, Anderies JM, Abel N. From metaphor to measurements: resilience of what to what? Ecosystems. 2001;4(8):765–781.doi:10.1007/s10021-001-0045-9
32. Milestad R, Darnhofer I. Building Farm Resilience: the prospects and challenges of organic farming. Journal of Sustainable Agriculture. 2003;22(3):81–97.doi:10.1300/ j064v 22n03_09
33. Folke C. Resilience: the emergence of a perspective for social-ecological systems analysis. Global Environmental Change. 2006;16: 53–267. doi:10.1016/j.gloenvcha.2006.04.002
34. Gunderson LH. Ecological resilience in theory and application. Annual Review of Ecological Systems. 2000;31(1):425–439. doi:10.1146/annurev.ecolsys.31.1.425
35. Walker B, Holling CS, Carpenter SR, Kinzig A. Resilience, adaptability and transformability in socio-ecological systems. Ecology and Society. 2004;9(2):5. Available from: http:// www.ecologyandsociety.org/vol9/iss2/art5/

36. Holling CS. The resilience of terrestrial ecosystems: local surprise and global change. In: sustainable development of the biosphere. Clark WC, Mumn ME, editors. Cambridge University Press; 1986. p. 292–317.
37. Aaker DA, Mascarenhas B. The need for strategic flexibility. Journal of Business Strategy. 1984;5(2):74–82. doi:10.1108/eb039060
38. Dedieu B, Chia E, Leclerc E, Moulin CH, Tichit M. editors. L'élevage en mouvement. Flexibilité et adaptation des exploitations d'herbivores. Editions Quae. Update Sciences & Technologies; 2008. 294 p.
39. Madelrieux S, Dedieu B, Dobremez L, Girard N. Patterns of work organisation in livestock farms: the ATELAGE approach. Livestock Science. 2008;121(1):28–37. doi:10.1016/j.livsci.2008.05.014
40. Kaine GW, Tozer PR. Stability, resilience and sustainability in pasture-based grazing systems. Agricultural Systems. 2005;83(1):27–48. doi:10.1016/j.agsy.2004.03.001
41. Astigarraga L, Ingrand S. Production flexibility in extensive beef farming systems. Ecology and Society. 2011;16(1):7. Available from: http://www.ecologyandsociety. org/vol16/iss1/art7/
42. Blanc F, Bocquier F, Agabriel J, D'Hour P, Chilliard Y. La composante animale de la flexibilité des systèmes d'élevage. In: Dedieu B, Chia E, Leclerc B, Moulin CH, Tichit M, editors. L'élevage en mouvement: flexibilité et adaptations des exploitations d'herbivores. Editions Quae; 2008. p. 73–94.
43. Ollion E, Ingrand S, Delaby L, Trommenschlager JM, Colette-Leurent S, Blanc F. Assessing the diversity of trade-offs between life functions in early lactation dairy cows. Livestock Science. 2016;183:98–107.doi:10.1016/j.livsci.2015.11.016
44. Gaillard C, Ingrand S, Blanquet J, Devevey C, Dumont R. Relations entre pratiques de finition et performances à l'abattage de vaches de réforme Simmental Française: intérêt du caractère mixte de la race. Inra Productions Animales. 2007;20:327–336. Available from: http://www6.inra.fr/productions-animales/2007-Volume-20/Numero- 4-2007/Relations- entre-pratiques-de-finition-et-performances
45. Magne MA. Modéliser le système d'information des agriculteurs. Le cas des éleveurs de bovins allaitants [thesis]. Université Montpellier II; 2007. Available from: http://metafort.irstea.fr/wp-content/uploads/2015/11/MAMagne.pdf
46. Vavra M. Sustainability of animal production systems: an ecological perspective. Journal of Animal Science. 1996;74:1418–1423. Available from: http://oregonstate.edu/dept/eoarcunion/sites/default/files/publications/VJAS_74_1418.pdf
47. Tichit M, Ingrand S, Moulin CH, Cournut S, Lasseur J, Dedieu B. Analyser la diversité des trajectoires productives des femelles reproductrices: intérêts pour modéliser le fonctionnement du troupeau en élevage allaitant. Inra Productions Animales. 2004;17(2):123–132. Available from: http://www6.inra.fr/productions- animales/2004-Volume-17/Numero-2-2004/Analyser-la-diversite-des-trajectoires-productives-des-femelles-reproductrices
48. Griffon M. Pour des agricultures écologiquement intensives. Des territoires à haute valeur environnementale et de nouvelles politiques agricoles. Les leçons inaugurales du groupe ESA;2007. 73 p. Available from: http://www.agrobiosciences.org/ article.php3?id_article=2276

CHAPTER 3

Livelihood Activities and the Role of Livestock in Smallholder Farming Communities of Southern Zambia

Gelson Tembo[1*], Alice Tembo[2], Fusya Goma[3], Evans Kapekele[4], Jairos Sambo[1]

[1]*Department of Agricultural Economics, University of Zambia, Lusaka, Zambia*

[2]*Ministry of Mines, Energy and Water Development, Luanshya, Zambia*

[3]*Veterinary Department, Ministry of Agriculture and Livestock, Lusaka, Zambia*

[4]*Disaster Management and Mitigation Unit (DMMU), Lusaka, Zambia*

ABSTRACT

This study represents one of the first comprehensive mixed-methods studies on the livelihood activities and the position of livestock in the hierarchy of these activities among households in live- stock-rearing communities of southern Zambia. The results identify livestock as the most important source of income and the second most important source of livelihood in general. Seasonality analysis also shows that livestock is very important in mitigating against hunger during lean periods between November and March. The study also identifies the richer households as the ones that benefit the most from livestock as a source of income. Policy and other interventions need to recognize not only the relative standing of livestock in these communities but also the fact that the livestock sub-sector is not the same to all rural households.

Keywords: Livestock; Livelihoods; Income; Shocks; Mixed-Methods; Zambia

1. INTRODUCTION

In many Southern African Development Community (SADC) countries, chronic vulnerability and poverty are entrenched and exacerbated by the ever-present risk of extreme climatic (drought and floods), economic and policy shocks. For most of these shocks and risks, food insecurity has been perceived mainly in terms of food crop availability and accessibility. The role of livestock, which touches upon the livelihoods of approximately 60 percent of the people in Southern Africa, is not fully appreciated. This neglect of the role livestock plays is partly due to lack of solid empirical evidence on the actual contribution livestock make to livelihoods and the survival strategies that are employed during times of shocks. Understanding the role of livestock in more specific terms could facilitate development of informed policy and other intervention strategies aimed at further strengthening the livestock-based livelihood systems. In Zambia, this knowledge is especially critical in livestock-based livelihood systems of Southern Province, where droughts and seasonal hunger are frequent and, often, rural people have to turn to their livestock as a coping strategy. Of the country's 10 provinces, Southern Province also stands out as one with the largest and most diversified smallholder livestock sub-sector. Unfortunately, there is a glaring dearth of empirical evidence on the underlying livelihood systems and the position of livestock in the hierarchy of livelihood activities. Elsewhere, studies have shown that livestock could impact nutritional status [1] [2] , income [1] [3] - [7]) and rural poverty [4] [8].

The study presented in this paper represents one of the first comprehensive mixed-methods treatments of the role of livestock as a source of livelihood and income, and in risk management against shocks. This study will assist policy makers and stakeholders in mainstreaming livestock intervention strategies in Zambia by understanding the livelihood systems and defining the role that livestock plays among livestock-owning smallholder households. In the remainder of the paper, we first present the methods and procedures used in Section two, followed by results and discussions in Section three. Summary and conclusion are presented in Section four.

2. METHODS AND PROCEDURES

2.1. Sample Design and Sampling

This study uses data from a mixed-methods survey of livestock-rearing communities in three districts of the Southern Province of Zambia, conducted during the 2006/2007 agricultural season. Southern Province is comprised of 11 districts, half of which are located in the low-rainfall (less than 800 mm annual rainfall) region, agro-ecological region (AER) I. This group of districts can be further sub-divided into valley districts and those that are located in the sandy plateau in the south-western part of the province. We call both these sub-categories of districts "hot-spot" districts, owing to their high poverty and vulnerability levels. The other six districts belong to medium-rainfall (800 - 1200 mm) region, AER II, here referred to as non-hotspot districts.

One district was randomly selected from each of these strata—one from each of the hot-spot sub-categories (Sinazongwe from the valley stratum; and Kazungula from the sandy plateau stratum), and one from the non-hotspot stratum (Namwala). Within each selected district and prior to primary data collection, all the veterinary camps or communities were stratified into four distinct groups based on relative productive potential (high or low) and market access (high or low). This was accomplished in close consultation with knowledgeable key informants such as the District Veterinary Officer (DVO) and his/her staff. One community was then randomly selected from each of the four camp/community strata.

2.2. Community Mapping and Wealth Ranking

In each selected community, a number of complementary quantitative and qualitative research techniques were used to collect the required data and information. Participants were drawn from within the livestock-rearing communities. The community participatory assessment process begun with a social mapping exercise to characterize the community's institutional, resource and asset context as well as to categorize the individual households based on their degree of vulnerability to food insecurity. This was done using participatory wealth ranking techniques, in which the community assigned the wealth status on each and every member of the community through a participatory process. Through this process, households were grouped into three strata—1) food secure (FS) or non-poor (NP), 2) food insecure (FI) or poor (P), and 3) extremely food insecure (EFI) or extremely poor (EP). For operational purposes, a household was categorized as non-poor if it had enough to eat throughout the year, i.e.

from harvest to harvest; as poor if it normally had enough food to last from harvest up to Christmas but not between Christmas and the next harvest; and as extremely poor if it had a longer period of food shortages, often experiencing severe food shortages even before Christmas.

2.3. Focus Group Discussions and Activity Rankings

Within each wealth stratum, focus groups were convened, each comprising a random sample of 8 - 10 individuals. In addition to being a powerful tool for collecting data that is enriched by purposeful use of interaction [9] - [12], Focus Group Discussions (FGDs) also offer considerable advantages in terms of cost per informant. The number of participants per FGD was determined in conformity with recommendations of some sections of the literature. Similarly, the number of FGDs per district was arrived at taking recommendations in the literature into account. Reference [13] , for example, suggests an optimal number of FGDs of 4 - 12 while [14] contends that data generated after about 10 FGDs are likely to be largely redundant.

During the focus group discussions a number of standard participatory tools were used to identify the issues and livelihood activities, and to determine the relative importance of the identified alternatives. These included tools such as participatory scoring and pair-wise ranking. Relative importance and seasonality of livelihood and income-generating activities were derived by using analytical calendars with scoring. Analytical calendars are matrices with each row representing a competing activity and each column representing the month. The activities in the analytical calendar are arrived at by the communities through a participatory process. Once this is done, two types of scoring are then undertaken. Vertical scoring, usually by distributing a fixed number of grains (say, 100), is done to determine the relative importance of each competing activity, overall and without taking seasonality into account. The more the grains assigned to the activity the more important the community considers the activity to be. The second type of scoring involves, for each activity, distributing the grains across the 12 months of the year. The months in which the activity is regarded to be important are assigned more grain. During seasonality analysis, the horizontal (monthly) scores are weighted by the relative importance of the activity/row as determined by the vertical scores. That is, the weight for each row entry is equal to the vertical score presented as a percentage of the total number of grains (e.g. 100). Unlike the unweighted seasonality (or row) scores, the weighted seasonality scores can be compared not only within activity but also across the activities.

2.4. In-Depth Interviews

During the community mapping and after all the households in the community had been identified, a very short questionnaire or listing form was administered to each of the identified households. A total of 309 households were listed and interviewed across all the communities in the three districts. While this information was important in its own right, the resultant list with wealth strata was also used as a sampling frame from which households were randomly selected for in-depth interviews. Using a semi-structured questionnaire, each in-depth household interview deciphered basic socio-economic characteristics and livelihood activities of each household. Care was taken to ensure representation of each of the wealth strata in these household case studies. A total of 56 household interviews were completed using a semi-structured questionnaire.

2.5. Analysis

Data collected through the wide spectrum of methods and instruments discussed above were analyzed using a wide range of techniques. These included post-interview brainstorming and collation of notes (for qualitative information), and quantitative analysis of rank/score data and scale data collected using pre-designed tables and semi-structured questionnaires. The quantitative analyses were accomplished using Microsoft Excel and the Statistical Package for Social Sciences (SPSS).

3. RESULTS AND DISCUSSION

The study area represents diverse livelihood systems comprising varying combinations of agricultural and non-agricultural activities. To help understand the intra- and inter-community variations in livelihood activities, the study used focus group discussions consisting of fairly homogeneous groups of community members. Four focus group discussions were carried out in each district for each food security stratum, one in each of the four sample communities. Each of these focus groups was asked to list the activities that constitute their livelihood system. Commonly listed activities included field crop production, livestock sales, gardening, fishing, wage and/or salaried employment (piece work), gathering and sale of forest products (grass, charcoal, honey, wild fruits, wood, etc.), crafts (carpentry, mat weaving, black-smith, knitting, etc.), beer brewing, hiring out animal draught power (ADP) for ploughing and transportation, selling animal products (milk, meat, eggs, etc.), and trading (sell of groceries and other necessities, commonly referred to as tu nthemba). The fact that smallholder livelihood systems are highly diversified is well documented in the literature [5]. Small-scale farmers have these diversified livelihood systems to help them cope during times of uncertainties which may include crop and market failures.

3.1. Livelihood Activities

Table 1 shows the numbers in each district of focus groups that indicated performing each livelihood activity, disaggregated by food security status. In columns (1) through (9), the entries range from zero (0), if none of the focus groups (per district) identified the livelihood activity in question as one of their means of survival, to four, if all the four focus groups mentioned the activity. Column (10) presents the total number of focus groups interviewed that identified the activity as an important source of livelihood, out of the 36 focus groups that were interviewed during the course of the study.

Field crop production is the most prevalent source of livelihood in the study area, indicated by 35 (or 97 percent) of the 36 focus groups (Column 10), followed by livestock (78 percent), gardening (72 percent), piece work (53 percent), and fishing (42 percent). The importance of field crop production and livestock in smallholder farming communities has been echoed throughout much of Africa (see, for example, [5] on Tanzania). Geographical and inter-strata variations in the composition of livelihood activities were also quite apparent. Field crops and livestock were mentioned by all the food security groups in the non-hotspot district (Namwala). Among the hotspot districts (Kazungula and Sinazongwe), only the extremely food insecure in one of the four Kazungula communities did not identify field crop production as their source of livelihood. The relative prevalence of livestock as a means of survival in the hotspot districts varies across food security categories, mentioned the least among extremely food insecure groups. Gardening, while being quite important in the hotspot districts, is not so important in Namwala, where only food insecure and extremely food insecure households use it as a source of livelihood.

Fishing and piece work are most prevalent in Sinazongwe District, where at least three-quarters of the interviewed focus groups mentioned them, and virtually non-existent among food secure households in the non- hotspot district, Namwala. Half of the food insecure and extremely food insecure focus groups in Namwala identified fishing as important. However, the importance of fishing has been put in check by the fact that the activity is rather specialized and the fact that fish have become progressively rarer. In all three districts, piece work is most common among the extremely food insecure.

Table 1. Diversity of livelihood activities across districts and food security status.

Activity actually mentioned by the focus group	Hotspot districts: Kazungula			Hotspot districts: Sinazongwe			Namwala (Non-hotspot district)			Total
	Food secure	Food insecure	Extremely food insecure	Food secure	Food insecure	Extremely food insecure	Food secure	Food insecure	Extremely food insecure	
	(1)	(2)	(3)	(4)	(5)	(6)	(7)	(8)	(9)	(10)
	-----Number of focus groups undertaking the livelihood activity[a]-----									
Field crops	4	4	3	4	4	4	4	4	4	35
Livestock sales	4	3	0	3	4	2	4	4	4	28
Gardening	3	4	2	4	3	4	0	2	4	26
Fishing	1	1	0	3	3	3	0	2	2	15
Piece work	0	2	3	3	3	4	0	1	3	19
Forest products	0	2	0	3	1	2	1	1	2	12
Crafts	1	2	0	1	1	2	0	1	2	10
Beer brewing	1	2	1	0	0	0	1	0	0	5
Hiring out ADP	1	0	0	0	1	0	0	2	0	4
Selling milk	1	1	0	0	0	0	1	1	0	4
Trading	2	2	4	1	2	1	0	0	1	13

[a]Out of four communities per district (Columns 1 - 9), or out of a total of 36 communities for the study (Column 10).

Trading is most common in Kazungula District, which is not surprising being a border town next to the country's tourist capital, Livingstone. Because of the nature of the terrain and inadequate land in Sinazongwe, many communities have resorted to fishing considering the fact that there are no restrictions involved as to what time of the year fishing is allowed or who should/shouldn't fish. This is unlike in Namwala, where the main rivers such as Namwala and Kafue are only open for fishing over the period March through to November.

Table 2 presents the proportions of the interviewed focus groups that ranked each activity among the top three most important activities at the time of the survey (2006) and 10 years before (1996), disaggregated by district and food security category. Table 2 also confirms the importance of crop agriculture and livestock in the study area with 81 percent and 53 percent of all the focus groups ranking them among the top three, respectively (Column 4).

As expected, the relative importance of these two activities is higher as we go higher on the food security scale—towards the food secure stratum. More than 90 percent of the food secure, for example, ranked field crop production among the top three most important livelihood activities. The relatively lower weight placed on agriculture and livestock by the food insecure and extremely food insecure groups means that these groups place relatively higher weight on other activities. Gardening, for example, seems to be a significant preoccupation of the food insecure segments of the society, ranked highly by at least 58 percent of the groups. Piece work and trading are also quite important among these groups with about half the interviewed FI and EFI focus groups ranking them in the top three.

Compared to agriculture-based activities, piece work and trading are often regarded inferior and adopted as means for coping with harsh realities of food insecurity. Thus, the waning importance of field crops and livestock and growing importance of piece work over the 10 years prior to the survey (compare Column 4 with Column 8) suggest an increase in levels of vulnerability. The importance of livestock as a source of livelihood went down by 21 percent (from 66.7 percent to 52.8 percent) while gardening increased by as much as 29 percent (from 38.9 percent to 50.0 percent) between 1996 and 2006. However, the fact that crops and livestock are important sources of livelihood shows that agriculture is key in these communities. While crops of various types are very important in day to day life during normal years, livestock have from time to time played the role of a life saver during bad seasons.

Cattle were the most prevalent animals, especially in the non-hotspot district, Namwala (Figure 1(a)).

Table 2. Relative importance and trends in livelihood activities in Kazungula, Namwala and Sinazongwe Districts of Zambia, 2006 and 1996.

Livelihood activity	At time of survey (2006)				10 years before (1996)			
	Food secure	Food insecure	Extremely food insecure	Total	Food secure	Food insecure	Extremely food insecure	Total
	(1)	(2)	(3)	(4)	(5)	(6)	(7)	(8)
	-----------------% of focus groups ranking the activity among top 3 most important-----------------							
Field crops	91.7	75.0	75.0	80.6	91.7	91.7	83.3	88.9
Livestock	75.0	50.0	33.3	52.8	91.7	75.0	33.3	66.7
Gardening	25.0	66.7	58.3	50.0	16.7	50.0	50.0	38.9
Fishing	25.0	8.3	8.3	13.9	33.3	0.0	8.3	13.9
Piece work	8.3	16.7	50.0	25.0	8.3	25.0	33.3	22.2
Forest products	0.0	8.3	8.3	5.6	0.0	8.3	8.3	5.6
Beer brewing	8.3	8.3	0.0	5.6	16.7	16.7	8.3	13.9
Crafts	8.3	16.7	16.7	13.9	8.3	16.7	16.7	13.9
Hiring out oxen	0.0	8.3	0.0	2.8	0.0	8.3	0.0	2.8
Selling milk	8.3	0.0	0.0	2.8	8.3	0.0	0.0	2.8
Trading	16.7	8.3	50.0	25.0	0.0	0.0	50.0	16.7

Namwala had about 100,000 cattle, which is almost twice as much as in the other two districts (Kazungula and Sinazongwe). Sinazongwe had the lowest cattle population, of about 52,000, while Kazungula had about 60,000.

Goats and poultry (small livestock) were most important in Sinazongwe with a goat population of more than 50,000 and that of poultry about 32,000. Sinazongwe had 7 times the goat population as did the closest district, Namwala. Sheep and pigs were considered important in Sinazongwe with the population higher than that of the other districts. Of all the three districts, Kazungula had the least populations of goats, poultry, and other relatively minor livestock.

The relative importance of cattle is also evident in the qualitative assessment, with most of the communities assigning a rank of 1 (Figure 1(b)). Goats are the second most important livestock species reared, followed by poultry (Figure 1(b)). The importance of cattle and goats among the food security groups is also evidenced in a study by [3] in the rural areas of South Africa. Sheep, donkeys, and pigs are the least important species carrying a mean rank of more than three (3) among all the food security groups.

3.2. Income Generation

Field crops and livestock are the two most important sources of income, together representing almost half of the income earned by rural households. These results are consistent with [4] , who found that field crops generated most of the livelihood incomes across social groups in West Africa (i.e. very poor, poor, relatively rich). When animal products and services (e.g. hiring out animal draught power) are considered, the contribution of the two sources to income extends to about 60 percent. While field crop production received a higher score as a livelihood activity in a more general sense than did livestock rearing, livestock and their products represent a much more important source of income. When only live animals and meat sales are considered, livestock generate 26 percent of all income earned by the households, compared to field crops' 21 percent. When animal products (eggs, milk, skins, manure, etc) are taken into account, the contribution of livestock swells to 39 percent, which is almost double the contribution of field crops. The dominance of livestock as an income earner holds both in the total sample (Column 7 in Table 3) as well as among most individual food security sub-samples. This is consistent with the national picture in which livestock is argued to contribute at least 50 percent of agricultural gross domestic product (GDP) [15].

Generally, small scale farmers grow most of their field crops for consumption and only sell a few when need arises. On the contrary, livestock is rarely eaten but rather sold to solve household problems (school fees, clothes, hummer mill services, health expenses, etc.). In a non-hotspot district, one was considered wealthy depending on the number of cattle one owned. Cattle are sold for other various purposes such as building a house, dowry payments, and settling conflicts (e.g. murder, adultery, pregnancy, theft) and exchanged for other assets including vehicles. After livestock and field crops, piece work is the next most important income source (15 percent), followed by a group of activities that we have grouped into the category "other" (13 percent). It is important to note that the contribution of livestock to income is largest among food secure households with an even larger prominence in the non-hotspot district. Piece work, on the other hand, is clearly most prominent among the extremely food insecure and most visible in the hotspot districts. Compared to all other food security groups, food insecure households appeared to have the most diversified set of income sources, with a significantly larger weight on numerous other minor sources collectively referred to as 'other' than any other group of households.

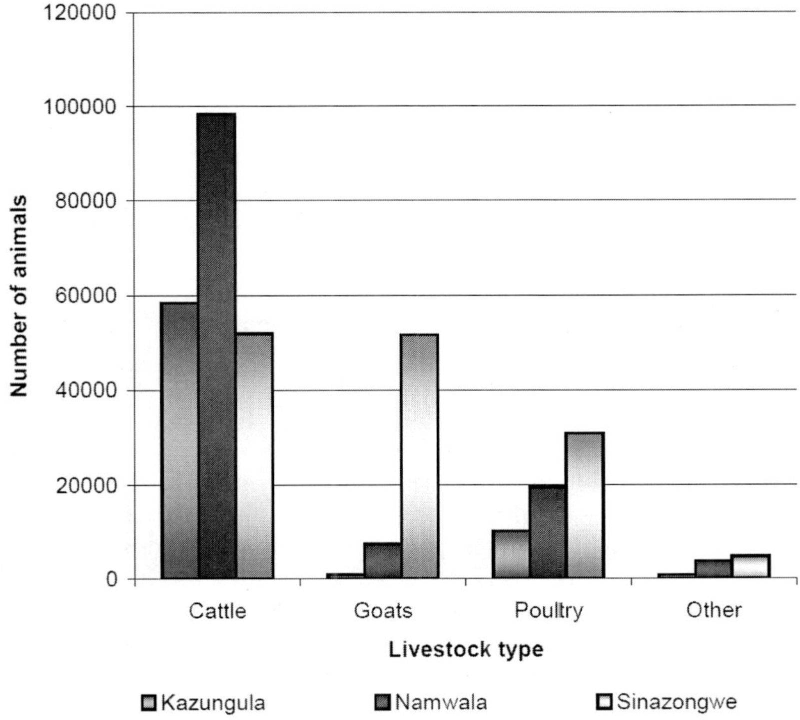

(a)

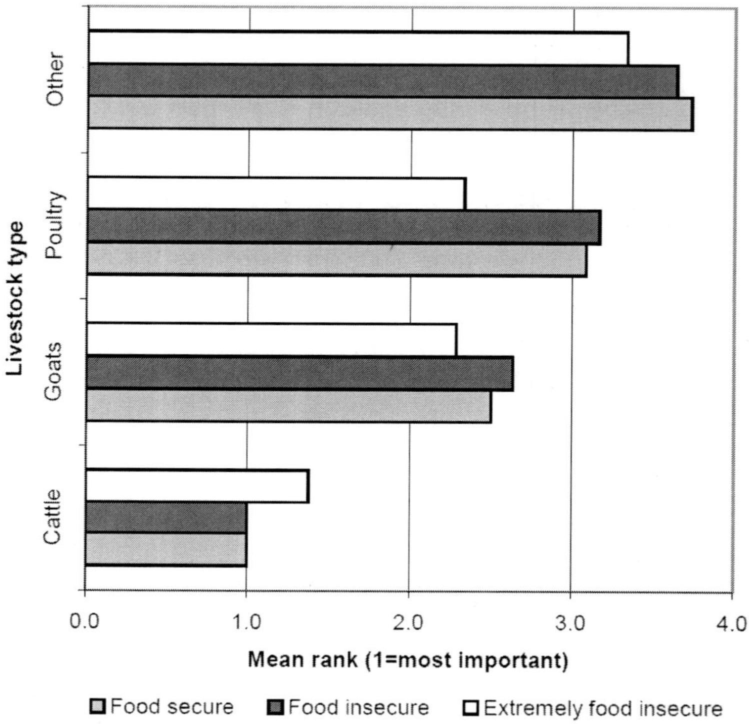

(b)

Figure 1. Numbers of livestock species and their relative importance to livelihoods, September 2006. Sources: Livestock numbers from Department of Veterinary and Livestock Development (DVLD). (a) Livestock population; (b) Rank of the livestock species (1 = most important).

Table 3 presents a total annual picture, which greatly masks the dips and peaks along the year and the households' abilities to mitigate the harder periods. Figure 2 shows the seasonality of the four most important income sources as depicted by weighted monthly scores—field crops (Figure 2(a)), livestock sales (Figure 2(b)), piece work (Figure 2(c)), and other sources (Figure 2(d))—and how the seasonal patterns differ by the household's food security status. Field crops are most important during and just after harvest, in the months of April through September. But the exact length of this crop income boom period differs markedly by food security status. While the extremely food insecure sell their crops almost immediately after harvest, the relatively more food secure households (FI and FS) are able to delay their sales and hence stretch the period further.

Table 3. Sources of income and their relative importance across food security status and district types, September 2006.

Source of income	Hotspot districts			Non-hotspot district			Total
	Food secure	Food insecure	Extremely food insecure	Food secure	Food insecure	Extremely food insecure	
	(1)	(2)	(3)	(4)	(5)	(6)	(7)
	-----------------Mean score (out of 100)-----------------						
Field crops sales	20.4	13.9	25.4	25.3	27.0	17.5	20.9
Livestock sales[a]	27.4	24.9	20.1	43.3	26.8	25.0	26.2
Other animal products[b]	12.5	3.1	3.0	10.3	5.3	12.0	7.1
Hiring out ADP	11.3	7.1	2.3	3.4	4.8	3.0	5.9
Piece work	10.9	14.3	26.8	0.0	5.3	19.3	14.6
Gardening	6.3	13.6	9.8	0.0	6.5	7.3	8.3
Fishing	3.5	6.9	1.5	0.0	3.5	5.0	3.7
Other[c]	7.9	16.3	11.3	17.7	21.0	11.5	13.3

[a]Livestock are sold both whole and as meat. [b]Other animal products mentioned by the communities included milk, eggs, skins, and manure. [c]Other income sources include hammer mill revenue (service and crop byproducts), remittances, selling of forest products, crafts, beer brewing, trading, traditional healing, guest house services, and cash transfers from NGOs.

One could speculate that this is because the more food secure households: 1) are less desperate to sell their crop at harvest as they have other means of survival, 2) have relatively larger harvests that need to be sold over an extended period of time, and/or 3) have better technical know-how for storing crops. Whatever the reason, the fact that the poorer households sell their produce at and/or immediately after harvest implies that they receive the lowest prices for them, a fact that could further strengthen the poverty vicious cycle, continually widening the gap between the poor and the richer segments of the society.

Figure 2(b) and Figure 2(c) seem to present a picture opposite to that presented in Figure 2(a). Livestock sales and piece work both are most prominent the further we are from the harvest period—October through February. While the general pattern of the two figures is the same across wealth strata, it appears the more food secure the household is the larger the weight on livestock sales. This pattern is more clearly defined during the first quarter of the calendar year. Also, as expected, piece work was much more heavily weighted by extremely food insecure focus groups.

Figure 2(d) further reinforces the argument that the food insecure households have a much more diverse set of income sources, placing significantly higher weights on the category "other" throughout the year. Compared to the other three income sources in Figure 2, this latter category does not have any special seasonal pattern.

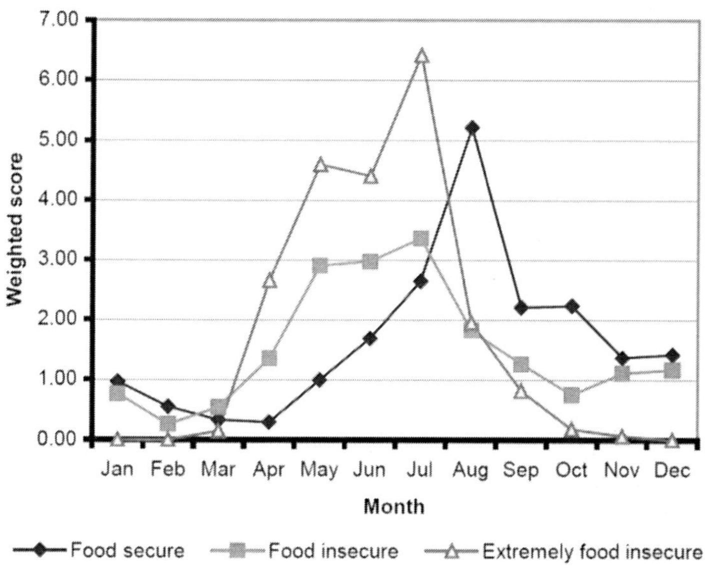

(a)

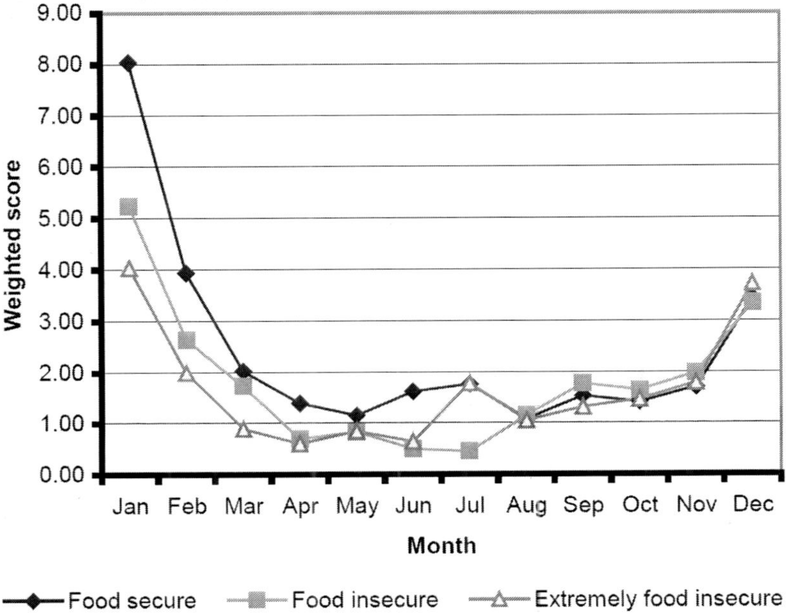
9.00
8.00
7.00
6.00
5.00
4.00
3.00
2.00
1.00
0.00
Weighted score
Jan Feb Mar Apr May Jun Jul Aug Sep Oct Nov Dec
Month
Food secure
Food insecure
Extremely food insecure

(b)

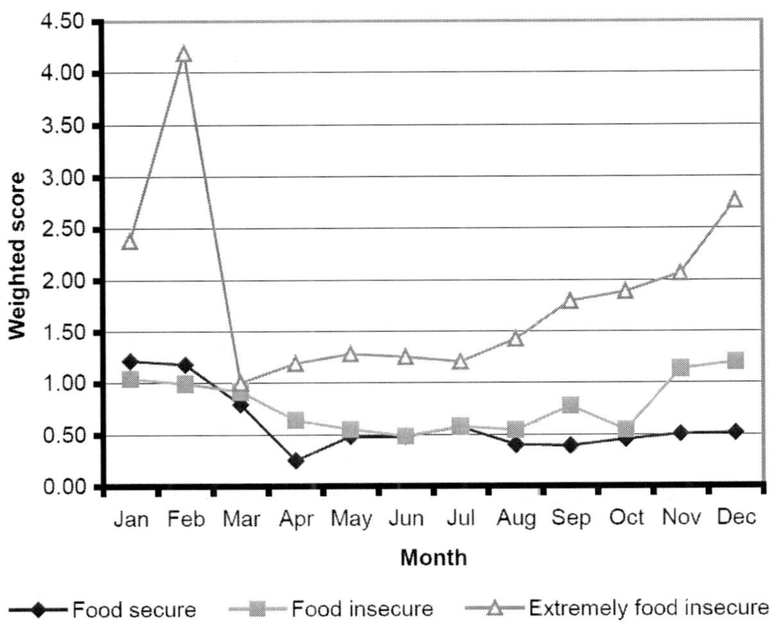
4.50
4.00
3.50
3.00
2.50
2.00
1.50
1.00
0.50
0.00
Weighted score
Jan Feb Mar Apr May Jun Jul Aug Sep Oct Nov Dec
Month
Food secure
Food insecure
Extremely food insecure

(c)

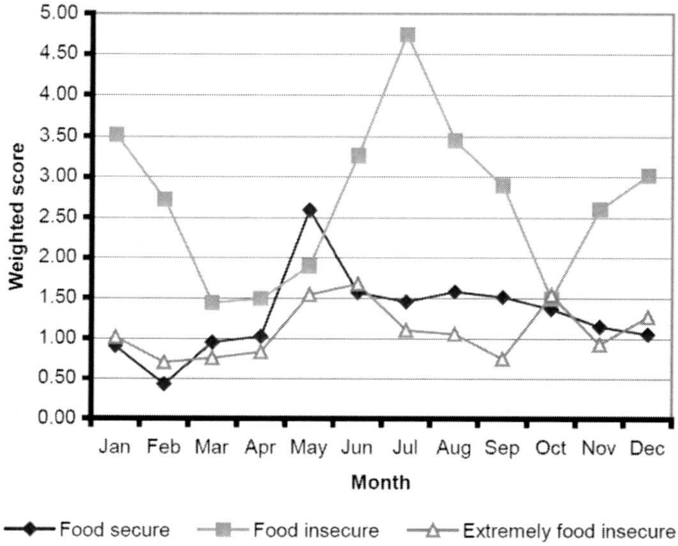

(d)

Figure 2. Seasonality of the four most important income sources, differentiated by food security status, 2006. (a) Field crops; (b) Livestock sales; (c) Piece work; (d) Other.

Livestock sales and piece work are high at the beginning and end of the year partly because this period coincides with the festive season in addition to being the time that parents prepare school fees as their children move into higher levels of education. Moreover, this is also the lean period of the year during which food stocks will have run out in most households.

4. SUMMARY AND CONCLUSION

This study uses qualitative research techniques to identify the livelihood activities and to determine the position of livestock in the hierarchy of those activities among households in livestock-rearing communities of Southern Zambia. Seasonality analysis was also done to identify the shifts in relative importance of the activities across the different times of the year. The results indicate that, when livelihood is broadly defined, livestock rearing is second only to field crop production both in terms of prevalence (number of communities citing it) and relative importance (as indicated by weighting exercises). However, livestock and livestock-related products and services are by far the most important source of income. Livestock as a source of income is especially important among the food secure households, who earn more than half of their income through sales of livestock, livestock products (milk, eggs, meat, manure) and livestock-related services (hiring out of animal drought power), compared to less than 40 percent among other food security groups. That livestock is important as sources of livelihoods and, especially, income identifies the need to identify and implement interventions that could enhance livestock ownership and productivity. While food secure households are more commercial in approach, less food secure households have their sustenance and subsistence in mind. This inherent dichotomy in approach to livelihood needs to be reflected in intervention design and implementation.

ACKNOWLEDGEMENTS

Gelson Tembo, Alice Tembo, Fusya Goma, Evans Kapekele, Jairos Sambo (ILRI) and the Food and Agriculture Organization of the United Nations (FAO). However, the views expressed in this paper do not necessarily represent the official positions of ILRI and/or FAO. All errors in interpretation are the authors' own.

REFERENCES

1. Biradar, N., Desai, M., Manjunath, L. and Doddamani, M.T. (2013) Assessing Contribution of Livestock to the Live- lihood of Farmers of Western Maharashtra. Journal of Human Ecology, 41, 107-112.
2. Kassa, H., Ayalew, W., Habte, G.Z. and Gebre, M.T. (2003) Enhancing the Role of Livestock Production in Improving Nutritional Status of Farming Families: Lessons from a Dairy Goat Development Project in Eastern Ethiopia. Livestock Research for Rural Development, 15. http://www.lrrd.org/lrrd15/6/kass156.htm
3. Chaminuka, P., Udo, H.M.J., Eilers, K.C.H.A.M. and Zijpp, A.V.D. (2014) Livelihood Roles of Cattle and Prospects for Alternative Land Uses of Wildlife/Livestock Interface in South Africa. Land Use Policy, 38, 80-90. http://dx.doi.org/10.1016/j.landusepol.2013.10.007
4. Mburu, S., Zaibet, L., Fall, A. and Ndiwa, N. (2012) The Role of Working Animals in the Livelihoods of Rural Com- munities in West Africa. Livestock Research for Rural Development, 24. http://www.lrrd.org/lrrd24/9/mbur24156.htm
5. Lwelamira, J., Binamungu, H.K. and Njau, F.B. (2010) Contribution of Small Scale Dairy Farming under Zero-Graz- ing in Improving Household Welfare in Kayanga Ward, Karagwe District, Tanzania. Livestock Research for Rural Development, 22. http://www.lrrd.org/lrrd22/2/lwel22031.htm
6. Mabe, L.K., Antwi, M.A. and Oladele, O.I. (2010) Factors Influencing Farm Income in Livestock Producing Commu- nities of North-West Province, South Africa. Livestock Research for Rural Development, 22. http://www.lrrd.org/lrrd22/8/mabe22142.htm
7. Dovie, D.B.K., Shackleton, C.M. and Witkowski, E.T.F. (2006) Valuation of Communal Areas Livestock Benefits, Rural Livelihoods and Related Policy Issues. Land Use Policy, 23, 260-271. http://dx.doi.org/10.1016/j.landusepol.2004.08.004

8. Holmann, F., Rivas, L., Urbina, N., Rivera, B., Giraldo, L.A., Guzman, S., Martinez, M., Medina, A. and Ramirez, G. (2005) The Role of Livestock in Poverty Alleviation: An Analysis of Colombia. Livestock Research for Rural Devel- opment, 17. http://www.lrrd.org/lrrd17/1/holm17011.htm
9. Kitzinger, J. (1994) The Methodology of Focus Groups: The Importance of Interaction between Research Participants. Sociology of Health and Illness, 16, 103-121. http://dx.doi.org/10.1111/1467-9566.ep11347023
10. McLafferty, I. (2004) Focus Group Interviews as a Data Collecting Strategy. Journal of Advanced Nursing, 48, 187- 194. http://dx.doi.org/10.1111/j.1365-2648.2004.03186.x
11. Merton, R.K., Fiske, M. and Kendall, P.L. (1990) The Focused Interview: A Manual of Problems and Procedures. 2nd Edition, Free Press, New York.
12. Morgan, D. (1996) Focus Groups. Annual Reviews of Sociology, 22, 129-152. http://psycho.univlyon2.fr/sites/psycho/IMG/pdf/Morgan_fc_v2.pdf http://dx.doi.org/ 10. 1146/annurev.soc.22.1.129
13. Krueger, R. and Casey, M.A. (2009) Focus Groups: A Practical Guide for Applied Research. 4th Edition, Sage Publi- cations Limited, London.
14. Millward, L. (1995) Focus Groups. In: Breakwell, G., Hammond, S. and Fife-Schaw, C., Eds., Research Methods in Psychology, 4th Edition, Sage Publications Limited, London.
15. Daka, D. E. (2002) Livestock Sector in Zambia: Opportunities and Limitations. In: IAEA (International Atomic Energy Agency), Eds., Development and Field Evaluation of Animal Feed Supplementation Packages, Proceedings of the Fi- nal Review Meeting, Cairo, 25-29 November 2000. http://www-naweb.iaea.org/nafa/aph/public/reports-12.pdf

CHAPTER 4

Livestock Demand, Global Land Use Changes, and Induced Greenhouse Gas Emissions

Sunil P. Dhoubhadel[1]*, Farzad Taheripour[2], Matthew C. Stockton[3]

[1]*Department of Agriculture, Fort Hays State University, Hays, USA*

[2]*Department of Agricultural Economics, Purdue University, West Lafayette, USA*

[3]*West Central Research & Extension Center, University of Nebraska-Lincoln, North Platte, USA*

ABSTRACT

With increasing global population and income, the demand for livestock products continues to grow and is likely to grow in future as well. This raises concerns about the implications of such growth for the environment. A well-known Computable General Equilibrium model (CGE), GTAP- BIO, is used to project the growth of livestock output between 2004 and 2022 and to estimate the land use changes and associated Greenhouse Gas (GHG) emissions. Results indicate that the increased livestock output leads to considerable increase in pasture (about 45 million hectares) and decrease in forest area (about 44 million hectares) between 2004 and 2022. Estimated emissions associated with this change are about 20 billion tons of carbon dioxide equivalent (CO_2 e) during 2004-2022 or an annual average of 1.1 billion tons. Results also indicate that a significant portion of the emissions (about 11%) can be reduced if consumer preferences could be altered. In practice, this will require policy interventions that promote substitution of non-ruminant for ruminant meat products.

Keywords: Livestock Demand, Land Use, GHG Emissions, CGE Model

1. INTRODUCTION

With growing concern about the climate change, potential sources of Greenhouse Gas (GHG) emissions mitigation are being identified. Lately, the livestock sector has received considerable attention in this regard as the sector reportedly contributes significantly to human-induced GHG emissions [1] - [3].

The global demand for livestock products has grown over the years and is likely continue that growth in the future [4]. Increasing overall livestock output has implications for the environment. The sources of GHG emissions from the livestock sector can be segregated into two broad categories: 1) emissions that emanate from land use changes due to livestock production; typically when forest land is converted into pasture or cropland to accommodate increased livestock production, it results in increased GHG emissions due to land clearing and reduced long term carbon sequestration; 2) emissions which are attributed to non-land use change sources such as enteric fermentation, feed production and processing, manure handling, and processing and transportation of animal products.

Unlike the emissions estimates for non-land use sources, which have been well documented in the literature [1] [2], the emissions induced by land use changes due to livestock production have not been adequately addressed and are a contentious issue [5]. While [3] and [6] estimate about 2.4 to 2.6 billion tons carbon dioxide equivalent (CO_2 e) emissions per year as a result of land use change due to livestock production, these reports, however, do not provide a detailed explanation of the methods used to obtain these estimates. A FAO report authored by [2] estimates approximately 7.1 billion tons of CO_2 e emissions annually from the livestock sector and attribute about 0.65 billion tons CO_2 e per year (about 9.2% of the total emissions from the sector) to the land use changes. However, this estimate fails to account fully the emissions from land use change associated with livestock production at the global level as it mainly accounts for land use changes in Latin America due to soybean production in Brazil and Argentina and pasture expansion in Brazil, Chile, Nicaragua, and Paraguay. Given that the report does not account for land use changes in all regions of the world and all crops, it is likely that it underestimates the emissions from land use change caused by livestock production at the global level. Moreover, [2] 's emission estimates are based on the reference years 1996-2006, which need updating to account for the livestock production after 2006. A more recent estimate of GHG emissions from livestock sector by [1] does not include the emissions from land use changes. Given this gap in the literature, this work focuses on estimating GHG emissions induced by land use changes due to projected increase in livestock production between 2004 and 2022.

The objectives of this paper are threefold: 1) provide a baseline projection for regional livestock output growth between 2004 and 2022 at the global scale; 2) use the baseline projection estimates of livestock output to estimate the expected global land use changes by regions; and 3) estimate the GHG emissions associated with those changes. Considering the global scale of this work and the number of interactions among many economic sectors and regions, a Computable General Equilibrium (CGE) modeling framework is used.

2. THE MODEL

A modified version of the standard GTAP (Global Trade Analysis Project) model, GTAP-BIO, is used for this analysis. The GTAP model is a multi-sector, multi-region CGE model, which is widely used for policy, trade, and environmental analysis. GTAP models consumer (private households) preference using a Constant Difference of Elasticity (CDE) function, which allows income elasticities to differ across goods. Producer/firm production functions are modeled using nested Constant Elasticity of Substitution (CES) forms. Moreover, it assumes constant returns to scale in production and perfectly competitive markets. The Armington assumption is used for traded goods i.e. the imports are differentiated by country of origin. The general equilibrium nature of the model is imposed by applying Walras Law in the model, i.e. in equilibrium all markets clear (i.e. aggregate demand equals aggregate supply in all input and output markets, imports equal exports in the international market), all firms earn zero economic profits, and all consumers operate on their budget constraints, and global investments equal global savings. The standard GTAP model is fully described and discussed with underlying assumptions and equations in the book "Global Trade Analysis" [7], the discussion in this section is focused on the features of GTAP-BIO model that are pertinent to the objectives of this study.

The GTAP-BIO was originally developed and used to estimate the impact of increased biofuels production on global land use and associated GHG emissions [8] [9]. The GTAP-BIO explicitly models the competition among crops, livestock, and the biofuels sector in the land market. Besides modeling the land use, the model explicitly incorporates feed demand for livestock. The livestock feed demand includes not only conventional feed sources but also distillers' grains, a by-product of corn ethanol production, primarily used as a livestock feed ingredient (for details see [10]). Given that the GTAP-BIO model explicitly models land supply and the livestock feed demand in addition to retaining the general equilibrium features of the standard GTAP model, the model is very appropriate to estimate the growth of livestock demand and associated land use changes.

Figure 1 illustrates the land supply module in the GTAP-BIO model. This module determines the supply of land for forest, crop, and pasture production purposes in 18 Agro-Ecological Zones (AEZs) around the world. The elasticity of transformation, ETL1 determines the transformation of land for forest, crops and pasture, while ETL2 governs allocation of crop land among different crops. ETL3 is the elasticity of transformation of pasture land for meat and milk production. These elasticities are tuned according to the changes in land cover and harvested areas at the global scale [11].

3. EXPERIMENTS AND DATA SOURCES

To examine the impact of increased livestock output on global land use and resulting GHG emissions, two experiments are developed and applied using the GTAP-BIO model and GTAP database version 7, which corresponds to the reference year 2004.

1) Experiment 1: This experiment is carried out by using the forecasted changes (Figure 2) in GDP, population, capital, skilled labor, and unskilled labor between the years 2004 (baseline) and 2022 as exogenous shocks to the GTAP-BIO model for all GTAP regions (Table 1). The land use changes obtained from this experiment can be attributed to changes in demand for all goods and services in the regional economies including changes in livestock production.

2) Experiment 2: Experiment 1 is repeated, while the household and intermediate demands for livestock products held constant1. The difference in land use between this experiment and the first experiment can be attributed to the effect of the livestock demand on land use.

Two additional steps are followed. First, the resulting land use changes are introduced into the emissions factors developed by Plevin et al. [12] to calculate induced land use emissions due to changes in livestock outputs for the time period of 2004-2022. These authors have developed a comprehensive model, "the AEZ_EF v47", specifically designed to estimate GHG emissions associated with land use changes. The model considers various sources and sinks of GHG emission such as the above and below-ground live biomass, dead organic matter, soil organic matter, harvested wood products, non-CO_2 emissions (e.g. methane and nitrous oxide), and foregone sequestration in estimating the induced land use emissions. Moreover, the model is designed to fit well with the GTAP-BIO model such that the regions and AEZs in the AEZ_EF v47 model exactly match with the 19 regions and 18 AEZs in the GTAP model. This facilitates direct use of the land use change results from the GTAP model simulations into the AEZ_EF v47 model. Second, a series of simulations with changes in the substitution parameter of the GTAP-BIO model are tested to examine the sensitivity of the results.

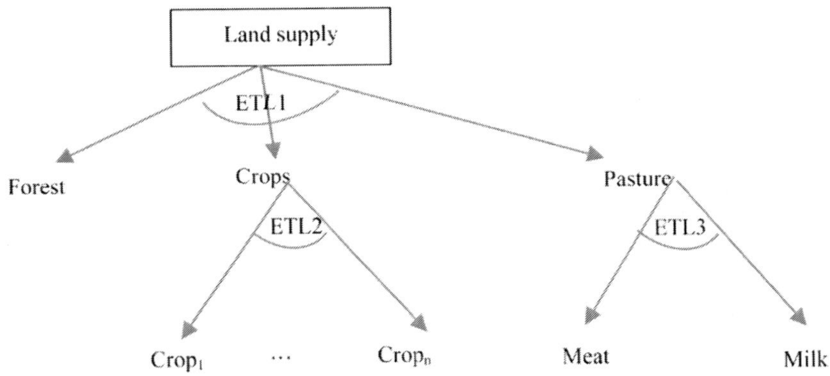

Figure 1. Land supply in the GTAP-BIO Model.

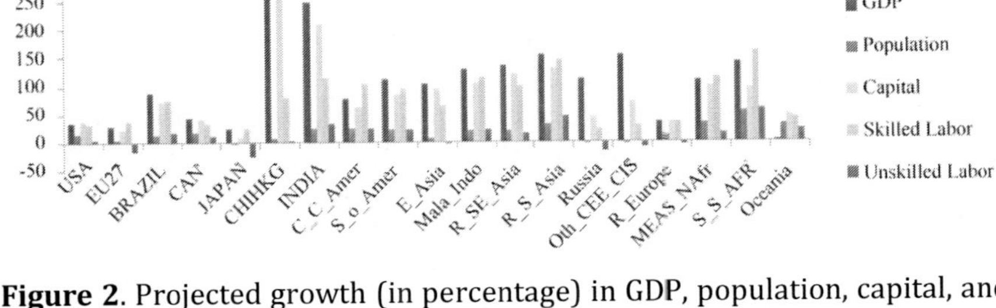

Figure 2. Projected growth (in percentage) in GDP, population, capital, and skilled and unskilled labor (2004-2022). Source: GDP, population and capital changes from CEPII baseline version 2.1 [13]. The data for skilled and unskilled labor are obtained from the baseline projection database prepared by [14].

Table 1. Definition of regions included in the GTAP-BIO Model.

GTAP-BIO regions	Description of the regions
USA	United States
EU27	European Union 27
BRAZIL	Brazil
CAN	Canada
JAPAN	Japan
CHIHKG	China and Hong Kong
INDIA	India
C_C_Amer	Central and Caribbean Americas
S_o_Amer	South and Other Americas
E_Asia	East Asia
Mala_Indo	Malaysia and Indonesia
R_SE_Asia	Rest of South East Asia
R_S_Asia	Rest of South Asia
Russia	Russian Federation
Oth_CEE_CIS	Other East Europe and Rest of Former Soviet Union
R_Europe	Rest of European Countries
MEAS_NAfr	Middle Eastern and North Africa
S_S_AFR	Sub Saharan Africa
Oceania	Oceania countries

The projected data for GDP, population, and capital (Figure 2) mainly come from the Centre d'Études Prospectives et d'Informations Internationales (CEPII) baseline database version 2.1 [13]. The projected changes in skilled and unskilled labor between the years 2004 to 2022 (Figure 2) are from the baseline projection database prepared by [14] for the GTAP model. Both the CEPII and [14] information are presented originally as a country level data and therefore are aggregated into the 19 GTAP regions. The percentage change in each of the five variables between 2004 and 2022 is calculated and then used as a shock to these variables in the model thereby simulating projected growth in livestock output and associated land use change.

4. RESULTS

4.1. Change in Livestock Output[2]

Figure 3 presents projected changes in the livestock output at the global scale for the time period of 2004 to 2022, obtained from the first experiment. The largest increase is observed for non-ruminant products (111%) followed by ruminant (84%) and dairy products (64%).

Figure 4 shows the projected increases in livestock output for each of the 19 regions. Among these regions, China-Hong Kong (CHIHKG) and INDIA are the regions with the greatest increases in livestock output. CHIHKG has more than a 200% increase in all three categories of livestock output with INDIA not far behind with nearly a 200% increase. This increase is mainly driven by the large simultaneous increases in income (GDP) and population (Figure 2) in these regions. Among livestock output categories, CHIHKG has the highest growth for ruminants, but INDIA has the largest increase in non-ruminants. The rest of South Asia (R_S_Asia), Other East Europe and the rest of the former Soviet Union excluding Russia (Other_CEE_CIS), Middle-Eastern and North Africa (MEAS_NAfr), and Sub-Saharan Africa (S_S_AFR) are the other regions with relatively high growth in livestock outputs.

4.2. Land Use Impacts

Table 2 and Figure 5 show the impact of livestock output growth on land use. The figures represent the difference in land use between the first and second experiments explained above. As is evident in the table and the figure, with the increase in livestock output, pasture area is expanded while forest land is reduced in all regions. However, the overall change in crop land has a mixed outcome. In most of the regions, cropland is reduced, while in regions such as Canada (CAN), CHIHKG, and Mala_Indo cropland expanded. Globally there is a net increase of about 44.5 billion hectares of pasture, with a decrease in cropland and forest of 1.1 billion and 43.3 billion hectares respectively (**Table 2**). Among the regions, CHIHKG emerges as the region of largest pasture expansion/deforestation followed by S_S_AFR, BRAZIL, South and Other Americas (S_o_Amer), and R_S_ Asia (**Table 2**). The expansion of pasture is small in the advanced economies such as the EU27, CAN, JAPAN, Oth_Europe, and Oceania.

4.3. Emissions Due to the Land Use Changes

As stated previously, the results on regional land use change obtained from the GTAP-BIO model simulations are used in combination with the "AEZ_EF v47" model to estimate the induced land use emissions as a result of regional changes in livestock demands for livestock outputs3. Table 3 presents the induced land use emissions as a consequence of changes in livestock demand between 2004 and 2022. Results indicate a major contribution of deforestation to the total emissions; particularly, the conversion of forest to pasture accounts for the majority of the emissions (Table 2 and Table 3). Conversely, the conversion of cropland to pasture reduces emissions. The regions with relatively higher emission are S_S_Afr, CHIHKG, BRAZIL, and S_o_Amer (Table 3). Mal_Indo, R_SE_ Asia, and R_S_Asia are other regions with major land use emissions. At the global level, the induced emissions are about 19.8 billion tons CO_2 e (Table 3). Since these emissions are the aggregated emissions for the 18 year period (2004-2022), the average annualized emissions due to the increase in livestock demand are about 1.1 billion tons CO_2 e per year. Given that the 0.65 billion tons CO_2 e per year estimate by [2] might be underestimated, the estimate of 1.1 billion tons CO_2 e per year obtained from this study is very plausible.

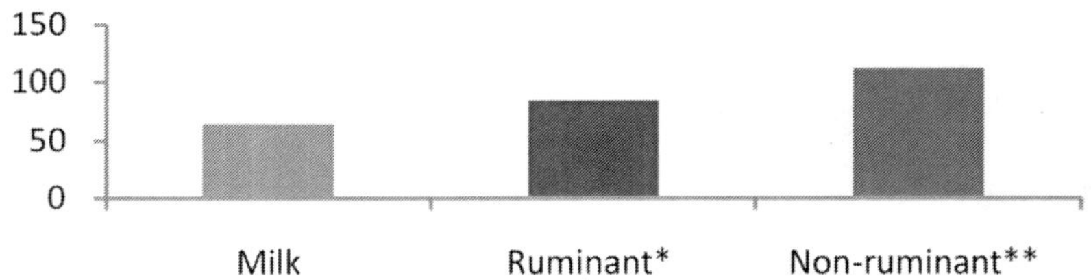

Figure 3. Projected growth (in percentage) in the global livestock output (2004-2022).

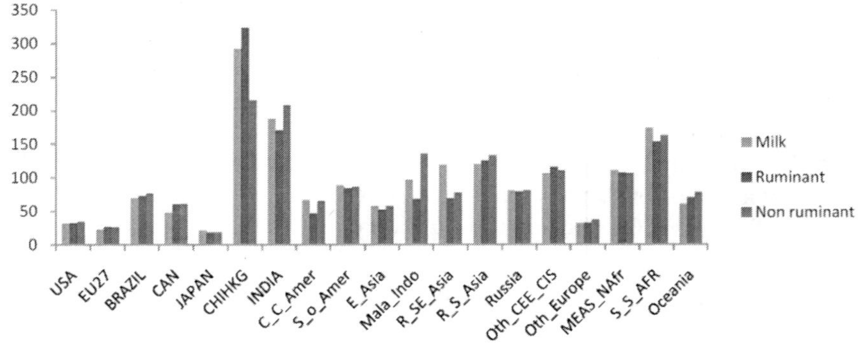

Figure 4. Projected growth (in percentage) in the regional livestock output (2004-2022).

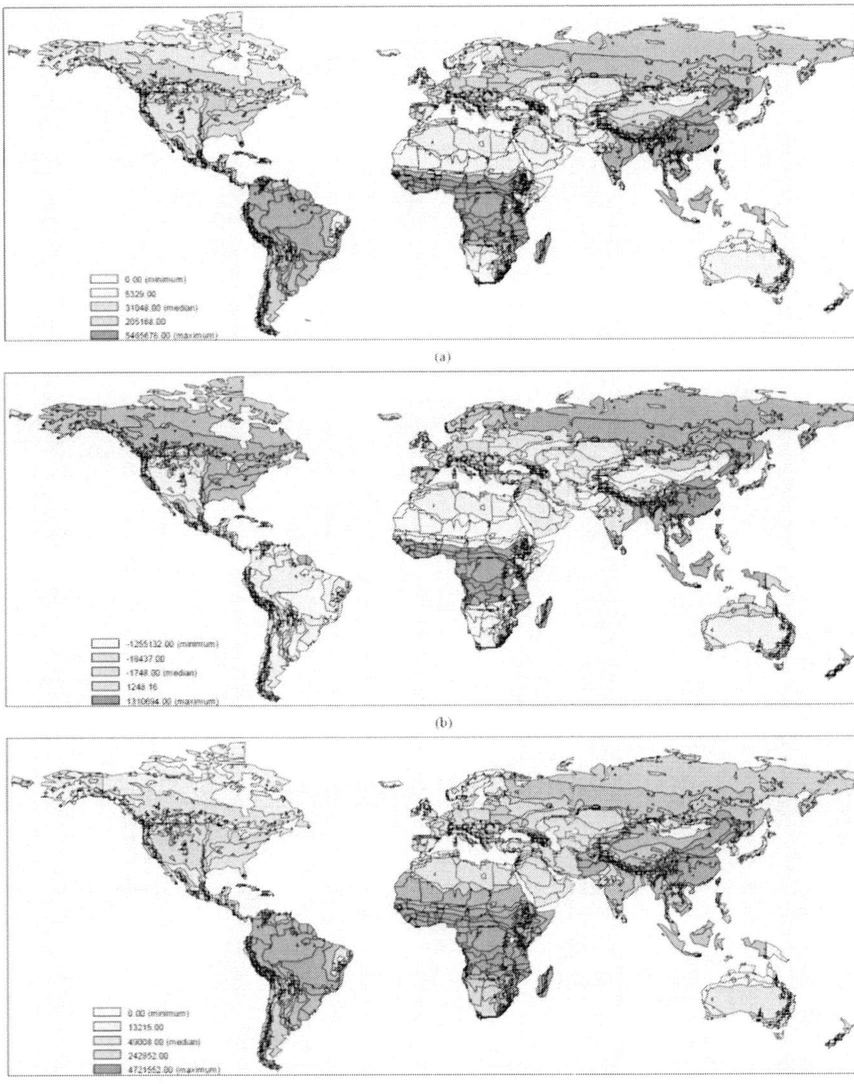

Figure 5. (**a**) Decrease in forest land due to increase in livestock demand (2004-2022); (**b**) Change in crop land due to increase in livestock demand (2004-2022). Negative values are for decrease in cropland and positive values indicate increase in cropland; (**c**) Increase in pastureland due to increase in livestock demand (2004-2022).

Table 2. Projected change in land use (in hectare) induced by change in the demand for livestock outputs (2004-2022).

GTAP-BIO regions	Land use changes		
	Forest	Cropland	Pasture
USA	(280,016)[a]	(26,032)	305,936
EU27	(121,680)	(7488)	129,212
BRAZIL	(4,204,000)	(430,460)	4,634,288
CAN	(113,376)	41,208	72,114
JAPAN	(7640)	(1861)	9503
CHIHKG	(18,803,440)	2,440,520	16,363,040
INDIA	(678,732)	(75,184)	753,938
C_C_Amer	(166,356)	(22,488)	188,832
S_o_Amer	(3,472,112)	(272,012)	3,744,208
E_Asia	(136,132)	(7582)	143,720
Mala_Indo	(429,544)	95,640	333,937
R_SE_Asia	(687,912)	(31,856)	719,812
R_S_Asia	(423,938)	(1,582,812)	2,006,784
Russia	(466,064)	(21,344)	487,464
Oth_CEE_CIS	(243,572)	(148,840)	392,352
Oth_Europe	(3368)	(1289)	4659
MEAS_NAfr	(13,054)	(80,708)	93,744
S_S_AFR	(13,071,280)	(907,216)	13,978,688
Oceania	(38,498)	(60,884)	99,456
World	(43,360,713)	(1,100,688)	44,461,687

[a]Values in the parenthesis are the negative changes.

4.4. Sensitivity Analysis

As with all models of this type, the results are a function of the magnitude of the parameters. Therefore, if the parameter values are altered, it is expected that the results would be altered. It is in this spirit that a sensitivity analysis is undertaken. The demand for livestock products comes directly from the preferences of those purchasing it, the final consumers. The elasticity associated with the substitution among livestock products is relaxed by making them more elastic to incorporate some change in consumer preferences over the time. The increase in elasticity value in effect amounts to an increase in price sensitivity for any individual product type.

This increase in elasticity is accomplished by altering the substitution parameter (SUBPAR) in the model. The SUBPAR parameter in the model helps determine how easily goods are substitutable in consumption. The SUBPAR values for livestock products in the model are decreased by 50% for all regions compared to the baseline case. The decrease in SUBPAR increases own and cross price compensated elasticities.

Table 3. Total induced emission (in 1000 tons CO_2 e) from land use changes due to changes in livestock output (2004- 2022).

GTAP-BIO Regions	Land Conversion Sequences[b]				Total
	F-to-C	C-to-P	C-to-CP	F-to-P	
USA	19,716	−3774	−66,315	62,612	12,239
EU27	404	−1128	0	25,155	24,431
BRAZIL	0	−77,551	−163,062	2,907,306	2,666,693
CAN	21,047	−276	0	19,294	40,064
JAPAN	505	−702	0	1469	1272
CHIHKG	1,783,421	−33,138	0	4,898,689	6,648,973
INDIA	98,341	−16,468	0	277,917	359,791
C_C_Amer	0	−5465	0	92,577	87,112
S_o_Amer	164	−63,217	0	1,815,169	1,752,116
E_Asia	216	−637	0	14,834	14,414
Mala_Indo	255,131	−9428	0	250,385	496,087
R_SE_Asia	17,938	−11,469	0	406,877	413,346
R_S_Asia	179,956	−83,821	0	103,653	199,788
Russia	19,241	−7282	0	44,155	56,114
Oth_CEE_CIS	0	−11,919	0	47,571	35,653
Oth_Europe	1	−260	0	437	177
MEAS_Nafr	0	−6064	0	4972	−1,092
S_S_Afr	382,294	−113,408	0	6,678,720	6,947,606
Oceania	2174	−3193	0	15,212	14,193
World	2,781,398	−437,208	−229,377	17,688,554	19,768,978

[b]F = forest, C = cropland, P = pasture, and CP = cropland-pasture (land that can alternate between crop cultivation and pasture).

Figure 6[4] and Figure 7 present the percentage point difference in the global and regional livestock outputs when SUBPAR is reduced. Compared to the baseline simulations, with the reduced parameter values, the global livestock output increased for all categories. The increased price sensitivity in consumption to both own-price and cross-price changes significantly increased the global output for non-ruminants (**Figure 6**). At the regional level, INDIA, CHIHKG, MEAS_NAfr, Oth_CEE_CIS, R_S_Asia, Russia, S_O_Amer, and BRAZIL are the regions with the largest increase in non-ruminant output compared to the baseline scenario (**Figure 7**).

Table 4 shows the difference in land use change with higher price sensitivity. The results on the "A" columns are the baseline results, the "B" columns are the price sensitive results, and the "C" columns list the difference between two scenarios (C = B − A). With the increased price sensitivity, about 3.8 million hectares of forest are spared, while both cropland and pasture area decline by just over 3 million hectares and 767 thousand hectares, respectively. Even with the reduction in crop and pasture land, livestock output is higher at the global level for the price sensitive scenario compared to the baseline (**Figure 6**).

Table 4. Impact of higher substitution in livestock consumption on land use chage (in hectare).

GTAP-BIO regions	Land use change due to change in the demand for livestock output obtained from in the baseline scenarios (A)			Land use change due to change in the demand for livestock output obtained from the reduced SUBPAR parameters scenarios (B)			Difference in land use change due to reduced SUBPAR (C = B − A)		
	Forest	Cropland	Pasture	Forest	Cropland	Pasture	Forest	Cropland	Pasture
USA	(280,016)	(26,032)	305,936	(256,992)	(48,096)	305,104	23,024	(22,064)	(832)
EU27	(121,680)	(7488)	129,212	(114,896)	(20,176)	135,004	6784	(12,688)	5792
BRAZIL	(4,204,000)	(430,460)	4,634,288	(4,568,416)	(527,912)	5,096,464	(364,416)	(97,452)	462,176
CAN	(113,376)	41,208	72,114	(107,440)	32,292	75,206	5936	(8916)	3092
JAPAN	(7640)	(1861)	9503	(7618)	(2860)	10,475	22	(999)	973
CHIHKG	(18,803,440)	2,440,520	16,363,040	(15,675,056)	674,496	15,000,576	3,128,384	(1,766,024)	(1,362,464)
INDIA	(678,732)	(75,184)	753,938	(481,876)	(271,136)	753,042	196,856	(195,952)	(896)
C_C_Amer	(166,356)	(22,488)	188,832	(172,464)	(21,672)	194,184	(6108)	816	5352
S_o_Amer	(3,472,112)	(272,012)	3,744,208	(3,421,528)	(325,448)	3,746,944	50,584	(53,436)	2736
E_Asia	(136,132)	(7582)	143,720	(126,620)	(14,378)	140,992	9512	(6796)	(2728)
Mala_Indo	(429,544)	95,640	333,937	(102,092)	(196,864)	298,957	327,452	(292,504)	(34,980)
R_SE_Asia	(687,912)	(31,856)	719,812	(489,032)	(184,228)	673,275	198,880	(152,372)	(46,537)
R_S_Asia	(423,938)	(1,582,812)	2,006,784	(98,976)	(1,295,420)	1,394,360	324,962	287,392	(612,424)
Russia	(466,064)	(21,344)	487,464	(389,280)	(82,136)	471,360	76,784	(60,792)	(16,104)
Oth_CEE_CIS	(243,572)	(148,840)	392,352	(205,360)	(197,128)	402,688	38,212	(48,288)	10,336
Oth_Europe	(3368)	(1289)	4659	(3316)	(1515)	4840	52	(226)	181
MEAS_NAfr	(13,054)	(80,708)	93,744	(13,435)	(100,072)	113,520	(381)	(19,364)	19,776
S_S_AFR	(13,071,280)	(907,216)	13,978,688	(13,245,584)	(1,524,080)	14,769,792	(174,304)	(616,864)	791,104
Oceania	(38,498)	(60,884)	99,456	(34,076)	(73,100)	107,136	4422	(12,216)	7680
World	(43,360,713)	(1,100,688)	44,461,687	(39,514,057)	(4,179,433)	43,693,919	3,846,657	(3,078,745)	(767,767)

Table 5. Impact on induced emissions (in 1000 tons CO_2 eq.) with the reduced SUBPAR.

GTAP-BIO regions	Land conversion sequences					Emissions with the new SUBPAR (D)	Emissions in the baseline simulations (E)	Difference in emissions (D-E)
	F-to-C	C-to-F	C-to-P	C-to-CP	F-to-P			
USA	14.074	0	−4322	−74.787	59.772	−5263	12.239	−17.502
EU27	88	0	−2813	0	23.886	21.161	24.431	−3270
BRAZIL	0	0	−95.183	−187.099	3.158.970	2.876.688	2.666.693	209.995
CAN	17.645	0	−324	0	19.757	37.079	40.064	−2986
JAPAN	345	0	−879	0	1528	994	1272	−278
CHIHKG	768.930	0	−38.790	0	4.494.995	5.225.136	6.648.973	−1.423.837
INDIA	26.252	0	−28.145	0	226.234	224.342	359.791	−135.449
C_C_Amer	0	0	−5179	0	95.967	90.789	87.112	3677
S_o_Amer	3	0	−75.799	0	1.789.402	1.713.605	1.752.116	−38.510
E_Asia	114	0	−1191	0	13.351	12.273	14.414	−2141
Mala_Indo	0	0	−47.713	0	86.104	38.390	496.087	−457.697
R_SE_Asia	484	0	−37.947	0	298.009	260.546	413.346	−152.800
R_S_Asia	9134	0	−67.321	0	46.576	−11.611	199.788	−211.400
Russia	8194	0	−9383	0	39.032	37.844	56.114	−18.270
Oth_CEE_CIS	0	0	−17.162	0	40.151	22.989	35.653	−12.664
Oth_Europe	0	−12	−303	0	434	119	177	−59
MEAS_Nafr	0	0	−7369	0	5125	−2244	−1092	−1153
S_S_Afr	122.084	0	−137.012	0	6.958.418	6.943.490	6.947.606	−4115
Oceania	32	0	−3690	0	14.887	11.230	14.193	−2963
World	967.380	−12	−580.524	−261.885	17.372.597	17.497.556	19.768.978	−2.271.422

While most regions had a decline in deforestation under the price sensitive scenario, four regions had an increase, which includes BRAZIL, C_C_Amer, MEAS_NAfr, and S_S_AFR. The regions having the most reduction in deforestation are CHIHKG, INDIA, Mal_Indo, R_SE_Asia, and Rest of South Asia (R_S_Asia). Most of these regions are also the regions with the largest decrease in the cropland area (**Table 4**).

Table 5 compares total emissions from the baseline simulation to the emissions of the price sensitive scenario. At the global level, the total emissions with the reduced SUBPAR are about 17.5 billion tons of CO_2 e, approximately 2.3 billion tons less (about 11% less) than the baseline simulation. The 11% drop in emissions is mainly due to the decrease in deforestation and the increased conversion of crop land for pasture (Table 3 and Table 5). The regions with the substantial reduction in emissions are CHIHKG, Mala_Indo, R_SE_Asia, R_S_Asia, and INDIA (Table 5). The results indicate that increased price sensitivity in consumption of livestock products can lead to a substantial reduction in the GHG emissions from the sector.

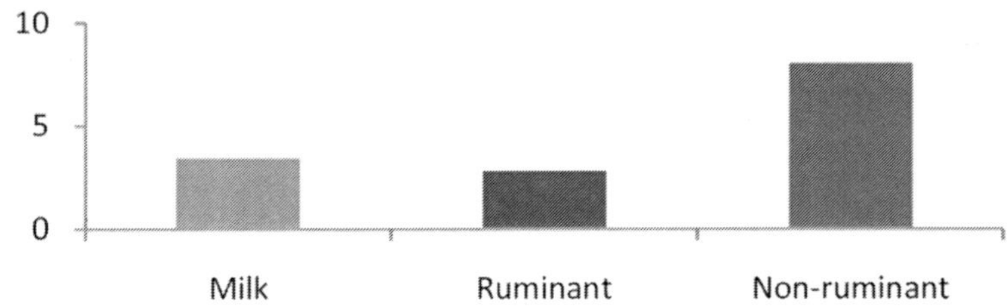

Figure 6. Percentage point difference in the global livestock output growth with the reduced SUBPAR compared to the baseline simulation.

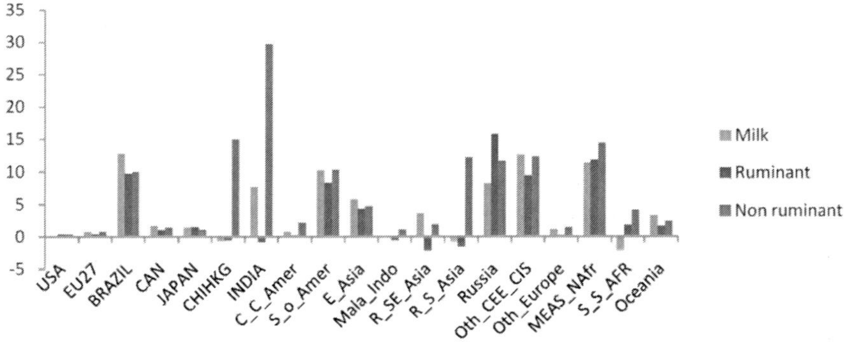

Figure 7. Percentage point difference in the regional livestock output growth with the reduced SUBPAR compared to the baseline simulation.

5. CONCLUSIONS

Based on regional projections of GDP, population, capital, and skilled and unskilled labor, demand for livestock outputs are forecasted for the period between the years 2004 to 2022. Globally, the demand for non-ruminant output increases the most. Regionally, this expansion is the most evident in fast-growing economies such as CHIHKG and INDIA.

Changes in land use as a result of the growth in livestock output are estimated to be large with a loss of forest amounting to over 43.3 million hectares, a reduction in crop land of about 1 million hectares, and an increase in pasture of over 44 million hectares. Given that the forests sequester more carbon than other land uses, clearing them results in significant GHG emissions. The change in land use due to increased livestock production increases emissions by about 20 billion tons of CO_2 e between 2004 and 2022 or about an average of 1.1 billion tons annually, which is about 15.5% of the total emissions (7.1 billion tons of CO_2 e) from livestock sector as estimated by [2] and about 2.2% of the total human-induced GHG emissions (49 billion tons of CO_2 e) as estimated by the Intergovernmental Panel on Climate Change for the year 2004 [15]. When consumer's responses to price changes are made more elastic i.e. more sensitive to price changes and willing to substitute more readily among livestock products, there is more increase in demand for livestock output especially the non-ruminant products with significant reduction in deforestation and GHG emissions.

The results from this study indicate that there is a potential for significant reduction in GHG emissions from livestock sector through policy interventions that target the consumption pattern of the private households. For this purpose, the intervention should encourage increased substitution among livestock products. In particular, the policies should encourage consumers to substitute non-ruminant for ruminant meat products.

ACKNOWLEDGEMENTS

The authors are grateful for the financial support from the University of Nebraska-Lincoln and Fort Hays State University.

REFERENCES

1. Caro, D., Davis, S. and Bastianoni, S. (2014) Global and Regional Trends in Greenhouse Gas Emissions from Livestock. Climatic Change, 126, 203-216. http://dx.doi.org/10.1007/s10584-014-1197-x
2. Gerber, P., Steinfeld, H., Henderson, B., Mottet, A., Opio, C., Dijkman, J., Falcucci, A. and Tempio, G. (2013) Tackling Climate Change through Livestock—A Global Assessment of Emissions and Mitigation Opportunities. Food and Agriculture Organization of the United Nations (FAO), Rome.
3. Steinfeld, H., Gerber, P., Wassenaar, T., Castel, V., Rosales, M. and de Haan, C. (2006) Livestock's Long Shadow: Environmental Issues and Options. Food and Agriculture Organization of the United Nations (FAO), Rome.
4. FAO (2011) World Livestock 2011—Livestock in Food security. Food and Agriculture Organization of the United Nations (FAO), Rome.
5. Herrero, M., Gerber, P., Vellinga, T., Garnett, T., Leip, A., Opio, D., Westhoek, H., Thornton, P., Olsen, J., Hutchings, N., Montgomery, H., Soussana, J., Steinfeld, H. and McAllister, T. (2011) Livestock and Greenhouse Gas Emissions: The Importance of Getting the Numbers Right. Animal Feed Science and Technology, 166-167, 779-782. http://dx.doi.org/10.1016/j.anifeedsci.2011.04.083
6. Goodland, R. and Anhang, J. (2009) Livestock and Climate Change: What If the Key Actors in Climate Change Are Cows, Pigs, and Chicken? World Watch, November/December.
7. Hertel, T. (1997) Global Trade Analysis Project: Modeling and Applications. Cambridge University Press, Cambridge.
8. Hertel, T., Golub, A., Jones, J., O'Hare, M., Plevin, R. and Kammen, D. (2010) Effects of U.S. Maize Ethanol on Global Land Use and Greenhouse Gas Emissions: Estimating Market Mediated Responses. Bioscience, 60, 223-231. http://dx.doi.org/10.1525/bio.2010.60.3.8
9. Tyner, W., Taheripour, F., Zhuang, Q. and Baldos, U. (2010) Land Use Changes and Consequent CO2 Emissions Due to U.S. Corn Ethanol Production: A Comprehensive Analysis. Department of Agricultural Economics, Purdue University, Indiana.
10. Taheripour, F., Hertel, T. and Tyner, W. (2011) Implications of Biofuels Mandate for the Global Livestock Industry: A Computable General Equilibrium Analysis. Agricultural Economics, 42, 325-342. http://dx.doi.org/10.1111/j.1574-0862.2010.00517.x
11. Taheripour, F. and Tyner, W. (2013) Biofuels and Land Use Change: Applying Recent Evidence to Model Estimates. Applied Sciences, 3, 14-38. http://dx.doi.org/ 10.3390/app3010014
12. Plevin, R., Gibbs, H., Duffy, J., Yui, S. and Yeh, S. (2014) Agro-Ecological Zone Emission Factor (AEZ-EF) Model (v47): A Model of Greenhouse Gas Emissions from Land-Use Change for Use with AEZ-Base Economic Models. GTAP Technical Paper No. 34. Global Trade Analysis Center, Purdue University, Indiana.
13. Foure', J., Benassy-Que'ere', A. and Fontangne', L. (2012) The Great Shift: Macroeconomic Projections for the World Economy at the 2050 Horizon. Centre d'études Prospectives et d'Informations Internationales (CEPII), France. http://www.cepii.fr/francgraph/doctravail/resumes/wp_resume.asp?annee=2012&ref=3&NoDoc=4179
14. Chappuis, T. and Walmsley, T. (2011) Projections for World CGE Model Baselines. GTAP Research Memorandum No. 22. Center for Global Trade Analysis, Purdue University, Indiana.
15. IPCC (2007) Climate Change 2007: Synthesis Report. Contribution of Working Groups I, II and III to the Fourth Assessment Report of the Intergovernmental Panel on Climate Change, Intergovernmental Panel on Climate Change, Geneva.

END NOTES

1. Livestock output demand from the government is also another source. Given that the share of government's demand in the total demand for the livestock output is negligible, this category of demand is not considered in the analysis.
2. The GTAP-BIO results (except for the land use change) are presented in proportionate change, hence, the results on livestock output changes are in percentage change rather than the absolute change.
3. For this purpose regional land use change results similar to the ones presented in **Table 2** and **Table 4** but disaggregated to the 18 AEZ levels of the GTAP-BIO model are used in the "AEZ_EF V47" model.
4. **Figure 6** shows the percentage point difference in global livestock output growth for price sensitive scenario compared to the baseline scenario. This figure is not directly comparable to **Figure 3**, which shows percentage change in global livestock output for the baseline scenario.

CHAPTER 5

Livestock metabolomics and the livestock metabolome: A systematic review

Seyed Ali Goldansaz[1,2], An Chi Guo[2,3], Tanvir Sajed[3], Michael A. Steele[1], Graham S. Plastow[1], David S. Wishart[2,3*]

1 Department of Agriculture, Food and Nutritional Sciences, University of Alberta, Edmonton, Alberta, Canada

2 Department of Biological Sciences, University of Alberta, Edmonton, Alberta, Canada,

3 Department of Computing Science, University of Alberta, Edmonton, Alberta, Canada

ABSTRACT

Metabolomics uses advanced analytical chemistry techniques to comprehensively measure large numbers of small molecule metabolites in cells, tissues and biofluids. The ability to rapidly detect and quantify hundreds or even thousands of metabolites within a single sample is helping scientists paint a far more complete picture of system-wide metabolism and biology. Metabolomics is also allowing researchers to focus on measuring the end-products of complex, hard-to-decipher genetic, epigenetic and environmental interactions. As a result, metabolomics has become an increasingly popular "omics" approach to assist with the robust phenotypic characterization of humans, crop plants and model organisms. Indeed, metabolomics is now routinely used in biomedical, nutritional and crop research. It is also being increasingly used in livestock research and livestock monitoring. The purpose of this systematic review is to quantitatively and objectively summarize the current status of livestock metabolomics and to identify emerging trends, preferred technologies and important gaps in the field. In conducting this review we also critically assessed the applications of livestock metabolomics in key areas such as animal health assessment, disease diagnosis, bioproduct characterization and biomarker discovery for highly desirable economic traits (i.e., feed efficiency, growth potential and milk production). A secondary goal of this critical review was to compile data on the known composition of the livestock metabolome (for 5 of the most common livestock species namely cattle, sheep, goats, horses and pigs). These data have been made available through an open access, comprehensive livestock metabolome database (LMDB, available at http://www.lmdb.ca). The LMDB should enable livestock researchers and producers to conduct more targeted metabolomic studies and to identify where further metabolome coverage is needed.

1. INTRODUCTION

Metabolites are sometimes referred to as the "canaries" of the genome [1]. Just as canaries for coalminers served as sensitive indicators of problems in coal mines, metabolites can be exquisitely sensitive indicators of problems in the genome (as well as the transcriptome or proteome). Metabolites are effectively the end products of complex interactions occurring inside the cell (the genome) and events, exposures or phenomena occurring outside the cell or organism (the environment). As a result, the comprehensive measurement of metabolites (via metabolomics) allows one to determine interactions between genes and the environment. In other words, metabolomics allows researchers to obtain a highly sensitive and more complete description of the phenotype [2; 3]. This metabolic readout of the phenotype is often called the "metabotype" [4]. Recent advances in both analytical chemistry and metabolite data analysis techniques are now making metabolomics far more accessible to a wider range of research disciplines. Indeed, metabolomics is now routinely used in biomedical research (for biomarker discovery and disease mechanism research), food and nutritional analysis, crop characterization and environmental monitoring [5; 6; 7; 8]. As a result, the field of metabolomics has experienced very rapid growth with just two papers published on the subject in 1999 to more than 2400 in 2015.

However, unlike in other areas of agriculture research where metabolomics is widely used in crop trait selection, pesticide monitoring, crop breeding or crop evaluation [9; 10; 11; 12], the application of metabolomics to livestock research is somewhat less widely used or appreciated. This is surprising given the potential of metabolomics to address many important questions in livestock and animal science. In particular, the power of metabolomics to non-invasively detect subtle phenotypic changes, innate phenotypic propensities and dietary responses makes it an ideal tool for livestock research, breeding and assessment [13; 14; 15; 16; 17; 18; 19]. Recently, there have been a number of papers in livestock metabolomics that have generated compelling results showing how metabolomics and metabolite-based phenotyping (metabotyping) can help farmers, veterinarians, livestock researchers and the livestock industry. These include papers demonstrating how metabolomics can be used to predict feed efficiency and residual feed intake (RFI) [20], ascertain disease propensity [21; 22; 23], evaluate dietary responses to different feeds [24; 25], assess carcass merit [26; 27; 28], fertility [29], milk quality [30; 31], determine bioproduct content [32] and ascertain other important economic or breeding traits associated with livestock.

Fast, effective, and quantitative phenotyping is critical for farm trials dealing with animal selection and breeding. Many traditional phenotypic measurements such as those related to animal feed consumption and RFI are expensive, time consuming and require specific recording equipment [20]. Others, such as carcass trait evaluation, may require animal slaughter, which obviously eliminates the potential breeding value of the animal. Similarly for reproductive traits, animals have to reach a stage of maturity and sexual activity to allow measurement of related traits. Metabolomics allows many of these trait measurements to be conducted earlier, more routinely, non-invasively and often at a lower cost than current techniques [33; 4]. However, metabolomics is not without its challenges. Metabolomic experiments must be carefully designed as diet and other variables such as sex, diurnal variations and sampling time can profoundly affect results. Likewise, metabolomic technologies, such as gas chromatography (GC), mass spectrometry (MS) and nuclear magnetic resonance (NMR) spectroscopy are not yet widely available in many livestock research facilities. Furthermore, there continues to be a significant shortage of data resources that could facilitate the interpretation of livestock metabolomic data.

Given the many applications of metabolomics in both the livestock industry and livestock research as well as the diversity of journals in which livestock metabolomics is often published, we felt it was important to conduct a thorough, systematic review of the field. By consolidating the results from diverse journals and different studies into a single review paper, we believed this content would provide a more complete picture of both the strengths and the weaknesses of livestock metabolomics. In conducting this review we sought answers to 4 key questions: 1) What are the most common applications of metabolomics in animal science and where are they trending?, 2) What are the preferred metabolomics technologies in livestock metabolomics and how are they evolving?, 3) What are the most obvious gaps or weaknesses in livestock metabolomics relative to other fields of metabolomics research? and 4) What are the known or measured metabolites for the 5 major livestock species (i.e., bovine, ovine, caprine, equine, and porcine) in different tissues and biofluids? This metabolite compilation, which we have called the livestock metabolome database or LMDB (available at http://www.lmdb.ca), is intended to help lay a more solid foundation in terms of data resources that would make livestock metabolomic studies much easier to perform, analyze and compare. The LMDB catalogues all metabolite compounds that have ever been identified and reported in the 5 livestock species (for multiple biofluids and tissues), along with concentration ranges, compound descriptions, chemical structures, reference NMR and MS spectra and other information associated with each metabolite for both healthy and a variety of abnormal physiological conditions.

2. MATERIALS AND METHODS

In compiling this review and assembling the livestock metabolome database, we used a combination of web-accessible data mining tools along with manual curation to survey 2313 peer reviewed journal articles covering the period from 1930 to 2015. From this initial set of articles, we reduced the number further to cover published livestock papers reporting the measurement or characterization of ≥8 metabolites for any of the 5 major livestock species (i.e., bovine, ovine, caprine, equine, and porcine). This reduced the target number of peer-reviewed manuscripts to a total of 149. The livestock species selected for this review were based on their global population, economic impact and use in agricultural systems [34; 35]. Details regarding the keyword selection, search engines and databases, journals and search strategy are given below and summarized in the preferred reporting items for systematic reviews and meta-analysis (PRISMA) checklist (S1 Table) and flow chart (Fig 1).

2.1 Keyword selection

As noted above, this review is focused on 5 main livestock species including cattle, sheep, goats, horses and pigs. Therefore, a combination of keywords was selected to target those specific animals and to identify the associated metabolomics studies. Keywords were divided into 3 main groups: 1) animal species, 2) sample types, and 3) metabolomic methods. Selected keywords for animal species included the name of the species and its various derivatives or synonyms, i.e., bovine, cattle, cow, calf, *Bos taurus*, etc. To target metabolomics papers in animal science, a broad range of metabolomics keywords were identified and used. These included different variations of the term "metabolomics" (such as metabolomics, metabonomics, metabolite profiling, metabolite fingerprint, chemical profile, chemical analysis, chemical composition, etc.) to target publications prior to and after 1999, as well as the names of various analytical platforms (i.e., NMR, mass spectrometry, liquid chromatography, gas chromatography-MS, etc.). Moreover, a wide variety of sample types such as different body fluids (i.e., serum, blood, plasma, urine etc.) and different organs or tissues were selected to further enrich the keyword search.

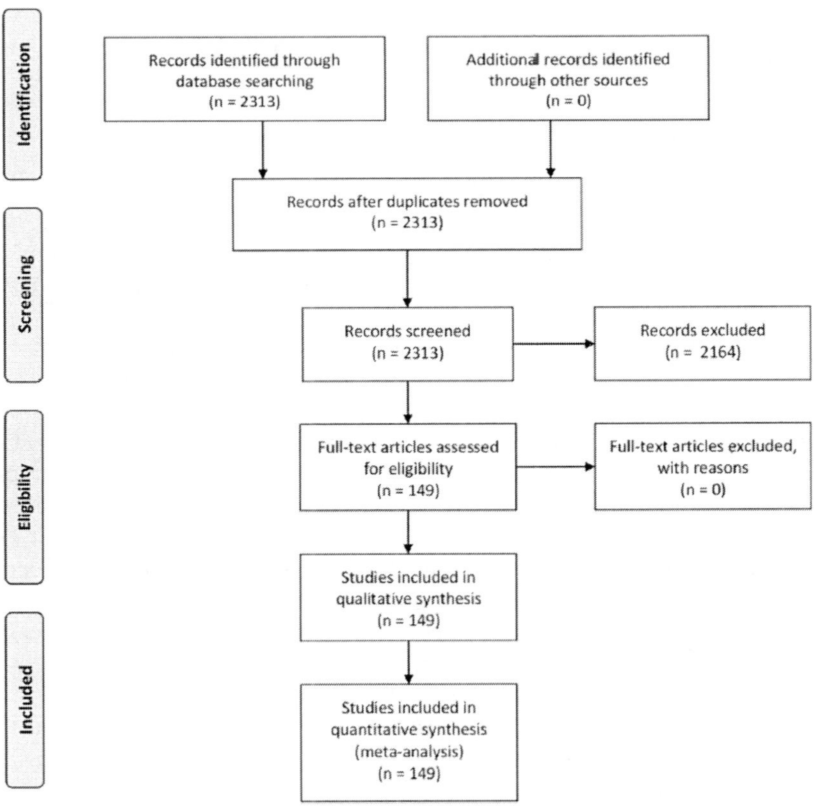

Figure 1. PRISMA diagram. The preferred reporting items for systematic reviews and meta-analysis (PRISMA) flow diagram identifies the total number of articles initially surveyed, the number of articles included and excluded for this systematic review. From: Moher D, Liberati A, Tetzlaff J, Altman DG, The PRISMA Group (2009). Preferred Reporting Items for Systematic Reviews and Meta-Analyses: The PRISMA Statement. PloS Med 6(7): e1000097. doi:10.1371/journal.pmed1000097. **FOR MORE INFORMATION, VISIT** www.prisma-statement.org.

2.2 Search engines and databases

An initial comparison among many open access search engines showed that most search results are similar regardless of the search engine used. Therefore, (https://scholar.google.ca/) was selected as the primary literature search engine. In addition, a number of agriculture-specific databases such as Agricola and AGRICULTUREnetBASE were also used. Other databases included Scopus, the Web of Science, ScienceDirect (http://www.sciencedirect.com/) and PubMed (http://www.ncbi.nlm.nih.gov/pubmed). Settings for all search engines and databases were adjusted to increase search efficiency and filter irrelevant results.

2.3 Search methods and selection criteria

Different keywords were combined to target metabolomics papers in the field of animal science. For example, "cattle", "cattle serum" or "cattle milk" was accompanied with "metabolomics", "chemical composition" or "metabolite profiling". Consequently, each combination of the keywords in the search engines generated a long list of results. These included various types of publications (full papers or abstracts) that contained any or all of the used keywords. A manual review was performed on all retrieved publications. Typically, the first 3–5 pages of the search results from the aforementioned search engines were manually reviewed to select for articles of interest. Among the papers identified as worth pursuing, research papers, abstracts or textbooks that showed relevance in their title or abstract were selected. In addition to papers reporting experimental results, res that included specific metabolite data sets were also selected. Among the selected manuscripts, only those papers that reported ≥8 metabolites were chosen for this review. The threshold of 8 or more as the minimum number of metabolites was based on a *post hoc* analysis of the retrieved papers and the need to optimize both metabolite coverage and the time devoted to manual analysis. We also determined that this selection cut-off allowed us to cover most, if not all, of metabolites reported in papers with <8 metabolites. Based on these criteria, a total number of 149 manuscripts covering all 5 animal categories were selected for this review. Selected publications were carefully read to extract and annotate a set of 10 pieces of information including: 1) metabolite names; 2) tissue or biofluid origin; 3) quantified values (concentration) if any; 4) experimental conditions; 5) animal breed; 6) sample size; 7) analytical platform; 8) field of research, 9) physiological condition (disease or state of health), and 10) Pubmed/DOI references.

2.4 Compilation of the livestock metabolome database

In compiling the data for this livestock metabolome database or LMDB (http://www.lmdb.ca), all reported concentrations were transformed into a standardized concentration unit (micromolar; μM) and each entry was associated with an abbreviated description of the experimental context, the sample type, and the methodologies used for the metabolomic analyses. In identifying a metabolite for inclusion in this study the compound had to: 1) have a molecular weight <1500 daltons; 2) it could not be a peptide, protein or oligonucleotide; 3) it had to correspond to a reasonably unique chemical entity (triglycerides and amino acids are not unique chemical entities, but LysoPC-16:2 is sufficiently unique) and 4) it had to be identified with a structurally interpretable name. This literature-based effort generated 1070 metabolites from 149 peer-reviewed papers, abstracts or textbooks. Metabolites extracted from these manuscripts were systematically categorized into the LMDB. Nearly all metabolites extracted were linked to a standard Human Metabolome Database (HMDB) identifier [36; 37; 38] which provides a freely-accessible comprehensive description of each metabolite. A brief description of experimental data for each metabolite was also extracted from the articles and included in the database including information on the analytical platform, experimental conditions and field of research. A PubMed and/or DOI id was also associated with each metabolite to provide a link to the article reporting that metabolite. Additional data on each metabolite, including structure, synonyms, chemical classifications, physicochemical data, reference NMR, GC-MS or LC-MS spectra and links to other databases were obtained through an in-house annotation tool called DataWrangler. All of this information was used to construct the on-line version of the LMDB (http://www.lmbd.ca). The LMDB was prepared using a Ruby-on-Rails [39] framework, modeled after other on-line species-specific metabolomic databases prepared in our laboratory. Details regarding their construction, required operating systems, browser compatibility and hardware requirements can be found elsewhere [36; 40; 41].

3. RESULTS AND DISCUSSION

3.1 Growth and trends in livestock metabolomics research

Based on the data collected from our literature survey, it is clear that the majority of metabolomics studies among all livestock categories have been conducted in cattle (Fig 2) with a total of 76 articles (50% of the selected articles) focusing on various fields of bovine research and assessment.

Metabolomics studies on pigs and sheep came second and third with 28% and 12% of the selected articles, respectively. The least studied group were horses with only 5 (3%) reported equine metabolomic studies. As might be expected, most livestock metabolomic studies focused on issues related to animal health, nutrition and production (65%). These studies are obviously useful for characterizing bioproduct quality, identifying biomarkers or understanding animal responses to different stressors. However, we were surprised to see relatively few efforts focused on metabolomic characterization of healthy animals with the aim of identifying baseline values for different metabolites in different biofluids or tissues. In fact, only 16 studies (10%) of this kind were reported. These "referential surveys" are foundational and are often needed before biomarker studies could/should be undertaken or fully understood.

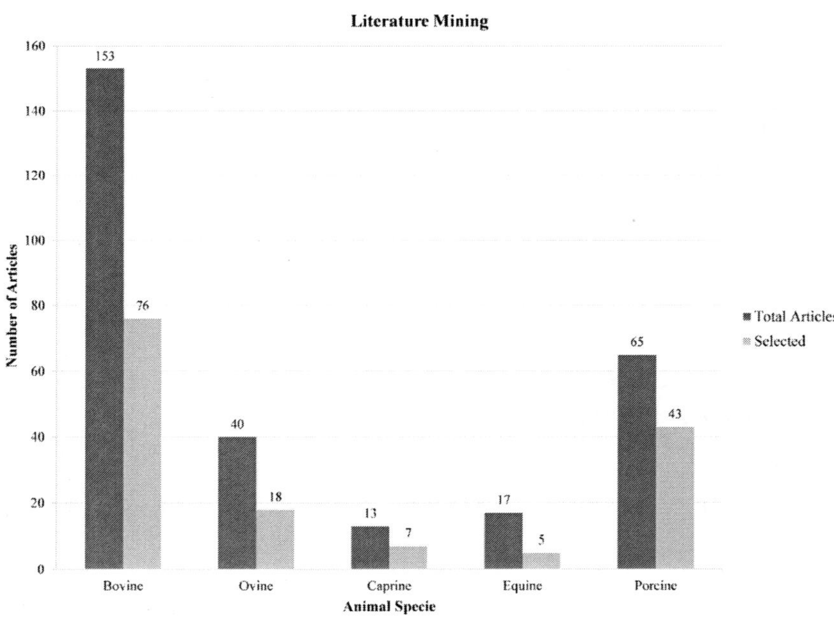

Figure 2. Literature mining. Total number of livestock metabolomics articles considering only articles that reported ≥8 metabolites resulted in selection of 149 manuscripts for this review.

As noted earlier, most metabolomic studies of cows, sheep, goats and pigs appear to be directed towards disease detection, production and bioproduct assessment, feed efficiency determination and reproduction. In contrast, the primary focus for equine metabolomics has been on drug discovery and doping detection, specifically for thoroughbred horses [42].

Given the large sums of money directed to horse racing, this is not unexpected. However, compared to the widespread applications of metabolomics in other livestock species for other purposes, it is clear that equine metabolomics is being under-utilized. Certainly, equine metabolomics could be used to select more desirable traits and higher value or higher performing animals, similar to what is being done for bovine metabolomics. Likewise, metabolomics could serve as a diagnostic or prognostic tool for improving equine health and disease resilience (as it has for essentially all other livestock species).

Temporal categorization of all 149 published studies showed that the majority of livestock metabolomics papers were published after 1999. Less than 9% (13 articles) of the selected papers were published prior to 1999 while, ~91% (136 articles) of the papers were published thereafter. The earliest paper in our collection dates from 1930. It is noteworthy that the term "metabolomics" was not coined until 1998 [43; 44] therefore, metabolomics studies prior to this date had to be identified using other keywords such as "chemical composition", "biochemical profiling", etc. Based on our observations, it is clear that interest in livestock metabolomics is growing rapidly, especially over the last couple of years. Our data indicates that from 2000–2010 just 29 articles (19%) were published in this field, while from 2011–2015 a total of 107 (72%) articles were published. In terms of percentage growth, the most rapidly expanding subfield appears to be caprine and equine metabolomics with a growth rate of 100% over the past 5 years. In terms of overall growth, the most significant changes were in bovine metabolomics with the number of papers growing from just 10 prior to 1999 to 49 in 2011–2015. The most recent additions to the field of livestock metabolomics are studies focused on goats (starting in 2014) and horses (starting in 2007).

3.2 Trends and gaps in livestock metabolomics applications

We found that livestock metabolomics studies can be categorized in 7 main areas (Table 1). These include animal health, animal nutrition, animal production, animal reproduction, animal physiology (mainly analysis of different biofluids), animal products (products originating from livestock such as milk, meat, yogurt, etc.), and human health (livestock models used for human health studies). This general categorization was based on a *post hoc* analysis of the types of articles where we manually assessed article keywords, subject headings, journal titles and the general focus of each article. Most of these categorizations (such as animal reproduction, human health and animal health) were relatively simple to make. For instance, the category "animal reproduction" obviously refers to articles using metabolomics to study reproduction in livestock. Likewise, the category "animal health" refers to articles using metabolomics to study livestock health or disease while "human health" refers to application of metabolomics to study human disease using various livestock models. Other categories proved to be somewhat more ambiguous. For instance, the field of "animal products" typically contains metabolomics investigations related to food, nutrition and human consumption of animal products, such as meat and cheese. On the other hand, "animal production" is focused on investigating the associated biochemical profile with each animal product. In some cases, we had to be fairly strict with our definitions. For instance, we limited "animal physiology" to include only those articles focused on analyzing various biofluids or characterizing the metabolite composition of specific biofluids, organs and tissues.

Table 1. Categorical comparison. Selected livestock metabolomics articles of 5 major livestock species were categorized based on the area of research, i.e., animal health, animal nutrition, animal production, animal reproduction, human health, animal physiology and animal products. It is noteworthy that articles in the area of human health mainly reflected animal models being used to study human related health issues.

	Bovine	Ovine	Caprine	Equine	Porcine
Animal Health	30	6	2	4	10
Animal Nutrition	10	6	5	0	14
Animal Production	22	3	2	2	11
Animal Reproduction	2	1	0	0	3
Human Health	6	4	2	0	14
Animal Products	16	1	0	0	2
Animal Physiology	13	2	0	1	0

Among the seven different categories, animal health (52) and animal production (40) had the most metabolomics articles published for the largest number of animal groups (Table 1). However, this varied depending on the livestock species being studied. In human health research, porcine metabolomic studies covered the majority of articles (14 articles) compared to all other livestock categories. This is not unexpected, given the comparable physiology of pigs to that of humans [45]. In the category of animal products, bovine-based studies had the most articles published (16 articles) relative to all other groups. Some of the more interesting applications of metabolomics found in our survey include the use of metabolomics for quality control of animal products [46; 47], evaluating nutritional value and impact of various feed sources on animal health and products [25], investigating disease biology by using animal models of human disease [48; 49], investigation of potential metabolite biomarkers of animal disease [22; 23], assessment of production traits [50; 51], reproductive performance [29], and general metabolome characterization [52; 53].

In terms of gaps in the existing literature, it is perhaps most useful to use bovine metabolomic studies as the "gold standard" by which to compare other livestock species. While metabolomics is routinely being used to understand the biology or diagnose a few common bovine production diseases (including acidosis, mastitis, milk fever) we found no metabolomic studies looking at common diseases in sheep or goats (such as brucellosis, campylobacteriosis, pneumonia, Q fever), in horses (equine flu, equine herpes, equine sleeping sickness, anemia, laminitis, azoturia), or in pigs (respiratory diseases, swine dysentery, parvovirus). Indeed, we found only 22 metabolomic studies focused on the health of sheep, goats, pigs and horses, compared to 30 metabolomic studies for cattle alone. Of these 22 non-bovine studies, most were focused on metabolic, growth and neurodegenerative disorders.

Livestock metabolomics studies also appear to be missing a number of opportunities currently being pursued in human biomedical research. One of particular note is the use of metabolomics to predict (as opposed to diagnose) or detect subclinical forms of disease. While disease diagnosis is useful, often it is too late or too costly to perform useful veterinary interventions. Detecting diseases before they manifest or predicting them before they occur allows inexpensive prophylactic or preventative measures to be taken. In human metabolomic studies, the identification of disease prediction biomarkers is becoming increasingly common [54; 55; 56; 57; 58]. This is because metabolic changes appear to precede significant physiological changes, possibly because metabolites play an important signalling role to activate later stage (i.e. symptomatic) physiological responses [59; 60]. However, we could only find 2 papers (limited to cattle) that focused on disease diagnosis/prognosis or (sub)clinical detection of diseases [21; 61]. A similar approach could also be used towards the prediction of later-life production traits on the basis of early-life metabolic fingerprints. This, too, is an area of interest in the field of human metabolomics, where later-life health is being predicted on the basis of early-life metabolic fingerprints [62; 63; 64]. Obviously the reliable prediction of economically important traits is an important tool for livestock management and strategic planning.

Metabolomics is already being used in the evaluation and/or prediction of production traits such as residual feed intake (RFI), carcass merit, reproductive performance and metabolic disorders for cattle. However, there is a surprising dearth of similar studies regarding evaluation or prediction of production traits for sheep, goats and pigs. Metabolomics potentially offers a unique opportunity for indirect, inexpensive marker-assisted measurement of these economical traits. This can be achieved through non-invasive sample collection of readily accessible biofluids such as blood, urine, milk and saliva. In most cases, the standard measurement or prediction of some traits such as RFI and carcass merit requires labour intensive, invasive, costly and time consuming measurements [65]. Metabolomic studies regarding the prediction of RFI in beef cattle have already been very promising with a reported initial prediction accuracy of 95% [20; 66]. Metabolomic data, when coupled with genomic data, appear to increase the accuracy of trait prediction [67]. This combination potentially allows one to screen for individual animals with superior traits that could be used for breeding stock. Given the positive results already seen for cattle, the application of these metabolomic concepts to other livestock species is certainly worth investigating. Overall it appears that there is still a considerable body of useful metabolomic work that could be pursued with most other livestock species by simply applying or extending what has already been done in bovine metabolomics.

3.3 Trends and gaps in sample size

Nearly 50% of the selected articles for all animal species used ≤30 animals or samples (from an even smaller number of animals) to conduct their metabolomics analysis. Other sample size categories shown in Table 2 account for ~10% of the peer-reviewed livestock metabolomics literature. The maximum number of samples reported from the selected papers were: 1587 (bovine), 163 (ovine), 80 (caprine), 36 (equine), and 506 (porcine). It is noteworthy that sample size does not always reflect the total number of animals used in the study. For instance, longitudinal studies typically collect multiple samples from a relatively small number of animals over an extended period of time. Relative to many reported human metabolomic studies [55] or rodent model studies [68] the number of samples and the number of subjects (i.e. animals) used in most livestock metabolomics studies is generally quite small. Indeed, many human and rodent model studies routinely measure 100s to 1000s of samples. This difference in sample size likely reflects the relatively high cost of performing large animal studies as well as the somewhat limited funding available to agriculture research relative to medical research.

However, it is important to note that the smaller sample sizes in livestock metabolomics also mean that statistical significance and "power" of the published results is also somewhat less than many human-subject or model organism studies. This represents a significant gap for livestock metabolomics and requires either study sizes to be increased or more effort being directed to conducting validation studies on similar-to-largely sized cohorts for confirmation previously reported results. Indeed, we found only one bovine metabolomic study reporting either independent cross validation (using a different animal cohort) or independent follow-up validation of any newly identified biomarkers or interesting metabolite findings [20]. On the other hand, follow-up validation studies are becoming routine in human metabolomic studies [69; 70; 71]. Clearly, this is a gap in livestock metabolomics that must be filled if metabolomic findings are going to be translated to practical pen-side or on-farm applications.

Table 2. Sample size

	Bovine	Ovine	Caprine	Equine	Porcine
≤30	25	6	4	5	30
31-50	12	2	0	0	7
51-100	9	4	3	0	1
>100	16	2	0	0	2
Undetermined	13	3	0	0	3

Sample size reported in livestock metabolomics papers were divided into 5 categories with papers using ≤30 samples, 31–50, 51–100, or those that have not mentioned the number of samples used in the analysis.

Another consistent problem detected in the published livestock metabolomics literature is incomplete reporting. We found that 13% of all published livestock metabolomics papers did not report the number of samples used in their research. Providing information on sample size is an essential scientific measurement and reflects on the quality and reliability of published papers. Failure to report sample sizes along with failure to provide information on the numbers of animals or animal replicates indicates a major flaw in manuscript preparation and scientific work.

3.4 Trends and gaps in biological sample types

As can be seen in Fig 3 and Table 3, a total of 30 different sample types have been used for livestock metabolomics analyses. The most commonly used sample types include milk, plasma, serum, urine and ruminal fluid. These biofluids account for 78% of the total sample types reported. Milk and plasma are the most commonly used samples in bovine metabolomics manuscripts. Among all other animal groups, plasma was the most widely examined sample type (Table 3), reflecting perhaps the ease of collection but also its potential utility as a proxy reporter for all of the organs in the body [72]. Some of the least frequently used samples include cerebrospinal fluid, colostrum, semen, adipose tissue, kidney and kidney perfusate, feces, amniotic fluid, bile and liver (Table 3). The relatively low number of papers reporting data on tissue metabolomics likely reflects the challenges and costs of animal culling especially for larger livestock, sample collection, and the need to rapidly perform metabolic quenching via liquid nitrogen (immediately after surgery or necropsy) to obtain useful tissue samples for metabolite analysis [73; 74].

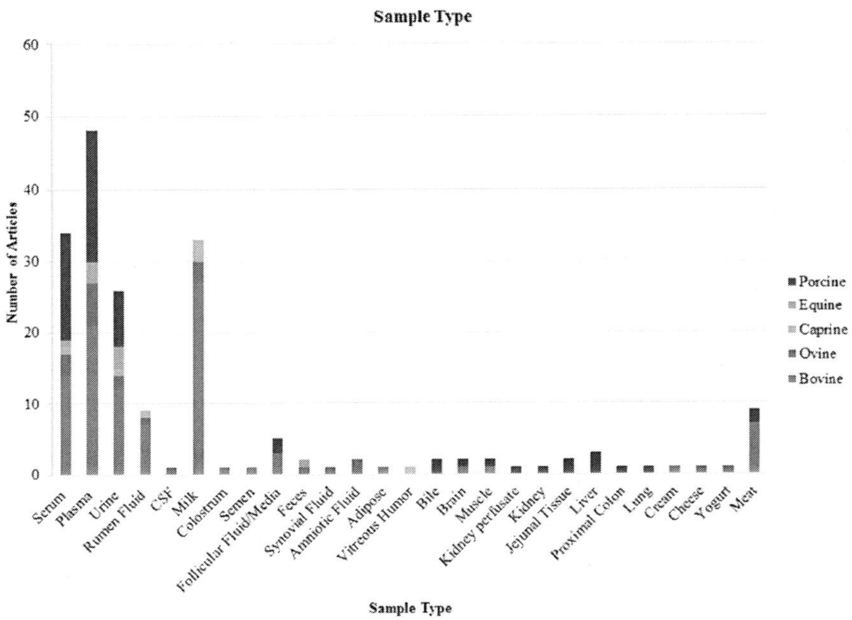

Figure 3. Sample types. Different varieties of samples and animal products have been analyzed in livestock metabolomics studies.

While studies on bovine milk are quite prevalent, there are essentially very few studies on sheep or goat milk (Table 3). Given the importance of goat and sheep milk in the global agrifood economy, it is surprising that only a total of 6 papers have been published on goat/sheep milk metabolites. One notable study, however, is that of Park and colleagues [75] who used LC-MS to identify/quantify 82 metabolites in sheep and goat milk. This paper reports a number of other macronutrient milk constituents including fat, protein, minerals and vitamins. In another more recent study, the effect of a specific grazing patterns and their associated dietary effect on goat milk was evaluated [76]. These authors used GC-MS techniques to identify and quantify 25 milk metabolites.

Table 3. Sample Types

	Bovine	Ovine	Caprine	Equine	Porcine
Adipose	1	0	0	0	0
Amniotic Fluid	0	2	0	0	0
Bile	0	0	0	0	2
Brain	0	1	0	0	1
Cerebral-Spinal Fluid	0	1	0	0	0
Cheese	0	1	0	0	0
Colostrum	1	0	0	0	0
Cream	1	0	0	0	0
Feces	0	1	0	1	0
Follicular Fluid/Media	3	0	0	0	2
Jejunal Tissue	0	0	0	0	2
Kidney	0	0	0	0	1
Kidney Perfusate	0	0	0	0	1
Liver	0	0	0	0	3
Lung	0	0	0	0	1
Meat	7	0	0	0	2
Milk	27	3	3	0	0
Muscle	1	0	0	0	1
Plasma	21	6	0	3	18
Proximal Colon	0	0	0	0	1
Rumen Fluid	7	1	1	0	0
Semen	1	0	0	0	0
Serum	14	3	1	1	15
Synovial Fluid	0	1	0	0	0
Urine	12	2	1	3	8
Vitreous Humor	0	0	1	0	0
Yogurt	0	1	0	0	0

Different varieties of samples have been used in livestock metabolomics analyses as identified by the number of published articles per sample per livestock specie.

Similar trends are also seen in other biofluid or sample types, with bovine samples or bovine-related papers dominating. For instance, there are a number of metabolomic studies on bovine ruminal fluid, plasma and urine, but very few studies on these biofluids for sheep, goat, horses or pigs (43 for all 4 species and 3 sample types). Likewise, metabolomics studies on colostrum and semen are limited to cattle only with one study each. Interestingly, some of the less-frequently used sample types such as cerebrospinal fluid, synovial fluid, amniotic fluid, bile and vitreous humor are limited to the less frequently studied livestock species (sheep, goat and pig). What is also quite striking is the dearth of fecal metabolomic studies among all livestock species (Table 3). With the growing interest in the microbiome and the clear role that gut (and rumen) microflora play in animal health, we were surprised by the complete absence of metabolomic papers on bovine fecal samples.

Given the importance of beef, sheep and goat meat, it is also surprising to see how little metabolomic data has been collected on meat samples. Indeed, only a total of 9 papers provided data on relatively small number (140) of meat metabolites. The most comprehensive meat metabolomics study was reported by Castejón et al. [32]. These authors profiled meat exudate using NMR to explore the effect of storage time on metabolite composition. They reported a total of 60 different metabolites. Overall, these data suggest that the livestock metabolomic literature is characterized by a significant under-representation of some important sample types, including milk, meat, fecal/rumen, semen samples and cerebrospinal fluid. These "gaps" in our knowledge and "gaps" in the published literature represent clear opportunities for livestock researchers to pursue.

With regards to the number of metabolites detected, quantified and/or reported among the different sample types, we found that the broadest level of coverage was for milk, plasma and serum (Table 4). Ruminal fluid, urine, feces and meat samples had slightly lower levels of coverage while the rest of the sample types reported in Fig 3 typically report <60 metabolites each. It is instructive to compare these livestock metabolite numbers to data reported for human metabolites identified in similar kinds of sample types. For instance, the most comprehensive human milk metabolomics paper reports just 129 identified metabolites [77], which is >3X lower than what has been reported in the livestock milk. The total number of metabolites reported for plasma/serum in humans is 4229 [72], which is significantly more than what is reported for livestock plasma/serum (with 759). Likewise, the total number of human urine metabolites has been reported to be 445 [2], which is more than twice that found in the urine of livestock species. Given their genomic similarity, our expectation is that the number of metabolites measurable in livestock for each of the biofluids should be comparable to the number of metabolites measured in humans. Currently, the Human Metabolome database recognized as the most comprehensive metabolomics database contains

>40,000 metabolites derived from various human biosamples [78]. As a result, this suggests there is still a significant gap to be filled with regard to the depth and breadth of metabolome characterization in livestock.

Table 4. Metabolite coverage.

	Milk	Plasma	Serum	Ruminal Fluid	Urine	Feces	Meat
Number of Metabolites	422	408	351	248	177	158	75

The number of metabolites detected, quantified and/or reported among the commonly used sample types in the livestock metabolomics publications up to 2016 (counting publications that reported >8 metabolites).

3.5 Trends and gaps in analytical instrumentation and methodologies

Metabolomics uses a wide variety of analytical instruments that vary in terms of their sensitivity and breadth of coverage. Nuclear magnetic resonance (NMR) continues to be among the most commonly used analytical platforms in metabolomics [79]. It is often chosen for its reliability and utility in absolute quantitation however, NMR is relatively insensitive and is limited to measuring substances in micromolar to millimolar (μM-mM) concentrations (Fig 4). Mass spectrometry (MS) platforms (especially LC-ESI-MS) can detect metabolites at nanomolar (nM) to picomolar (pM) concentrations, allowing a much higher number of metabolites to be detected. However, MS instruments are prone to frequent breakdowns and, relative to NMR, it is often difficult to quantify chemical concentrations via MS techniques. Gas chromatography-MS (GC-MS) is less sensitive than liquid chromatography (LC)-MS, but is generally more robust and more reproducible. As a result, GC-MS can sometimes be used to identify and quantify the metabolome with higher precision and reproducibility than either NMR or LC-MS.

Each of the 149 livestock metabolomics papers was carefully analyzed to identify which analytical platforms (NMR, LC-MS, GC-MS) were used more frequently to conduct metabolomic analyses. In certain studies, more than one platform was used so, we simply counted the frequency that each technique or technology was used in each study. Interestingly, the most commonly used metabolomics platform for all animal categories is NMR spectroscopy, accounting for 28% of all livestock metabolomics studies. Following closely behind NMR, in terms of frequency, is LC-MS with 25% of all studies using this analytical platform. It is noteworthy that the LC-MS category includes ultra performance liquid chromatography (UPLC)-MS, high-performance liquid chromatography (HPLC)-MS, and direct flow injection (DFI)-MS. Gas chromatography-MS is the third most prevalent (15%) analytical platform used in livestock metabolomics studies. The more limited use of GC-MS is typical of other metabolomic disciplines as well.

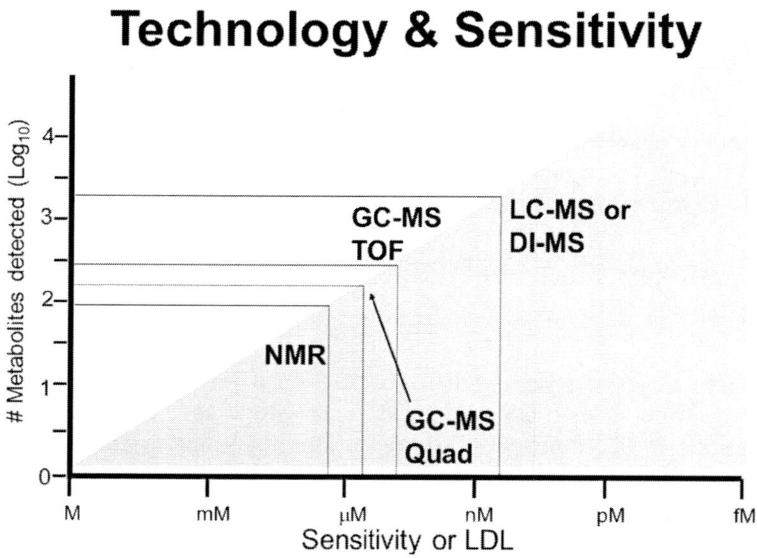

Figure 4. Relative sensitivity of metabolomics platforms. Nuclear magnetic resonance (NMR), gas chromatography-mass spectrometry (GC-MS), and liquid chromatography (LC)-MS are the commonly used metabolomics platforms with varying detection limits.

Other, less conventional or more targeted, methodologies account for the remaining 27% of the technologies used in livestock metabolomics studies. These methods include, but are not limited to, infrared spectroscopy (FTIR), silicic acid column chromatography, immunoassays, the Kjeldahl method (for organic nitrogen measurement), ELISAs, and miscellaneous, lab-specific methods. Relative to other fields of metabolomics, livestock metabolomics appears to use NMR spectroscopy somewhat more and LC-MS somewhat less. This may simply reflect the availability of instrumentation or the preferences of major research groups in livestock metabolomics. Certainly, sample abundance and supply is not a significant issue in livestock metabolomics so, the use of tools that require higher-volumes, but offer more quantitative results (such as NMR) is not unexpected. However, NMR is not the most sensitive technique and certainly if livestock metabolomics researchers wish to extend their coverage of the livestock metabolome, they will certainly need to makes use of more LC-MS methods.

Another gap that was noted in livestock metabolomics research is the near complete absence of ICP (inductively coupled plasma)-MS studies to measure metal ion levels in tissues and biofluids. Indeed, only 2 studies used ICP-MS, with the most complete characterization being conducted by Saleem et al. [52] who reported the identification and quantification of 20 metals in bovine ruminal fluid. The importance of metal ions as micronutrients for animal health and animal productivity cannot be underestimated [80; 81]. Therefore, it is surprising that so little metal ion data has been collected or analyzed in livestock metabolomic studies. It was also noted that the use of fluxomics [82] or the measurement of metabolite flux using stable isotopes is completely absent in livestock metabolomics studies. Fluxomics is particularly useful in understanding metabolic sinks and sources. It is also useful for assessing nutrition and metabolic efficiency—topics, which are obviously important in livestock research. However, to conduct metabolic flux analysis, isotopically labeled (^{13}C or ^{2}H) feed needs to be used. Given the size of most livestock animals (relative to rats and mice) and the need for significant quantities of expensive, isotopically labeled feed, fluxomic studies are likely too difficult and costly to perform [83]. Likewise, the use of imaging mass spectrometry or IMS (which is becoming very popular in human metabolomics studies) was completely absent in livestock studies. Imaging mass spectrometry is particularly useful for analyzing tissues and for understanding the metabolic changes that take place during tissue development or tissue transformation [84; 85].

A good metabolomics study should use more than one analytical platform, and ideally as many different (orthogonal) platforms as possible to broaden the metabolite coverage. In our analysis we found that 69% of the published studies used just 1 platform (either NMR, HPLC-UV, LC-MS, GC-MS or ICP-MS), 15% used 2 platforms and only 3% used 3 or more analytical platforms. The remaining 13% of studies used relatively non-conventional platforms or assays (immunoassays, FT-IR, etc.). The most comprehensive metabolomic analysis was a study that used 5 different platforms (NMR, HPLC-UV, LC-MS, GC-MS and ICP-MS) to characterize the bovine ruminal fluid metabolome [52]. Looking through the more recent studies, there is a general trend towards using more than one platform and a growing trend towards using LC-MS techniques over NMR methods. However, the surprisingly high number of livestock metabolomic studies that still use only a single platform also represents a significant issue that the field must remedy. Certainly the trend in human metabolomic studies is to use at least 2 and often 3 or more different analytical platforms [2].

Another gap that was identified from this literature analysis was the general lack of integration of other omics techniques (proteomics, transcriptomics or SNP measurements) with reported livestock metabolomic studies. Indeed, only 5 papers (3 bovine and 2 swine metabolomics studies) used metabolomics in conjunction with genomics or proteomics. One paper of note was an investigation that used genomics and metabolomics to evaluate RFI (residual feed intake) from cross breeds of dairy and beef cattle [66]. This group of researchers used metabolomics and phenotypic data to support their genomics investigations and identified two genes (*TP53* and *TGFB1*) that were strongly associated with cellular functions driving feed efficiency. In another study by Lu and colleagues [50], the effect of genetic polymorphisms on dairy milk characteristics was evaluated using a combination of metabolomics and proteomics. This paper identified alterations in triglyceride composition and reported changes in the milk metabolome and proteome of dairy cows with the K232A (lysine to alanine substitution) polymorphism in the well-studied *DGAT1* gene. Given the growing trend towards systems biology research and the more "holistic" interpretations of multi-omics data in other fields of life science, the near absence of multi-omics studies represents an important gap in livestock metabolomics (and omics) research.

3.6 Trends and gaps in metabolite quantification

The majority of livestock metabolomics publications are non-quantitative or semi-quantitative (yielding relative quantification) while 28.18% of published studies provide fully (absolute) quantitative data. The metabolites tracked in this review were categorized in two main groups: 1) quantified and 2) non-quantified metabolites. Any metabolite that was associated with an absolutely quantified value (millimolar, micromolar, nanomolar, mg/mL, ug/mL, etc.) in a given sample type was placed in the quantified category. The non-quantified group consists of either metabolites with no quantified value or ones that have only relative quantification (i.e. reported as a fraction or a percentage). Over all livestock species and all sample types, we found a total of 404 quantified metabolites and 666 non-quantified. The majority of both quantified and non-quantified metabolites are lipids and lipid-like molecules. Temporal trends in metabolite quantification show that proportionally fewer livestock metabolomics papers are providing quantitative data. For instance, 69% of papers published prior to 1999 had quantitative data, while 34% of papers from 1999–2010 and just 21% from 2011–2015 generated quantitative metabolite data.

Overall, livestock metabolomics still has an impressive proportion (~28%) of publications that report absolute concentration values. In contrast, most other fields of metabolomics quantify metabolites far less frequently [72].

Nevertheless, the steady decline in the proportion of livestock papers providing quantitative metabolomic data is not a good sign. The importance of absolute quantification in metabolomics cannot be over-emphasized. As a branch of analytical chemistry focusing on small molecule characterization, there is more than 100 years of history and a plethora of tools, standards and protocols designed specifically for absolute metabolite quantification [79]. Absolute quantification allows facile comparisons of readings between animals, research staff, platforms, laboratories and countries. Acquiring quantified values also allows one to determine normal and abnormal ranges for disease diagnosis, prediction as well as other relevant production measures. Obtaining quantified data and recognizing normal physiological concentrations is also a requirement in biomarker discovery [86; 60]. Indeed, absolute quantification and the existence of normal and abnormal ranges is the foundation to the entire field of clinical chemistry. In livestock metabolomics, having a "normal" quantified range specific for each animal species or breed is critical for defining referential "healthy" conditions. Likewise, being able to quantify specific changes in an animal's metabolome allows one to identify "abnormal" conditions such as overt disease, malnutrition, pregnancy difficulties, and most importantly subclinical conditions for which no obvious clinical indicators are visible [51; 87].

3.7 Trends and gaps in metabolite coverage

Based on our analysis of the literature and the definition of a metabolite given earlier, the majority of livestock metabolomics studies report ≤50 metabolites (79% of the total selected metabolomics publications) while the other 21% report >50 metabolites. The largest number of metabolites (or features) reported in a single paper was 647 [51] covering multiple biofluids for bovine samples while, the fewest reported was 8 (in a variety of papers from all different livestock species). As with metabolite quantification, there is a trend for more recent livestock metabolomics papers to report a greater number of metabolites. For instance, papers published prior to 1999 averaged 29 metabolites per study, those from 1999–2010 averaged 44 metabolites per study while, papers from 2011–2015 averaged 63. Among the later publications, the recent bovine study conducted by Sun et al. [51] who investigated potential biomarkers of milk production and quality using GC-time-of-flight/MS analyses of rumen fluid, milk, serum and urine claimed to detect the highest number of metabolites (i.e., 647). However, careful reading of the manuscript shows that they only formally identified 123. The remaining "metabolites" were unidentified MS peaks or features. In ovine metabolomic studies, Parveen and colleagues [88] reported 168 out of 205 detected metabolites using GC-MS to investigate sheep plasma and feces. Clark et al. [89] reported 97 metabolites out of the 571 detected features in caprine serum using a combination of both GC-MS and LC-MS. In equine metabolomics, the highest number of metabolites identified was from a study conducted by Escalona and colleagues [53] with 102 metabolites identified via NMR analysis of plasma, urine and fecal water. A porcine metabolomics study by Metzler-Zebeli et al. [90] reported 104 out of 132 detected serum metabolites using LC-MS.

Overall, our analysis shows a total of 1070 non-redundant or unique metabolites have been detected and/or quantified in the livestock metabolomics literature. Bovine studies covered the majority of detected metabolites (i.e., 768 different compounds) over multiple sample types. Porcine and ovine studies have the next highest number of detected metabolites with 412 and 285 different metabolites, respectively. Caprine and equine studies reported 167 and 109 different metabolites, respectively. The most frequently detected metabolites with >100 separate entries for different animals, biofluids or conditions include: alanine (124 times), valine (112 times), isoleucine (105 times), glycine (101 times), and lactate (101 times). In addition, 26 other metabolites were reported 50–100 times. Metabolites reported more than once and <50 times add to 560 while, 479 metabolites are reported only once.

It is important to provide some context to these numbers, especially with regard to metabolome studies reported for other animal or model species. The estimated size of the mammalian metabolome is >100,000 molecules [36; 2] and the total number of metabolites so far reported and/or theoretically expected to be in the human metabolome or HMDB is just over 42,000 [78]. While the number of expected or theoretical metabolites is large, the actual number of experimentally identified (and/or quantified) metabolites is actually quite small. For instance, based on the HMDB, the number of experimentally identified metabolites in the human metabolome is 3821 [78], in the *E. coli* metabolome it is 891 [41] and in the yeast metabolome it is 625 [40]. Among the different livestock species, it is clear that the coverage of the bovine metabolome is quite extensive and is approaching or even exceeding that of other model organisms. However, there is an obvious gap in terms of the coverage of other livestock species with caprine and equine metabolomes being very poorly characterized. Much more work is needed on goat and horse metabolomes to bring them up to the level seen in the bovine metabolome.

3.8 Trends and gaps in animal breeds

While we have largely focused on examining metabolomics data for different livestock species, we also noticed some interesting trends with regard to the choice of specific breeds in each livestock species. Similar to other fields of bovine research, the majority (45%) of bovine metabolomic studies use either pure- or cross-bred Holsteins. A smaller amount (11%) of other studies used cross breeds to investigate various aspects of the bovine metabolome. Other common bovine breeds used in metabolomic studies include Charolais (7%) and Jersey (3%). In ovine metabolomic studies, the main breed reported is Suffolk (19%) while other breeds (i.e., Sarda) are reported only once or twice. For caprine metabolomics studies, the preferred breeds have been Norwegian (22%) with other breeds such as Saanen and Alpine being reported only once. Likewise, among equine and porcine metabolomic studies, Standardbred horses (33%) and Landrace sows (22%) were most frequently used. Interestingly, no breed information was provided in 18%, 33%, 22%, 17%, and 9% of the bovine, ovine, caprine, equine, and porcine metabolomics manuscripts, respectively. It is surprising that this essential information is not provided in the manuscripts. This suggests the reporting standards found in livestock metabolomics manuscripts still needs improvement.

Based on the above statistics, one of the more obvious gaps in current livestock metabolomics research is the limited variety of breeds being used in most metabolomic studies. The vast majority of the published research appears to be focused on just one or two main breeds i.e., Holstein in cattle, Suffolk in sheep, Standardbred in horses. Evidently, assessing breed differences and their potential impacts on the metabolome has not been a priority for most livestock researchers. However, it is important to remember that the existence of dozens of livestock breeds is a consequence of centuries of selection for very unique phenotypic qualities—some of which are likely determined by their metabolism or metabolome. Different breeds will be characterized by specific production or metabolic parameters and these may be fundamentally different between breeds. While the composition of mammalian (and livestock) metabolomes is likely to be highly similar, metabolite concentrations are expected to differ substantially between different breeds. Identifying the unique aspects affiliated with each breed's metabolome is therefore, an important component of livestock metabolomics that should be considered in future studies. This is particularly true for purebred and breeding stock herds that are limited to very few animals/herds worldwide. Breeding stock animals provide most of the genetic background found in most commercial herds, which means they have a significant influence on the metabolome associated with their progeny. We were also surprised by the very limited research on the neonatal livestock metabolome. Indeed, we found only 16 neonate metabolomic studies, with 1 study focused on calves, 4 on lambs, 1 on kids, 10 on piglets and no studies on colts or foals.

3.9 Trends and gaps in biomarker discovery

One of the strengths of metabolomics lies in its utility for biomarker discovery [91]. Because metabolites can be more easily, cheaply and routinely quantified than most other biological molecules, they are ideal for use in biomarker panels. Indeed, metabolite biomarkers continue to be developed and used in clinical applications at a much greater rate than genes or proteins [78]. In surveying the papers compiled for this review, we found a total of 11 livestock metabolomics papers that proposed candidate biomarkers. This included 5 papers in animal health, 1 in animal nutrition, 2 in animal production, 1 in animal reproduction and 2 for animal models of human health. These studies were limited to cattle, sheep and pigs with no metabolomic biomarker studies being reported for goats or horses. Of these papers, we observed that most reported fewer than 30 candidate biomarkers, with the lowest number being 2 [61]. A few reports used higher number of metabolites, i.e., 64, as part of a statistical model to increase the accuracy of prediction [92; 93]. The majority (55%) of metabolomic biomarker papers did not provide any quantitative data, but rather reported only relative metabolite trends (up or down relative to some indeterminate standard). This means that only 5 papers, all from the bovine group, effectively provided useful or verifiable biomarker data. Furthermore, only a single paper [20] reported follow-up validation studies where the initially discovered biomarkers were subsequently validated on a separate cohort of samples.

Based on our data, most biomarker studies were conducted with relatively small sample sizes with the majority of studies being done on fewer than 100 animals. The largest biomarker study was one conducted on 321 animals (1587 samples), which investigated prognostic biomarkers of ketosis in dairy cows using NMR spectroscopy [61]. Overall, the quality of biomarker studies done for livestock metabolomics is not particularly good, especially given the standards expected of human biomarker studies [91].

Nevertheless, among the reported biomarker studies, we did find some very interesting and compelling results. One example is a biomarker study of RFI and other feed efficiency traits in beef steers [20]. In this study, NMR spectroscopy was used to identify and quantify plasma metabolites associated with RFI, initially in a discovery population and subsequently in the validation cohort. Karisa et al. [20] reported 3 candidate biomarkers of RFI that significantly ($P<0.05$) account for >30% of the phenotypic variation for this trait. Other metabolites were proposed to be associated with average body weight, average feed intake, dry matter intake and average daily gain. In another interesting study, predictive biomarkers of transition diseases in dairy cows were investigated [21]. This study monitored only 12 dairy cows over four time points during the transition (pre- and post-calving) period. Blood samples were drawn to quantify the metabolome changes associated with various periparturient diseases post-calving. Using direct flow injection (DFI)-MS, Hailemariam and colleagues [21] profiled 120 blood metabolites of which 3 were suggested as candidate biomarkers for transition diseases, with a sensitivity and specificity of ≥85%. Another study reported by Gray et al. [94] looked into biomarkers associated with vaccine efficacy. Using UPLC-MS metabolomic measurement of plasma derived from Holstein male calves, Gray and colleagues [94] found 12 metabolites that were altered post-vaccination. These biomarkers are being proposed as a newer, more efficient route to optimise vaccination and to make vaccine formulation and benchmarking much more efficient and targeted. This paper emphasizes on the importance of disease prevention and vaccination procedures in livestock, especially in using new technologies such as metabolomics to enhance evaluation of vaccine efficacy.

Identification of biomarkers will not only improve disease diagnosis but also allow the opportunity for disease prediction prior to manifestation of clinical signs. For example, if a metabolic disorder can be predicted well before (sub)clinical manifestation, farmers can make informative decisions with regards to their management, feeding, housing, etc., to change the cascade of biological events leading to that disease. Predictive attempts of such can make a significant financial and sustainability difference by maintaining production quantity and quality, saving on costs associated with treatment, veterinary visits, preventing animal culling and thus, maintaining longevity.

4. THE LIVESTOCK METABOLOME DATABASE

In assembling the material for this review, we identified a total of 1070 metabolites that have been detected and/or quantified in livestock metabolomic studies of cattle, sheep, goats, horses and pigs. This information has been systematically categorized into LMDB with all of the metabolites being fully described including information about the degree or quality of quantification (i.e., quantified, non-quantified) and the source sample types for each livestock species. All of the metabolites with quantitative data had their concentrations converted into a standardized concentration unit (i.e., μM) to improve consistency. In addition to the chemical data and source information, an abbreviated description of the experimental context for each metabolite was extracted from the articles and included in the online database (http://www.lmdb.ca). This information includes data on the analytical platform(s), experimental conditions, field of research, and animal breed used in acquiring the metabolomic data. All metabolites are linked to a standard HMDB (http://www.hmdb.ca/) identification number, which provides a freely-accessible and detailed description of the metabolite. A PubMed and/or DOI identifier is also associated with each metabolite entry, which provides a literature reference or a direct link to the article reporting that metabolite for readers who are interested in further details.

Only those metabolites that had reasonably complete descriptions (i.e., unique chemical names, sample types, source information, etc.) were included in the online database. A number of metabolites or "features" were identified during the review process but not included in the LMDB. These include those compounds that have either not been characterized at all (no chemical name, no data on sample types), or not fully characterized (unknown or undefined chemical structure). This collection of 415 "unknown" metabolites will be added to the LMDB once we can obtain sufficient structural and sample source information on them. Among the metabolites entered into the LMDB, 404 compounds were quantified and 666 were not. On a species level there were 768 bovine metabolites, 285 ovine metabolites, 167 caprine metabolites, 109 equine metabolites, and 412 porcine metabolites. Detailed descriptions of each compound are provided in the LMDB "metabocard" pages. Likewise, structural images, molecular formulas, names and synonyms, chemical classification/taxonomy information, physicochemical data (molecular weights, pI's, pKa's, boiling/melting points), referential spectral data (both experimental and theoretical NMR, MS/MS and EI-MS spectra), links to other online databases and full reference (authors, journals, volumes, etc.) information is also provided. The LMDB has been designed so that it can be easily browsed and it supports searches through standard text queries as well as via structure, mass, and spectral queries. Most of the information in the LMDB is hyperlinked to other resources within the LMDB, allowing for a more convenient and compact route to access the data. The LMDB is available at http://www.lmdb.ca. This database will be constantly updated with more metabolites and more detailed metabolite descriptions as more research in livestock metabolomics is published.

By assembling the LMDB and making this information freely available through both the web and this manuscript, we hoped to create a referential resource that other livestock researchers could readily use. Our past experience in assembling and maintaining the Human Metabolome Database (HMDB) clearly showed how useful a centralized, on-line resource could be in the field of human metabolomics [38]. Therefore, our expectation with the LMDB is that it will have a comparable impact on the field of livestock metabolomics. Indeed, we believe that establishing a comprehensive repository that stores and categorizes livestock metabolome information into a standardized format will be critical for future livestock research. It will also be important for identifying potential livestock disease biomarkers, improving animal selection (via metabolomic assays), enhancing animal nutrition and understanding novel biochemical mechanisms arising from various physiological perturbations. With more and more livestock metabolomics papers appearing each year and the continued growth in metabolite coverage, it will be challenging to maintain the LMDB. However, without even attempting to create the LMDB we suspect that livestock metabolomics would continue to lag behind the metabolomics activities seen in other areas (i.e. human, plant crops, microbes, food/beverage studies) and would face significant hurdles in the coming years trying to catch up.

5. CONCLUSION

Metabolomics is less than 15 years old, yet it has already delivered some remarkable achievements. This includes significant improvements in the ability to identify many environmental contaminants and toxins [95], significant advances in food and nutrient characterization [7; 6], the identification of many novel biomarkers for disease risk including risk markers for diabetes [54], heart disease [96] and cancer [58] as well as promising leads for a variety of drugs and therapies [97]. Metabolomics is also well-positioned to provide some important advances in both livestock research and the livestock industry, especially as it relates to livestock health, breeding and production. A number of examples were highlighted in this review including metabolome discovery for normal metabolite composition and concentrations [52; 53], identification of biomarkers of transition diseases [21] as well as production traits in dairy [51] and beef cattle [20] with the goal of introducing prognostic strategies in animal health as well as increasing prediction accuracies. Our observations also showed that a wide variety of biofuids have received attention for metabolomics research such as metabolic profiling of milk, plasma, serum, and urine, minimizing animal welfare concerns.

However, in order for livestock metabolomics to deliver on the promise and the excitement seen in other areas of metabolomics research, it is important to carefully assess what has been accomplished, what is known and what still needs to be done. The intent of this review was to provide a critical overview of the trends and gaps in livestock metabolomics research. Specifically, we sought answers to 4 key questions: 1) What are the most common applications of metabolomics in animal science and where are they trending?, 2) What are the preferred metabolomics technologies livestock metabolomics and how are they evolving?, 3) What are the most obvious gaps or weaknesses in livestock metabolomics relative to other fields of metabolomics research? and 4) What are the known or measured metabolites for the 5 major livestock species (i.e., bovine, ovine, caprine, equine, and porcine) in different tissues and biofluids? In addressing the first 3 questions we focused on areas relating to: 1) Animal Choices; 2) Research Applications; 3) Sample Size; 4) Sample Type; 5) Instrumentation and Methodologies, 6) Quantification; 6) Metabolite Coverage; 7) Animal Breeds, and 8) Biomarker Identification. In many cases we were able to identify some clear trends while at the same time identifying important shortcoming or areas where further improvements could be made. It was apparent that livestock metabolomics appears to be ahead with regard to metabolite quantification, the diversity of research applications and its efforts in biomarker identification. On the other hand, it was also clear that livestock metabolomics (especially with regard to sample size, instrumentation and metabolite coverage) was lagging somewhat further behind than human, microbial or plant crop metabolomics.

Based on our assessment of the shortcomings with current livestock metabolomics studies, it is clear that future metabolomics research should focus on expanding or extending metabolome discovery using healthy control animals, increase sample numbers, direct more effort towards metabolite quantification, perform more integrated multi-omics experiments, use a greater variety of analytical platforms or techniques (including ICP-MS, MSI and fluxomics), increase the breadth of metabolite coverage (by using more sensitive platforms, such as ESI-MS), investigate a greater and new varieties of biosamples such as semen, amniotic fluid, saliva and urine, extend the number and types of animal breeds used in metabolomic studies and be more conscientious in the design and implementation of biomarker studies.

Another important outcome of this study was the collection and consolidation of livestock metabolite information into a single, centralized resource (the LMDB). It became readily apparent in conducting this review that the livestock metabolomics literature is highly diffuse and that valuable information is being "lost" or is not readily available. By compiling the LMDB and making an on-line version of the database freely available, we hope it could serve as a hub for livestock researchers and the livestock industry to further advance the field of livestock metabolomics.

REFERENCES

1. Pearson H. Meet the human metabolome. Nature. 2007;446: 8. pmid:17330009
2. Bouatra S, Aziat F, Mandal R, Guo AC, Wilson MR, Knox C, et al. The human urine metabolome. PLoS One. 2013;8: e73076. pmid:24023812
3. Monteiro MS, Carvalho M, Bastos ML, Guedes de Pinho P. Metabolomics Analysis for Biomarker Discovery: Advances and Challenges. Curr Med Chem. 2013;20: 257–271. pmid:23210853

4. Fontanesi L. Metabolomics and livestock genomics: Insights into a phenotyping frontier and its applications in animal breeding. Anim Front. 2016;6: 73–79.
5. Moore ER, Kirwan J, Doherty MK, Whitfield PD. Biomarker discovery in animal health and disease: the application of post-genomics technologies. Biomark Insights. 2007;2: 185–196. pmid:19662203
6. Wishart DS. Metabolomics: application to food science and nutrition research. Trends Food Sci Tech. 2008;19: 482–493.
7. Kim S, Kim J, Yun EJ, Kim KH. Food metabolomics: from farm to human. Curr Opin Biotechnol. 2016;37: 16–23. pmid:26426959
8. Jalali A, Hatamie A, Safarpour T, Khajeamiri A, Safa T, Buazar F. Impact of pharmaceutical impurities in Ecstasy tablets: gas chromatography-mass spectrometry study. Iran J Pharm Res. 2016;15: 221–229. pmid:27610162
9. Simo C, Ibanez C, Valdes A, Cifuentes A, Garcia-Canas V. Metabolomics of genetically modified crops. Int J Mol Sci. 2014;15: 18941–18966. pmid:25334064
10. Summer LW, Lei Z, Nikolau BJ, Saito K. Modern plant metabolomics: advanced natural product gene discoveries, improved technologies, and future prospects. Nat Prod Rep. 2015;32: 212–229. pmid:25342293
11. Mahdavi V, Farimani MM, Fathi F, Ghassempour A. A targeted metabolomics approach toward understanding metabolic variations in rice under pesticide stress. Anal Biochem. 2015;478: 65–72. pmid:25766578
12. Mahdavi V, Ghanati F, Ghassempour A. Integrated pathway-based and network-based analysis of GC-MS rice metabolomics data under diazinon stress to infer affected biological pathways. Anal Biochem. 2016;494: 31–36. pmid:26582432
13. Fiehn O. Metabolomics—the link between genotypes and phenotypes. Plant Mol Biol. 2002;48: 155–171. pmid:11860207
14. Houle D, Govindaraju DR, Omholt S. Phenomics: the next challenge. Nat Rev Genet. 2010;11: 855–566. pmid:21085204
15. Duggan GE, Hittel DS, Sensen CW, Weljie AM, Vogel HJ, Shearer J. Metabolomic response to exercise training in lean and diet-induced obese mice. J Appl Physiol. 2011;110: 1311–1318. pmid:21270351
16. Jones DP, Park Y, Ziegler TR. Nutritional metabolomics: progress in addressing complexity in diet and health. Annu Rev Nutr. 2012;32: 183–202. pmid:22540256
17. May DH, Navarro SL, Ruczinski I, Hogan J, Ogata Y, Schwarz Y, et al. Metabolomic profiling of urine: response to a randomised, controlled feeding study of select fruits and vegetables, and application to an observational study. Br J Nutr. 2013;110: 1760–1770. pmid:23657156
18. Gilany K, Moazeni-Pourasil RS, Jafarzadeh N, Savadi-Shiraz E. Metabolomics fingerprinting of the human seminal plasma of asthenozoospermic patients. Mol Reprod Dev. 2014;81: 84–86. pmid:24254375
19. Minai-Tehrani A, Jafarzadeh N, Gilany K. Metabolomics: a state-of-the-art technology for better understanding of male infertility. Andrologia. 2015;48: 609–16. pmid:26608970
20. Karisa BK, Thomson J, Wang Z, Li C, Montanholi YR, Miller SP, et al. Plasma metabolites associated with residual feed intake and other productivity performance traits in beef cattle. Livest Sci. 2014;165: 200–211.
21. Hailemariam D, Mandal R, Saleem F, Dunn SM, Wishart DS, Ametaj BN. Identification of predictive biomarkers of disease state in transition dairy cows. J Dairy Sci. 2014;97: 2680–2693. pmid:24630653
22. LeBlanc SJ, Leslie KE, Duffield TF. Metabolic predictors of displaced abomasum in dairy cattle. J Dairy Sci. 2005;88: 159–170. pmid:15591379
23. Sundekilde UK, Poulsen NA, Larsen LB, Bertram HC. Nuclear magnetic resonance metabonomics reveals strong association between milk metabolites and somatic cell count in bovine milk. J Dairy Sci. 2013;96: 290–299. pmid:23182357
24. Saleem F, Ametaj BN, Bouatra S, Mandal R, Zebeli Q, Dunn SM, et al. A metabolomics approach to uncover the effect of grain diets on rumen health in dairy cows. J Dairy Sci. 2012;95: 6606–6623. pmid:22959937

25. Abarghuei MJ, Rouzbehan Y, Salem AZM, Zamiri MJ. Nitrogen balance, blood metabolites and milk fatty acid composition of dairy cows fed pomegranate-peel extract. Livest Sci. 2014;164: 72–80.
26. Weikard R, Altmaier E, Suhre K, Weinberger KM, Hammon HM, Albrecht E, et al. Metabolomic profiles indicate distinct physiological pathways affected by two loci with major divergent effect on Bos taurus growth and lipid deposition. Physiol Genomics. 2010;42A: 79–88. pmid:20647382
27. Karisa BK, Thomson J, Wang Z, Bruce HL, Plastow GS, Moore SS. Candidate genes and biological pathways associated with carcass quality traits in beef cattle. Can J Anim Sci. 2013a;93: 295–306.
28. Kuhn C, Weikard R, Widmann P. Metabolomics: a pathway for improved understanding of genetic modulation of mammalian growth and tissue deposition. Proc 10th World Cong Genet App Livest Prod. 2014;1–6.
29. Chapinal N, Carson ME, LeBlanc SJ, Leslie KE, Godden S, Capel M, et al. The association of serum metabolites in the transition period with milk production and early-lactation reproductive performance. J Dairy Sci. 2012;95: 1301–1309. pmid:22365212
30. Melzer N, Wittenburg D, Hartwig S, Jakubowski S, Kesting U, Willmitzer L, et al. Investigating associations between milk metabolite profiles and milk traits of Holstein cows. J Dairy Sci. 2012;96: 1521–1534.
31. Melzer N, Wittenburg D, Repsilber D. Integrating milk metabolite profile information for the prediction of traditional milk traits based on SNP information for Holstein cows. PLOS One. 2013;8: e70256. pmid:23990900
32. Castejón D, García-Segura JM, Escudero R, Herrera A, Cambero MI. Metabolomics of meat exudate: Its potential to evaluate beef meat conservation and aging. Anal. Chim. Acta. 2015;901: 1–11. pmid:26614053
33. Zhang A, Sun H, Wang P, Han Y, Wang X. Modern analytical techniques in metabolomics analysis. Analyst. 2012;137: 293–300. pmid:22102985
34. Robinson TP, Wint GRW, Conchedda G, Van Boeckel TP, Ercoli V, Palamara E, et al. Mapping the global distribution of livestock. PLoS One. 2014;9: e96084. pmid:24875496
35. Thornton PK. Livestock production: recent trends, future prospects. Philos Trans R Soc Lond B Biol Sci. 2010;365: 2853–2867. pmid:20713389
36. Wishart DS, Tzur D, Knox C, Eisner R, Guo AC, Young N, et al. HMDB: the Human Metabolome Database. Nucleic Acids Res. 2007;35: D521–526. pmid:17202168
37. Wishart DS, Knox C, Guo AC, Eisner R, Young N, Gautam B, et al. HMDB: a knowledgebase for the human metabolome. Nucleic Acids Res. 2009;37: D603–610. pmid:18953024
38. Wishart DS, Jewison T, Guo AC, Wilson M, Knox C, Liu Y, et al. HMDB 3.0—The Human Metabolome Database in 2013. Nucleic Acids Res. 2013;41: D801–807. pmid:23161693
39. Viswanathan V. Rapid web application development: a ruby on rails tutorial. IEEE Software. 2008;25: 98–106.
40. Jewison T, Neveu V, Lee J, Knox C, Liu P, Mandal R, et al. YMDB: The Yeast Metabolome Database. Nucleic Acids Res. 2012;40: D815–20. pmid:22064855
41. Sajed T, Marcu A, Ramirez M, Pon A, Guo A, Knox C, et al. ECMDB 2.0: A richer resource for understanding the biochemistry of E. coli. Nucleic Acids Res. 2016;44: D495–501. pmid:26481353
42. Peters RJB, Stolker AAM, Mol JGJ, Lommen A, Lyris E, Angelis Y, et al. Screening in veterinary drug analysis and sports doping control based on full-scan, accurate-mass spectrometry. Tren Anal Chem. 2010;29: 1250–1268.
43. Tweeddale H, Notley-Mcrobb L, Ferenci T. Effect of Slow Growth on Metabolism of *Escherichia coli*, as Revealed by Global Metabolite Pool ("Metabolome") Analysis. J Bacteriol. 1998;180: 5109–5116. pmid:9748443
44. Oliver SG, Winson MK, Kell DB, Baganz F. Systematic functional analysis of the yeast genome. Trends Biotechnol. 1998;16: 373–378. pmid:9744112
45. Nielsen KL, Hartvigsen ML, Hedemann MS, Lærke HN, Hermansen K, Knudsen KEB. Similar metabolic responses in pigs and humans to breads with different contents and compositions of dietary fibers: a metabolomics study. Am J Clin Nutr. 2014;99: 941–949. pmid:24477039

46. Cevallos-Cevallos JM, Danyluk MD, Reyes-De-Corcuera JI. GC-MS based metabolomics for rapos simultaneous detection of Escherichia coli O157:H7, Salmonella Typhimurium, Salmonella Muenchen, and Salmonella Hartford in ground beef and chicken. J Food Sci. 2011;76: M238–246. pmid:22417363
47. Regal P, Nebot C, Díaz-Bao M, Barreiro R, Cepeda A, Fente C. Disturbance in sex-steroid serum profiles of cattle in response to exogenous estradiol: a screening approach to detect forbidden treatments. Steriods. 2011;76: 365–375.
48. Merrifield CA, Lewis M, Claus SP, Beckonert OP, Dumas M, Duncker S, et al. A metabolic system-wide characterisation of the pig: a model for human physiology. Mol Biosyst. 2011;7: 2577–2588. pmid:21761043
49. Mickiewicz B, Heard BJ, Chau JK, Chung M, Hart DA, Shrive NG, et al. Metabolic profiling of synovial fluid in a unilateral Ovine model of anterior cruciate ligament reconstruction of the knee suggests biomarkers for early osteoarthritis. J Orthop Res. 2015;33: 71–77. pmid:25283885
50. Lu J, Antunes Fernandes E, Páez Cano AE, Vinitwatanakhun J, Boeren S, van Hooijdonk T, et al. Changes in milk proteome and metabolome associated with dry period length, energy balance, and lactation stage in postparturient dairy cows. J Proteome Res. 2013;12: 3288–3296. pmid:23738862
51. Sun H, Wang D, Wang B, Wang J, Liu H, Guan LL, et al. Metabolomics of Four Biofluids from Dairy Cows: Potential Biomarkers for Milk Production and Quality. J Proteome Res. 2015;14: 1287–1298. pmid:25599412
52. Saleem F, Bouatra S, Guo AC, Psychogios N, Mandal R, Dunn SM, et al. The bovine ruminal fluid metabolome. J Metabolomics. 2013;9: 360–378.
53. Escalona EE, Leng J, Dona AC, Merrifield CA, Holmes E, Proudman CJ, et al. Dominant components of the Thoroughbred metabolome characterised by 1H-nuclear magnetic resonance spectroscopy: A metabolite atlas of common biofluids. Equine Vet J. 2015;47: 721–730. pmid:25130591
54. Wang-Sattler R, Yu Z, Herder C, Messias AC, Floegel A, He Y, et al. Novel biomarkers for pre-diabetes identified by metabolomics. Mol Syst Biol. 2012;8: 615. pmid:23010998
55. Reinehr T, Wolters B, Knop C, Lass N, Hellmuth C, Harder U, et al. Changes in the serum metabolite profile in obese children with weight loss. Eur J Nutr. 2014;54: 173–181. pmid:24740590
56. Kordalewska M, Markuszewski MJ. Metabolomics in cardiovascular diseases. J Pharm Biomed Anal. 2015;113: 121–136. pmid:25958299
57. Mirsaeidi M, Banoei MM, Winston BW, Schraufnagel DE. Metabolomics: applications and promise in mycobacterial disease. Ann Am Thorac Soc. 2015;12: 1278–1287. pmid:26196272
58. Shajahan-Haq AN, Cheema MS, Clarke R. Application of metabolomics in drug resistant breast cancer research. Metabolites. 2015;5: 100–118. pmid:25693144
59. Mathew S, Krug S, Skurk T, Halama A, Stank A, Artati A, et al. Metabolomics of Ramadan fasting: an opportunity for the controlled study of physiological responses to food intake. J Transl Med. 2014;12: 161. pmid:24906381
60. Johnson CH, Ivanisevic J, Siuzdak G. Metabolomics: beyond biomarkers and towards mechanisms. Nat Rev Mol Cell Biol. 2016;17: 451–459. pmid:26979502
61. Klein MS, Buttchereit N, Miemczyk SP, Immervoll A, Louis C, Wiedemann S, et al. NMR metabolomic analysis of dairy cows reveals milk glycerophosphocholine to phosphocholine ratio as prognostic biomarker for risk of ketosis. J Proteome Res. 2011;11: 1373–1381. pmid:22098372
62. Moco S, Collino S, Rezzi S, Martin FJ. Metabolomics perspectives in pediatric research. Pediatr Res. 2013;73: 570–576. pmid:23314292
63. Morrow AL, Lagomarcino AJ, Schibler KR, Taft DH, Yu Z, Wang B, et al. Early microbial and metabolomic signatures predict later onset of necrotizing enterocolitis in preterm infants. Microbiome. 2013;1: 13. pmid:24450576
64. Mohamad N, Ismet RI, Rofiee M, Bannur Z, Hennessy T, Selvaraj M, et al. Metabolomics and partial least square discriminant analysis to predict history of myocardial infarction

of self-claimed healthy subjects: validity and feasibility for clinical practice. J Clin Bioinforma. 2015;5: 3. pmid:25806102
65. Moore SS, Mujibi FD, Sherman EL. Molecular basis for residual feed intake in beef cattle. J Anim Sci. 2009;87: E41–E47. pmid:18952728
66. Widmann P, Reverter A, Weikard R, Suhre K, Hammon HM, Albrecht E, et al. Systems biology analysis merging phenotype, metabolomics and genomic data identifies non-SMC condensing I complex, subunit G (NCAPG) and cellular maintenance processes as major contributors to genetic variability in bovine feed efficiency. PLoS One. 2015;10: e0124574. pmid:25875852
67. Karisa B, Moore S, Plastow G. Complicated relationships: a review of biological interaction networks and pathways in animal science. Springer Sci Rev. 2013b;1: 73–83.
68. Chen X, Xie C, Sun L, Ding J, Cai H. Longitudinal metabolomics profiling of Parkinson's disease-related α-synuclein A53T transgenic mice. PLoS One. 2015;10: e0136612. pmid:26317866
69. Alves RDAM, Dane AD, Harms A, Strassburg K, Maleki Seifar R, Verdijk LB, et al. Global profiling of the muscle metabolome: method optimization, validation and application to determine exercise-induced metabolic effects. J Metabolomics. 2015;11: 271–285.
70. Bussche JV, Marzorati M, Laukens D, Vanhaecke L. Validated High Resolution Mass Spectrometry-Based Approach for Metabolomic Fingerprinting of the Human Gut Phenotype. Anal Chem. 2015;87: 10927–10934. pmid:26451617
71. Ganna A, Fall T, Salihovic S, Lee W, Broeckling CD, Kumar J, et al. Large-scale non-targeted metabolomic profiling in three human population-based studies. J Metabolomics. 2016;12: 4.
72. Psychogios N, Hau DD, Peng J, Guo AC, Mandal R, Bouatra S, et al. The human serum metabolome. PLoS One. 2011;6: e16957. pmid:21359215
73. Kosmides AK, Kamisoglu K, Calvano SE, Corbett SA, Androulakis IP. Metabolomic Fingerprinting: Challenges and Opportunities. Crit Rev Biomed Eng. 2013;41: 205–221 pmid:24579644
74. Verma M, Khoury MJ, Ioannidis JPA. Opportunities and Challenges for Selected Emerging Technologies in Cancer Epidemiology: Mitochondrial, Epigenomic, Metabolomic, and Telomerase Profiling. Cancer Epidemiol Biomarkers Prev. 2013;22: 189–200. pmid:23242141
75. Park YW, Juarez M, Ramosc M, Haenlein GFW. Physico-chemical characteristics of goat and sheep milk. Small Rum Res. 2007;68: 88–113.
76. Steinshamn H, Inglingstadb RA, Ekebergb D, Mølmanna J, Jørgensen M. Effect of forage type and season on Norwegian dairy goat milk production and quality. Small Rum Res. 2014;122: 18–30.
77. Andreas NJ, Hyde MJ, Gomez-Romero M, Lopez-Gonzalvez MA, Villaseñor A, Wijeyesekera A, et al. Multiplatform characterization of dynamic changes in breast milk during lactation. Electrophoresis. 2015;36: 2269–2285.
78. Wishart DS, Mandal R, Stanislaus A, Ramirez-Gaona M. Cancer Metabolomics and the Human Metabolome Database. Metabolites. 2016;2: 6.
79. Wishart DS. Computational strategies for metabolite identification in metabolomics. Bioanalysis. 2009;1: 1579–1596. pmid:21083105
80. McLaughlin MJ, Parker DR, Clarke JM. Metals and micronutrients ± food safety issues. Field Crop Res. 1999;60: 143–163.
81. Singh R, Gautam N, Mishra A, Gupta R. Heavy metals and living systems: An overview. Indian J Pharmacol. 2011;43: 246–53. pmid:21713085
82. Winter G, Krömer JO. Fluxomics—connecting 'omics analysis and phenotypes. Environ Microbiol. 2013;15: 1901–16. pmid:23279205
83. Srivastava A, Kowalski GM, Callahan DL, Meikle PJ, Creek DJ. Strategies for Extending Metabolomics Studies with Stable Isotope Labelling and Fluxomics. Metabolites. 2016; 6(4): E32. pmid:27706078
84. Norris JL, Caprioli RM. Analysis of Tissue Specimens by Matrix-Assisted Laser Desorption/Ionization Imaging Mass Spectrometry in Biological and Clinical Research. Chem Rev. 2013;113: 2309–2342. pmid:23394164

85. Römpp A, Both J, Brunelle A, Heeren RMA, Laprévote O, Prideaux B, et al. Mass spectrometry imaging of biological tissue: an approach for multicenter studies. Anal Bioanal Chem. 2015;407: 2329. pmid:25575583
86. Torii Y, Kawano Y, Sato H, Sasaki K, Fujimori T, Kawada J, et al. Quantitative metabolome profiling reveals the involvement of the kynurenine pathway in influenza-associated encephalopathy. J Metabolomics. 2016;12: 84.
87. Ghazi N, Arjmand M, Akbari Z, Mellati AO, Saheb-Kashaf H, Zamani Z. 1H NMR- based metabolomics approaches as non- invasive tools for diagnosis of endometriosis. Int J Reprod Biomed (Yazd). 2016;14: 1–8.
88. Parveen I, Moorby JM, Fraser MD, Allison GG, Kopka J. Application of Gas Chromatography-Mass Spectrometry Metabolite Profiling Techniques to the Analysis of Heathland Plant Diets of Sheep. J Agric Food Chem. 2007;55: 1129–1138. pmid:17249687
89. Clark M, Murray JD, Maga EA. Assessing unintended effects of a mammary-specific transgene at the whole animal level in host and non-target animals. Transgenic Res. 2014;23: 245–256. pmid:24214495
90. Metzler-Zebeli BU, Ertl R, Klein D, Zebeli Q. Explorative study of metabolic adaptations to various dietary calcium intakes and cereal sources on serum metabolome and hepatic gene expression in juvenile pigs. J Metabolomics. 2015;11 545–558.
91. Xia J, Broadhurst D, Wilson M, Wishart D. Translational biomarker discovery in clinical metabolomics: an introductory tutorial. J Metabolomics. 2013;9: 280–299.
92. Osorio MT, Downey G, Moloney AP, Röhrle FT, Luciano G, Schmidt O, et al. Beef authentication using dietary markers: Chemometric selection and modelling of significant beef biomarkers using concatenated data from multiple analytical methods. Food Chem. 2013;141: 2795–2801. pmid:23871026
93. Imhasly S, Naegeli H, Baumann S, von Bergen M, Luch A, Jungnickel H, et al. Metabolomic biomarkers correlating with hepatic lipidosis in dairy cows. BMC Vet Res. 2014;10: 122. pmid:24888604
94. Gray DW, Welsh MD, Doherty S, Mansoor F, Chevallier OP, Elliott CT, et al. Identification of systemic immune response markers through metabolomics profiling of plasma from calves given an intra-nasally delivered respiratory vaccine. Vet Res. 2015;46: 7. pmid:25828073
95. Skelton DM, Ekman DR, Martinović-Weigelt D, Ankley GT, Villeneuve DL, Teng Q, et al. Metabolomics for in situ environmental monitoring of surface waters impacted by contaminants from both point and nonpoint sources. Environ Sci Technol. 2014;48: 2395–2403. pmid:24328273
96. Shah SH, Kraus WE, Newgard CB. Metabolomic profiling for the identification of novel biomarkers and mechanisms related to common cardiovascular diseases. Circulation. 2012;169: 1110–1120.
97. Corona G, Rizzolio F, Giordano A, Toffoli G. Pharmaco-metabolomics: an emerging "omics" tool for the personalization of anticancer treatments and identification of new valuable therapeutic targets. J Cell Physiol. 2012;227: 2827–2831. pmid:22105661.

CHAPTER 6

Livestock feed resources, nutritional value and their implication on animal productivity in mixed farming system in Gasera and Ginnir Districts, Bale Zone, Ethiopia

Mekuanint Gashaw*1 and Girma Defar2

[1]*Department of Animal Sciences, Debre Tabor University, School of Agriculture and Environmental Science, Debre Tabor P. O. Box. 272, Ethiopia.*

[2]*Department of Animal and Range Science, Maddawalabu University, School of Agriculture, Bale Robe, P. O. Box 247 Ethiopia.*

ABSTRACT

The study was conducted in Gasera and Ginnir districts of Bale mid to highland areas, Oromia regional state, southeast Ethiopia; to assess livestock feed resources, to determine major feed nutritional value and their implication on animal productivity. Stratified random sampling (poor, medium and wealth) was applied to identify the sampled unit of 156 households in the selected districts. The chemical compositions of sampled feeds were determined using wet chemistry at Holeta Agricultural nutrition laboratory. Livestock production systems in the districts were extensive type. The metabolizable energy (ME) of the major available roughage feeds was 956,094.71 and 980,392.51 MJ per annum in the Gasera and Ginnir districts, respectively. Crop residues (aftermath and barley straw, wheat straw and teff straw) contributed about 80.4 and 81.73% in Gasera and Ginnir, respectively of the total annual ME supply of roughage feeds where the remaining is from natural grasses. The DCP (digestible crude protein) amount estimated in sampled feed was not statistically different in the study districts (P>0.05). Crude protein (CP) and in vitro dry matter digestibility (IVDMD) were significantly higher (P < 0.05) for feed from natural grasses than crop residues and aftermath. To the contrary, neutral detergent fiber (NDF), acid detergent fiber (ADF) and acid detergent lignin (ADL) contents were significantly higher (P < 0.05) for crop residues and aftermath. Mean crude protien content of crop residues and aftermath was less than the critical level of CP (7% DM) required for optimum function of rumen microbes, implying the need for supplementation. On the other hand, the amount of dry matter available from major sampled feed sources indicated that, it satisfied the maintenance requirement of cattle; however, the bulkiness of the feed cannot indicate the quality of the utilizable nutrients. This implies supplementation with concentrates and improved forage feeds are compulsory to overcome energy and protein deficiency in both study districts, especially during dry periods for reasonable livestock production.

Keywords: Livestock, feed resources, chemical composition.

1. INTRODUCTION

Ethiopia believed to have the largest livestock population in Africa. According to the recent CSA's (2012/13) sample survey report, Ethiopia is endowed with 53.99, 25.5, 24.06, 9.01,0.92 and 50.38 million heads of cattle, sheep, goat, equines, camels and poultry, respectively, excluding non-sedentary populations of three zones of Afar and six zones of Somali Regions. This indicates that, Ethiopia holds large potential for livestock products (milk, meat, manure, skin and hide) production in particular from large and small ruminants mainly due to its large livestock population, the relatively favorable climate and disease free environment (Holloway et al., 2000) compared to other tropical countries. In addition, the country enjoys diverse topographic and climatic conditions and hence livestock production at different levels takes place across all agro-ecological zones.

Because of the low potential for crop production including absence of or limited irrigation technologies in Ethiopia and most countries in COMESA (Common Market for Eastern and Southern Africa), livestock remains a major source of income and food for the majority of rural people in mixed and pastoral farming systems. In this respect, livestock ownership, in terms of both quantity and quality is an important asset because of its multiple social, economic and cultural uses (UNESC, 2012). Like other developing countries, agriculture is the largest economic sector in Ethiopia accounting for about 46% of the national GDP, employment of 85% of the labor force and 90% of the poor depend on the sector for their livelihood (World Bank, 2008). According to MoARD (2007), the livestock sector in Ethiopia accounts for 16% (national), 27-30% (agricultural) GDP, respectively and 13% of the country's export earnings. Similarly, World Bank (2011) reported that the livestock subsector provides 16% of the total GDP and generates 14% of the country's foreign exchange.

Different authors have classified livestock production systems in Ethiopia based on the criteria that include degree of integration of livestock with crop production, level of input and intensity of production and agro-ecology. Accordingly, about five production systems have been defined; namely mixed crop-livestock farming, pastoral, agro-pastoral, intensive dairying and peri-urban dairying (MoA, 1998; Mohammed et al., 2004; Yitay, 2007). Across these different production systems, the product from livestock basically (milk, draught power, meet, manure and cultural value) have vital place where the livelihood of small holder is based on integrated agriculture.

The mixed-farming system provides 50% of the meat and 90% of the milk production from the world and employs 70% of the world's poor livestock producers (Thornton and Herrero, 2001). Highland livestock production in Ethiopia supports 85% of the total human population (Deressa et al., 2008; World Bank, 2011). The importances of livestock production in Ethiopia, in particular in the study area (Bale Zone) were used as source of milk production, meat production, draught power and cash income. In mid to high land area of Bale Zone, there are large number of livestock population. However, productivity of the sector is still poor because of lack of feed in quality and quantity among other factors, especially during dry season. Thus, this study was conducted in Bale mid to highland area to know available feed resources for livestock and finally to analyze the nutritional value of the feeds. At the end, this study was give baseline information for policy makers and farming community to take overcome problems regarding to livestock feed. Finally, this paper aimed to address the following specific objectives;

i) To identify major livestock feed resources in the study area

ii) To evaluate major livestock feed nutritional value in the study area and;

iii) To assess the implication of feed resources and feeding strategies on livestock productivity under the existing farming system in the selected districts.

2. MATERIALS AND METHODS

2.1 Description of the study area

The study was conducted in Bale Zone, Oromiya Regional State of Ethiopia. The Zone holds a total area of 63555 km2 of landmass having 3 town districts and 18 rural districts of which 9 falls under the mixed farming system and the rest 9 are pastoral (BoFED, 2012).

Gasera district is found 490 km south east of Addis Ababa in the highlands of Bale Zone. Pelvic vertisols and Eutric nitosols soil types are dominant and they have good agricultural potentials (Bale Zone Agriculture Office). The main economic stay of the district is crop-livestock farming system. Major crops cultivated in the districts are wheat, barley, teff, maize, faba bean and linseed. On the other hand, livestock species raised by the farmers include cattle, sheep, goat, donkey, horse, mule and poultry with a population of 115,848, 13,119, 17,284, 17,318, 5,756, 1,237 and 36,560, respectively (CSA, 2012).

Ginnir district on the other hand is located at 570 km southeast of Addis Ababa in the mid and highlands of Bale Zone. The soil type is Chromic and Pelvic Vertisols which are very fertile with fairly good Agricultural potential (Bale Zone Agriculture Office). Cattle has the biggest population among the livestock species with 190,023 heads followed by goat, sheep and donkey (CSA, 2012).

3. METHODOLOGY

3.1 Sampling procedures and sample size

A reconnaissance survey was conducted and discussions held with agricultural office of Bale zone as per the thematic research topics by Maddawalabu University to select the two mixed farming system districts for the study. From the mixed-farming system districts, two districts were selected based on the information of Zone Agriculture Office and thematic research need assessment of Maddawalabu University on existing problems related to crop-livestock farming, feed resources and other recurring issues.

Based on the information obtained from the respective district offices of agriculture, with no consideration of the lowland *kebeles* in each district, six focal kebeles proportionally three from each district was randomly selected for household interview and feed sample collection.

The sample size of households were determined using Arsham (2005): $N=0.25/(SE)^2$, where: N = sample size, and SE = standard error of the proportion. Assuming the standard error of 4% at a precision level of 5% and the confidence interval of 95%, 156 households practicing crop-livestock farming were selected using random sampling technique for interview. The number of households per *kebeles* was fixed based on the proportion of the households practicing crop-livestock farming.

Before start the survey, semi structured questionnaire was pre tested on sample households to evaluate appropriateness of the design, clarity and interpretation of the question by farmers, relevance of the questions, quality of the data received and time taken for an interview to make appropriate modification and corrections. The prepared questionnaire was translated in to local languages (Afan Oromo) and administered to the interviewee's by co-investigator and enumerators. One-day training was organized for enumerators on how to administer the questionnaire. Interview was done by the co-researcher together with the enumerators and development agents from the target *Kebeles*. The interview were carried out at the farmer's home to enable counter checking of the farmer's response with respect to the livestock population and species, availability of feed resources, livestock feeding system. Group discussion was organized in each *Kebeles* with eight to ten key informants and development agents including farmers who had many years' experience and knowledge of mixed farming to collect data about livestock system and problems were ranked by vote.

The feed samples for laboratory analysis were consider the major livestock feeds of the following different feeds used by the farmers to feed their animals in the districts namely; mixed natural grasses, mixed fallow land herbaceous forages (grasses), wheat straw, barley straw, emmer wheat straw, teff straw, pulse hallums (straw), wheat bran and homemade concentrates.

3.2 Data collection

A combination of primary and secondary data was used. For primary data generation; field visits, interviews, focal group discussions, feed sampling and laboratory analysis were the methods employed; and for secondary data the sources were published or unpublished reports from different organizations and literatures.

3.3 Sample collection

Major livestock feed samples (mixed natural grasses, mixed fallow land herbaceous forages (grasses), crop residues and concentrates) were collected for laboratory analysis based on farmers' response during interview.

3.4 Estimation of forage biomass yield

To determine the potential forage biomass yield and DM, representative samples of mixed natural grass and herbaceous vegetation were taken from protected areas in wet season based on pasture plants full growth to 50% flowering stage. For grazing pasture (1 m x 1 m) area, harvesting was done at the ground level (Zewdie, 2010). Fresh weights of harvested samples were taken immediately by using a spring balance of 20g precision. For further chemical analysis, a composite sample was taken from the bulk.

Dry roughage feeds (crop residue) samples were collected during dry season after crop was harvested and threshed. The quantity of available crop residues produced by farmers was estimated by applying grain to straw ratio as suggested by FAO (1987). Accordingly, for a ton of wheat, barley, emmer wheat and teff straw, a multiplier of 1.5 were used and for pulses haulm (straw) a multiplier of 1.2 were used. The quantity of potentially available crop residues for animal consumption were estimated by assuming 10% wastage (Adugna and Said, 1994) and the grazing potential of crop aftermath (stubbles) were estimated using a mean of 0.5 ton per ha (FAO, 1987). The amounts of grain yield obtained from the respective crops were quantified by interviewing the farmers and crosschecking it with the data recorded by development workers for any deviation.

3.5 Assessment of livestock feed requirement

The annual availability of feed was compared with the annual requirements of the livestock population. Livestock populations were converted in to Tropical Livestock Unit (TLU) according to Gryseels (1988). The DM requirements for maintenance was calculated based on daily DM requirements of 250 kg tropical cattle (an equivalent of 1 TLU). Nutrients supplied by each feed types were estimated from the total DM output and nutrient content of that feed on DM basis. The total nutrient requirements (DM, CP and ME) per day per TLU were estimated based on recommendations of Kearl (1982) and McCarthy (1986) for tropical livestock.

3.6 Chemical analysis of sample feeds

Chemical analyses of feed samples were performed at Holeta Agricultural Research Center's nutrition laboratory. The representative samples of major feed resources from the study districts were collected, bulked, dried, sub-sampled and ground to pass through 1 mm mesh sieve for determination of dry matter (DM), ash, nitrogen (N), neutral detergent fiber (NDF), acid detergent fiber (ADF), acid detergent lignin (ADL) and *in vitro* dry matter digestibility (IVDMD). The samples were pre-dried at 135°C for two hours in an oven dry to remove the moisture content in the feed before chemical analysis and analyzed for DM, OM, CP and ash contents following the methods described by AOAC (1990). *In vitro* dry matter digestibility IVDMD, NDF, ADF and ADL were determined according to the methods of Van Soest and Robertson (1985). The crude protein (CP) content was calculated as Nx6.25. Metabolisable Energy (ME) and Digestible Crude Protein (DCP) content of a particular feed were estimated from IVDOMD and CP contents, respectively based on the following equations.

ME (MJ/kg DM) =0.15*IVDOMD (g/kg) (MAFF, 1984)

DCP (g) = 0.929*CP (g) -3.48 (Church and Pond, 1982)

3.7 Statistical Analysis

Collected data was analyzed using appropriate software (SPSS, version 20). Descriptive statistics were employed to describe the various variables in the mixed crop-livestock production system.

4. RESULT AND DISCUSSION

4.1 Feed resources and feeding strategies

4.1.1 Available feed resources

The principal feed resources available to livestock in the study districts included natural grass from (communal grazing land, individual pasture land and fallow land, road side, cultivated land borders (alley) and marginal lands), crop residues, crop aftermath and homemade concentrates (HMC) and it is coincided with the report of Alemayehu and Sissay (2003) and Yeshitila (2007).

Table 1. Available annual major livestock feed resources in Gasera and Ginnir districts of Bale mid to highland areas.

Feed type	Gasera		Ginnir		Overall	
	DM (tons)	% (N=78)	DM (tons)	% (N=78)	DM (tons)	% (N=78)
Natural grass	4.86 ± 0.35^{a}	36.76	3.53 ± 0.47^{b}	34.64	3.96±0.41	35.70
Fallow land grass	0.89 ± 0.02^{a}	12.29	0.68 ± 0.03^{b}	13.60	0.74±0.03	12.34
Crop residue	8.35 ± 0.78^{b}	31.19	9.14 ± 0.86^{a}	33.06	8.77±0.82	33.12
Aftermath grazing	2.74 ± 0.26^{a}	17.70	2.78 ± 0.22^{a}	17.08	2.75±0.24	17.22
HMC	0.18 ± 0.01^{a}	02.06	0.13 ± 0.02^{b}	01.62	0.15±0.02	01.72
Total feed supply	16.02 ± 0.28^{a}	100	16.26 ± 0.32^{a}	100	16.37±0.30	100

^{ab} = means with different superscripts within a row are significantly different (*P < 0.05*); HMC= the mixture of partially milled cereal crops (barley, wheat, emmer wheat etc.) and pulse crops which are not used for human consumption. DM = dry matter; N = number of respondents.

Table 2. Estimated available major crop residues per household in Gasera and Ginnir districts of Bale mid to highland areas.

Crop residues	Gasera		Ginnir		Overall	
	Mean ± SE DM (tons)	%	Mean ± SE DM (tons)	%	Mean ± SE DM (tons)	%
Wheat	7.47 ±0.84[b]	39.10	8.31 ± 0.53[a]	42.08	7.76 ±0.42	40.59
Barley	2.07 ±0.33[a]	32.34	1.79 ± 0.41[b]	28.26	1.82 ±0.08	30.82
Emmer wheat	0.73 ±0.06[b]	9.44	0.78 ± 0.10[a]	9.12	0.76 ±0.01	09.28
Teff	0.54 ±0.04[b]	8.56	0.76 ± 0.08[a]	9.42	0.60 ±0.40	08.72
Pulses	0.28 ±0.02[a]	5.76	0.29 ± 0.05[a]	7.26	0.29 ±0.01	06.31
Maize	0.19 ±0.06[a]	5.40	0.14 ± 0.06[b]	3.76	0.16 ±0.07	04.28
Total residue supply	11.29 ±0.35[b]	100	12.08 ±0.23[a]	100	11.45±0.99	100

ab = means with different superscripts within a row are significantly different ($P < 0.05$); DM = dry matter; N = number of respondents; SE = standard error

Generally, crop residues from (barley, wheat, emmer wheat, teff, pulses and maize) were found to contribute more to the feed resource base in the study area (Table 2) which reconcile with the report of Tolera. et al., 2012. The extent of availability of different type of feed resources like natural grass, fallow land grass varied significantly ($P < 0.05$) between the districts (Tables 1) even though aftermath grazing and total feed supply is not significantly different. Thus, the reason for the difference could be attributed to the variation in land-use system in the districts, frost, crop rust, crop diseases and whereby yield of crop residue is related to area of land allocated for crop production and the crop grain yield.

Table 2 indicates that from the feed supplied by crop residues; 40.49% was barley straw, 30.82% wheat and the other 28.69% (almost one third of the total residue) were supplied by teff, emmer wheat, pulses and maize stover in the study districts. Even though higher yield of wheat straw is associated with the allocation of more land to wheat cultivation, primarily due to availability of improved seeds, productivity and encouraging price for this crop, its palatability as animal feed is very low. Pulses (field pea and faba bean) straws contributed only about 7.26% of the total crop residue while maize contributed 3.76%. This could be due to the low yield and small area of land allocated for the crops in the districts because of environmental and other related factors. The contribution of crop residues to the feed resource base in Ginnir was significantly higher ($P < 0.05$) than that in Gasera district (Table 2). Crop residues supplied about 61.29% of the annual DM available per household in Ginnir area (Table 8).

In Gasera district, pastureland including grazing of communal, individual and fallow lands provided 5.75 tons of the annual DM supply per household, which was significantly higher ($P < 0.05$) than that of Ginnir (4.21 tons) (Table 1). This indicates that livestock in Gasera district were relatively less dependent on crop residue for feed, whereby the rest of the feed was obtained from crop residues, crop aftermath and HMCs. Feed obtained from fallow lands was also significantly higher ($P < 0.05$) in Gasera (Table 1). This may be due to fallowing of the land, which is a common practice in the district. Feed from aftermath grazing was not significantly different ($P < 0.05$) in both districts. This might be associated with the land use system in the districts where more land is assigned to cultivation and crop grain yield that resulted in higher aftermath grazing potential.

The overall, feed DM did not significantly different ($P > 0.05$) for both Ginnir and Gasera districts (Table 1) which disagreement with the finding of Dawit et al. (2013). This is because of the available feed from crop residues in one district being compensated by that from pasture land in the other district. According to Sere and Steinfeld (1996), if more than 10% of the feed DM for livestock comes from crop subsector, the production system is categorized as mixed crop-livestock farming systems.

Table 3. Feeding priorities of crop residues and supplements for livestock practiced in Gasera and Ginnir districts of Bale mid to highland areas (in % of respondents).

Animal type	Districts	Crop residue (rank)					Supplement feed (rank)				
		N	1st	2nd	3rd	4th	N	1st	2nd	3rd	4th
	Gasera	78	64.5	25.4	7.0	3.1	56	42.0	41.1	14.8	1.1
Oxen	Ginnir	78	66.2	25.0	8.8	---	53	38.1	45.7	14.1	2.1
	Overall	156	65.3	25.2	7.9	3.1	109	40.0	43.4	14.4	1.5
	Gasera	78	32.3	48.0	16.5	2.2	61	29.5	33.5	20.0	17.0
Lactating cows	Ginnir	78	30.0	52.0	14.7	2.3	65	32.0	28.4	24.0	15.6
	Overall	156	31.1	50.0	15.5	2.2	126	30.5	30.5	22.0	16.3
	Gasera	67	11.4	23.5	48.0	17.1	70	45.0	40.0	15.0	---
Calves	Ginnir	71	12.0	27.6	44.4	16.0	66	48.5	39.0	12.5	---
	Overall	138	11.5	25.1	46.2	16.5	136	46.5	39.5	13.5	---
	Gasera	73	10.6	47.0	33.5	8.9	72	20.8	52.1	22.3	4.7
Weak animals	Ginnir	70	6.0	40.6	47.4	6.0	72	34.7	24.2	31.1	10.0
	Overall	143	8.3	43.5	40.5	7.5	144	27.5	38.1	26.5	7.3

Therefore, the production system in the districts is with the mentioned range which indicated that the system is classified as mixed crop-livestock production system.

4.2 Feeding strategy and seasonal feed supply

Feeding strategy depends on the nature of the farming system, objective of herding animals and the availability of feed resources in that area which is to be affordable by the farmers. Livestock of mixed type (cattle of different category, small ruminants and equines), except draft animals, were herded together on degraded communal pasturelands, roadsides and marginal lands. Draft animals and in some cases, lactating cows owned by different households in the locality were kept around crop lands herded or tethered (FGD and field observation). In line with this, farmers were ranked feeding priority of crop residues and supplements to the livestock types they owned (Table 3). About 66.2 and 64.5% of them gave priority to draught animals in feeding of crop residues in (due to disease, stress, feed shortage, etc) and calves were given comparable attention in feeding crop residue relative to oxen and lactating cows (Table 3). The reason for such type of management is probably to maintain body condition of draught animals, particularly during peak times of cropland ploughing.

4.3 Feeding calendar

Grazing was the major source of livestock feed throughout the year on available poor grazing lands, in the districts with significant contribution of other feed resources during the dry season listed in Table 4.

Crop residues were the main source of feed during the dry season from early February to June when pasture from grazing area not able to provide reasonable quantity of feed in the study districts. Teff and emmer wheat residues were equally preferred next to barley straw while pulses and wheat straw are the last choice when the other residues are not available at enough quantity. Therefore, residues from cereal crops were collected and stacked under shade around homestead commonly after threshing the crop in dry season. Farmers usually fed pulse straw directly on threshing field rather than after storage. This may be due to the small quantity of pulse crop production and the straw from these crops is sensitive to moisture to develop moulds whereas straws from pulse crops are less palatable, said the respondents. In general, the intensity of cultivation and crop production is generally higher in both districts.

Table 4. Available feed resources and feeding calendar in the study districts.

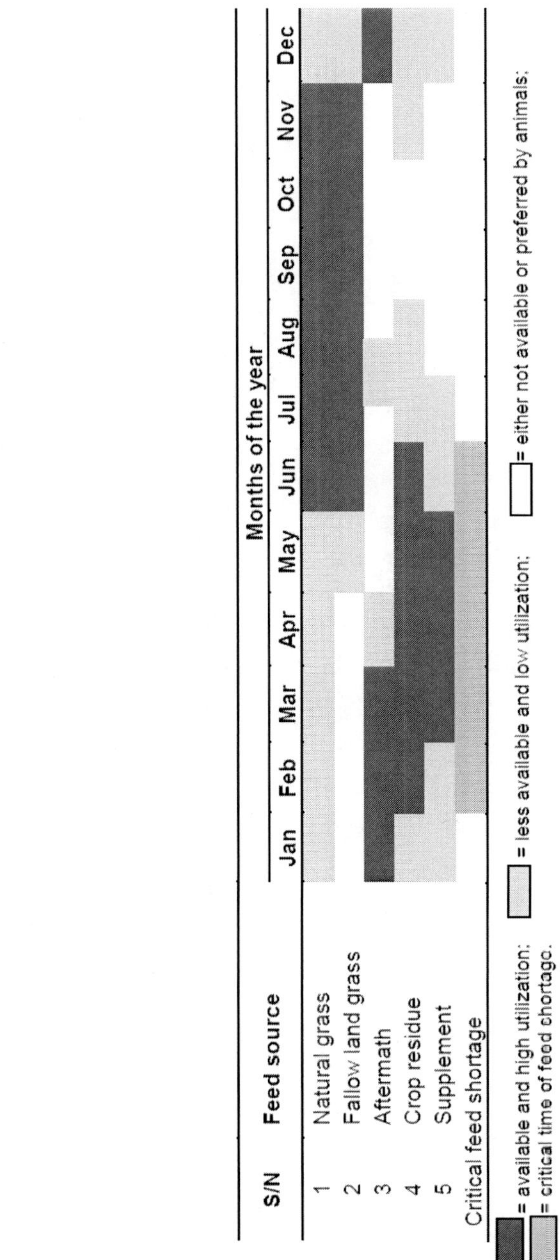

Table 5. Mean chemical composition and *in vitro* dry matter digestibility of major livestock feed resources in Gasera district of Bale mid to highland area.

Variables	DM	% of DM					
		Ash	CP	IVDMD	NDF	ADF	ADL
Natural grass	38.53^{b}	10.99^{a}	8.96^{b}	64.87^{b}	73.15^{b}	43.53^{b}	$7.29b^{c}$
Fallow land grass	25.06^{b}	8.96^{b}	10.21^{b}	69.55^{b}	71.56^{b}	42.61^{b}	5.65^{c}
Crop aftermath	94.62^{a}	9.12^{ab}	3.77^{c}	40.36^{c}	79.85^{a}	56.34^{a}	11.24^{a}
Wheat straw	93.55^{a}	9.14^{ab}	3.22^{c}	41.92^{c}	77.72^{a}	49.51^{a}	10.30^{ab}
Barley straw	92.89^{a}	9.67^{ab}	4.01^{c}	48.31^{bc}	75.26^{a}	48.45^{a}	8.97^{b}
Emmer wheat straw	93.52^{a}	9.72^{ab}	3.91^{c}	47.77^{bc}	76.72^{a}	49.82^{a}	9.01^{b}
Teff straw	93.13^{a}	9.01^{ab}	3.78^{c}	46.84^{bc}	74.17^{ab}	49.63^{a}	9.22^{b}
Pulses straw	93.64^{a}	8.91^{b}	7.33^{b}	54.44^{c}	66.23^{bc}	51.71^{a}	13.64^{a}
HMCs	88.12^{a}	6.52^{c}	16.13^{a}	86.05^{a}	40.55^{c}	28.47^{c}	5.11^{c}
Wheat bran	89.70^{a}	6.86^{c}	14.43^{a}	84.32^{a}	46.80^{c}	33.24^{c}	6.26^{c}

ab = means with different superscripts within a column are significantly different (*$P < 0.05$*); HMC= the mixture of partially milled cereal crops (barley, wheat, emmer wheat etc.) and pulse crops which are not used for human consumption DM = dry matter; CP = crude protein; DCP = digestible crude protein; IVDMD = *in vitro* dry matter digestibility; ME = metabolizable energy; NDF = neutral detergent fiber; ADF = acid detergent fiber; ADL = acid detergent fiber.

Livestock were allowed to graze aftermath of crops preferably that of barley twice in a year in both study districts as barley crop is planted twice a year. Aftermath grazing of cereal crops occurred after harvest from mid July to late August during short rainy season (mostly of barley and teff) and from November to April in Gasera and Ginnir districts. Initially, aftermath grazing is reserved and grazed privately for some period and then after it became accessible to the community in the locality. This is probably to exploit the potential of crop aftermath at individual level since land available for stubble grazing is of private resource. Moreover, livestock have to be herded closely to protect from crop damage until harvesting of food crops is over in dry season. Homemade concentrates were provided for the animals in dry season when feed shortage is more pronounced and to some animals any time under special cases (FGD).

4.4 Chemical composition of major feed resources

4.4.1 Dry matter and ash content

The DM content of the natural grass from grazing and fallow land is significantly lower ($P < 0.05$) than crop residues and concentrates in the study districts (Table 5). Ash content of natural pasture from grazing land and fallow land was 10.99 and 8.96% in Gasera while 9.89 and 9.14% in Ginnir district, respectively. Similarly, ash content of homemade concentrate and wheat bran were found to be 6.52 and 6.86% of the DM in Gasera district. Ash content of the concentrates was significantly different ($P < 0.05$). This is due to the feed type (cereal grain products and by-products) difference and the actual dominating organic components in the feed. Higher ash content observed in the study area for natural pasture indicates that it may satisfy mineral requirement of the animals for production at the time of availability of that particular forage.

Ash content was found to be 9.12, 9.14, 9.67, 9.72, 9.01 and 8.91% of the DM for Aftermath, wheat, barley, emmer wheat, teff and pulses residues, respectively in Gasera district (Table 5). Similarly, the ash content of Aftermath, wheat, barley, emmer wheat, teff and pulses residues was found to be 10.22, 9.34, 9.11, 9.31, 9.31 and 8.95% of the DM, respectively in Ginnir district (Table 6). The values for wheat straw and crop aftermath are more comparable than others. Overall the ash content of crop residues in the study districts is comparable to the results of Solomon (2004) and Yitaye (1999) in different parts of Ethiopian highlands. In contrast, the latter author obtained ash content of 14.61 % DM for barley which much more than the current result. This variation may be due to environmental factors, crop management practices and varietal difference of the crops under production.

Table 6. Mean chemical composition and *in vitro* dry matter digestibility of major livestock feed resources in Ginnir district of Bale mid to highland area.

Variables	DM	% of DM					
		Ash	CP	IVDMD	NDF	ADF	ADL
Grazing land grass	44.97^{b}	9.89^{a}	8.10^{b}	63.96^{b}	74.537^{b}	45.62^{b}	7.82^{c}
Fallow land grass	35.17^{b}	9.14^{ab}	9.62^{b}	69.84^{b}	71.88^{b}	43.55^{b}	6.31^{c}
Crop aftermath	94.71^{a}	10.22^{a}	3.44^{c}	40.78^{d}	78.65^{a}	55.77^{a}	11.77^{a}
Wheat straw	93.72^{a}	9.34^{ab}	3.62^{c}	42.22^{d}	78.25^{a}	52.34^{a}	11.00^{a}
Barley straw	92.53^{a}	9.11^{ab}	3.92^{c}	49.83^{bc}	74.78^{a}	47.17^{ab}	8.80^{b}
Emmer wheat straw	93.14^{a}	9.31^{ab}	3.88^{c}	48.15^{bc}	76.86^{a}	48.03^{ab}	8.93^{b}
Teff straw	93.13^{a}	9.31^{ab}	3.78^{c}	49.28^{bc}	73.01^{ab}	49.14^{ab}	9.20^{b}
Pulses straw	93.57^{a}	8.95^{b}	7.02^{b}	51.60^{c}	69.75^{bc}	50.59^{a}	12.72^{a}
HMCs	87.87^{a}	6.54^{c}	16.44^{a}	86.66^{a}	38.85^{c}	30.32^{c}	5.42^{c}

ab = means with different superscripts within a column are significantly different ($P < 0.05$); HMC= the mixture of partially milled cereal crops (barley, wheat, emmer wheat etc.) and pulse crops which are not used for human consumption DM = dry matter; CP = crude protein; DCP = digestible crude protein; IVDMD = *in vitro* dry matter digestibility; ME = metabolizable energy; NDF = neutral detergent fiber; ADF = acid detergent fiber; ADL = acid detergent fiber.

According to Preston and Leng (1984), mineral content of straw is generally low and imbalanced but this may be quite adequate for maintenance, while the requirements for minerals by the animal for production are increased many fold. Therefore, mineral content of straw in the study area is likely to be insufficient for livestock production.

4.4.2 Crude protein content

Feed obtained from natural grass (grazing land and fallow land) at Gasera had relatively higher CP content than that from Ginnir (Tables 5 and 6). The lowest CP content of grazing land grass was 8.96% DM in Gasera and 8.10% DM in Ginnir district. The value is in agreement with the reports of Yihalem (2004) for northern highlands of Ethiopia. High CP content observed in the present study district may be attributed to high legume proportion of natural pasture at the time of harvest. In general, feeds harvested from natural grass, can support maintenance requirement of ruminants since CP content is above 7.2% DM (ARC, 1980).

Mean CP content of crop aftermath, wheat, barley, emmer wheat, teff and pulses straw was 3.77, 3.22, 4.01, 3.91, 3.78 and 7.33% DM in Gasera district, respectively. On the other hand 3.44, 3.52, 3.92, 3.88, 3.78, and 7.02% DM values found to be the CP content of aftermath, wheat, barley, emmer wheat, teff and pulses straw in Ginnir district, respectively (Tables 5 and 6). Lower CP content in Ginnir area may be related to the longer period of time required for physiological maturity to harvest the crop which induces dilution of CP and enhances lignifications (McDonald et al., 1995). The mean CP content of pulses straw was found to be significantly higher ($P < 0.05$) than that of different crop straws (Tables 5 and 6). This is due to the high nitrogen content of the pulse crops as they are legumes. For wheat straw, however, CP content was found to be higher than the previous reports of 2.4% DM (Seyoum et al., 2001) and 2.7% DM (Gashaw, 1992) in the central highlands of Ethiopia. This may be due to varietal differences and variation in management practices as well as soil fertility gradient of the areas. For pulses straw mean CP content was 7.33 and 7.02% DM in Gasera and Ginnir districts, respectively (Tables 5 and 6).

All straws from different crops, except that of pulses had CP content of less than the critical level of CP for optimum microbial rumen function (Van Soest, 1982). However, it was mentioned in the previous section that out of available feed resources in the study area, crop residues and crop aftermath contributed significant amount of annual feed budgeting. This shows that supplementation by protein source is required for reasonable livestock production. Natural pastures were observed to have significantly higher ($P < 0.05$) CP content as compared to straws and crop aftermath in the districts (Tables 5 and 6). A minimum of 15% CP is required for lactation and growth of ruminants (Norton, 1982) and hence the entire crop residues in the study area are by far below the requirement for production, indicating the need for supplementation of animal reared for the same purpose is compulsory.

4.4.3 In vitro dry matter digestibility

In vitro DM digestibility aftermath and wheat straw was40.36 and 41.92% in Gasera district and 40.78 and 42.22% in Ginnir district, respectively (Tables 5 and 6). This is by far the lowest IVDMD of the major livestock feeds indicated in the table in the study districts. The percentage IVDMD for barley, emmer wheat, teff and pulses straw were comparable in both districts. The IVDMD was similar ($P < 0.05$) for barley, emmer wheat, teff and pulses straw, while the lowest was observed for wheat straw and aftermath among the crop residues. Lower IVDMD for aftermath and wheat straw in the area may be due to low leaf proportion for aftermath and prolonged time required for crop maturity for wheat compared to other cereals grown in the area that induces intensive lignifications. Mean IVDMD observed was within the range reported by Daniel (1988) and Gashaw (1992) for barley. Percent IVDMD observed was 64.87 and 69.55% in Gasera and 63.96 and 69.84% in Ginnir districts for grazing and fallow land, respectively. Among the crop residues, crop aftermath had the lowest (40.36 and 40.78%) IVDMD followed by IVDMD of wheat straw (41.92 and 42.22%) in Gasera and Ginnir districts, respectively.

Unlike CP content, mean IVDMD of pulse straw (54.44% and 51.60%) was statistically similar to that of cereal straws in both study districts. Variation of IVDMD was observed between crop aftermath and cereal straw rather than among cereal crops (barley, emmer wheat, teff and pulses except wheat straw) among straws of wheat, barley and emmer wheat. Crop residues with less than 50% OM digestibility reduce intake to 50 g/kg $LW^{0.75}$ or less (Mosi and Butterworth, 1985). Thus, crop aftermath and wheat straw had less than 50% to limit feed intake, suggesting the need for either improvement of the nutritive value of these feed resources or supplementation with good quality feed for reasonable livestock production.

4.4.4 Fiber content of the major feeds

Neutral detergent fibre ranged from 66.23 to 79.85% for roughage feeds (natural grasses and crop residues in Gasera and 69.75 to 78.65% in Ginnir districts (Tables 5 and 6). The highest and lowest NDF cont was recorded for crop aftermath and pulses straw, respectively in both districts of the study area. In general, NDF content is higher for crop aftermath and cereal residues, followed by natural grass and pulses straw with lowest content is from HMCs and wheat bran.

NDF contents of crop residues appeared to be close to the previous reports for central highlands of Ethiopia (Seyoum et al., 2001). Crop aftermath had significantly higher (P < 0.05) NDF content than pulses straw (Tables 5 and 6). However, variation in NDF content was not significant (P > 0.05) for natural pasture in the study districts. Generally, NDF content of roughage feeds with less than 45% was grouped as a high quality feed, while feed with NDF content of 45- 65 was categorized as medium quality feed by Singh and Oosting (1992). Therefore, all crop residues had NDF content above 65% to be categorized under low quality feed that could affect livestock production.

The NDF content of roughage feeds that ranged from 66.23-79.85% in the study districts reported was to be high enough to limit DM intake and digestibility (Alemu, 1982). According to Van Soest (1985), NDF content above 55% was reported to limit DM intake and hence, NDF content of the natural pasture and crop residues in the study area could affect feed intake, which directly limits productivity of animals.

The mean maximum ADF content was 45.62, 56.34 and 51.71% DM for natural pasture (grazing and fallow land), crop aftermath and crop residues, respectively from both districts of the study area (Tables 5 and 6). ADF values for natural pasture was within the range of the report of Yihalem (2004) that ranged from 39.97-52.23% DM for the northern highlands of Ethiopia. The same author also observed increase in ADF as stage of harvesting advanced. The ADF content was significantly lower (*P < 0.05*) for natural pasture than either cereal straws or crop aftermath. However, the difference was not significant (*P > 0.05*) for straw and stubble (crop aftermath). Mean ADF content ranges from 48.45 to 49.82% DM for wheat, barley, emmer wheat and teff straw, where the minimum value is for barley and the maximum is for emmer wheat in Gasera district, while in Ginnir district it ranges 47.17 to 52.34% DM with maximum value for wheat straw (Tables 5 and 6). Pulses mean ADF content was 51.71 and 50.59% DM for Gasera and Ginnir districts, respectively. The ADF value obtained for crop residues in this study was comparable to the results of Yitaye (1999).

Lignin (ADL) content values was found for natural grass (5.65-7.82%), crop aftermath (11.24-11.77%), cereal straws (8.80-11.00%), pulses straw (12.72-13.64%) and available concentrates (5.11-6.26%) DM in the study districts (Tables 5 and 6). Lignin content was higher for straw of pulses as compared to that contained in cereal straws. This may be associated with varietal, climate and crop management differences in the area. Lignin content was significantly lower (*P < 0.05*) for natural pasture than either crop residues or crop aftermath which is comparable with that of available concentrates in the districts.

According to McDonald et al. (1995), in young pasture containing only 50 g lignin/kg, 80% of the fiber (cellulose and hemicellulose) may be digested but in older herbage with 100 g lignin in one kg, the proportion of the fiber digested may be less than 60%. Thus, lignin contents of the crop residues and aftermath in the study districts were high enough (above 7%) to limit DM intake and livestock production at the end. Relatively natural pasture in some of the wet months may had less than 5% lignin content, perhaps not to influence feed intake. However, the content went beyond 5% at the end of rainy season. This is probably due to the positive correlation between lignifications and plant maturity (Reed et al., 1988), where maturity of the plant is accompanied by an increase in the fiber and lignin content of the plant.

4.5 Nutrient content estimation and implication on animal productivity

4.5.1 Estimation of major feed resources nutrient supply

Nutrient content of the major roughage feed resources per kg of feedstuff sampled in the Gasera and Ginnir districts are indicated in Table 7. The value of ME in natural grass (grazing and fallow land grass) ranged 9.59-9.73 MJ/kg DM for grazing land and 10.43-10.48 MJ/kg DM for fallow land grasses in the study districts. Content of ME in crop residues ranged between 6.05 for aftermath to 8.16 MJ/kg DM for pulses straw in Gasera and 6.11 for aftermath to 7.74 MJ/kg DM for pulses straw in Ginnir district. Seyoum and Zinash (1989) observed ME content of barley 7.8 MJ/kg DM, which was comparable with energy content of barley straw and other roughage feeds in the current study. For crop aftermath, energy concentration was the lowest value which was recorded in Gasera (6.05 MJ/kg DM).

Lower nutritive value of the crop residues in the study area may be related to the long period of time required for crop maturity because of erratic rainfall condition and other related factors. According to Preston and Leng (1984), the main sources of energy diet in crop residues are the cell wall constituents (cellulose and hemicellulose), which are not readily digestible (Tables 5 and 6). However, the same authors reported addition of a readily fermentable carbohydrate source to the roughage diets possibly increase the microbial biomass and rate of degradation of the fibers for continuous intake and supply of energy. Therefore, the ME content of feed in this study might not be adequate and readily available to animals since the overall intake of the roughage feeds is low due to high fiber content and low CP content.

Table 7. Mean ME and DCP content of major sampled feed resources in the Gasera and Ginnir districts of Bale mid to highland areas.

Feed type	ME ($MJkg^{-1}DM$)		DCP ($gkg^{-1}DM$)	
	Gasera	Ginnir	Gasera	Ginnir
Grazing land natural grass	9.73	9.59	79.76	71.76
Fallow land grass	10.43	10.48	91.37	85.89
Crop aftermath	6.05	6.11	31.54	28.48
Wheat straw	6.28	6.33	26.43	28.15
Barley straw	7.24	7.47	33.77	32.94
Emmer wheat straw	7.16	7.22	32.84	32.56
Teff straw	7.03	7.38	31.63	31.64
Pulses straw	8.16	7.74	64.61	61.73

DM = dry matter; DCP = digestible crude protein; ME = metabolizable energy.

Energy content of natural grass from fallow land in this study was similar to the energy content of similar feeds from the central highlands of Ethiopia and Sinana district of Bale highlands with 10.2 MJ/kg DM (Gashaw, 1992) and 10.80 MJ/kg DM (Solomon, 2004). Higher energy concentration of feed from natural grass (from grazing and fallow land) may be due to natural herbaceous legume mixed grass in the study districts particularly as observed in the fallow land grasses.

The DCP content of the natural grass (fallow and grazing land grasses) ranged 71.76 g/kg DM in Ginnir to 91.37 g/kg DM in Gasera for this particular study (Table 7). Mean DCP content of cereal straws was found within the range of 28.48 g/kg DM for crop aftermath to 61.73 g/kg DM pulses straw in Ginnir district while the DCP of feeds from Gasera district found within the range of 26.43 to 64.61 g/kg DM (Table 7). Digestible CP observed for pulses was the highest, which ranged from 61.73 g/kg DM (in Gasera) to 64.61 g/kg DM (in Ginnir). It was reported that nitrogen (N), but not energy, is the first limiting nutrient for animal performance (productivity) through the effect on feed intake and digestibility particularly during the dry season (Preston and Leng, 1984). In the study districts, the major feed resources during dry periods had low CP/DCP content (Tables 5 and 6) and, therefore, supplementation is required even to satisfy maintenance requirements of animals.

4.6 Annual nutrient yield estimation of major feed resources

Metabolizable energy available was 956,094.719 and 980,392.51 MJ per annum in the Gasera and Ginnir districts, respectively (Table 8). As the overall total ME supplied from the available feeds in the districts is comparable to each other, the same is true for the same feed items from the two districts that they contribute almost comparable percent of ME. The annual ME supplied by wheat straw was high proportion (41.42% vs 39.78% in Gasera and Ginnir, respectively) since wheat is the major crop produced in the area. However, its utilization is negligible compared to other crop residues as indicated by the respondents and FGD participants. This might be because of its poor nutrient content and low palatability by the animals inherently. Whereas annual available ME from grazing lands was higher in Gasera district as compared to Ginnir. This is related to variation in the land use system and chemical composition of the feed resources in the area.

Table 8. Annual average ME and DCP supply from sampled major feed resources in the Gasera and Ginnir districts of Bale mid to highland areas.

Feed type	ME (MJ)				DCP (gm)			
	Gasera	%	Ginnir	%	Gasera	%	Ginnir	%
Grazing land grass	108,333.00	11.33	95,309.83	9.72	929,101.10	17.70	713,183.88	13.83
Fallow land grass	79,160.98	8.27	83,826.46	8.55	693,474.54	13.21	687,099.20	13.32
Crop aftermath	97,955.67	10.24	101,794.18	10.38	510,664.77	9.73	474,484.20	9.19
Wheat straw	396,042.54	41.42	390,025.35	39.78	1,666,784.16	31.75	1,734,472.93	33.62
Barley straw	174,066.25	18.20	187,362.91	19.11	811,908.47	15.47	826,202.72	16.02
Emmer wheat straw	26,234.95	2.74	34,332.04	3.50	120,329.04	2.29	154,855.56	3.00
Teff straw	20,999.80	2.19	35,963.25	3.67	94,484.18	1.80	154,183.93	2.98
Pulses straw	53,301.52	5.57	51,778.49	5.28	422,035.75	8.04	413,036.66	8.00
Subtotal of districts	**956,094.71**	**100**	**980,392.51**	**100**	**5,248,782.00**	**100**	**5,157,519.08**	**100**
Overall total	**1,936,487.22**				**10,406,301.08**			

DCP = Digestible crude protein; ME = metabolizable energy.

In general crop residues (aftermath and straw) contribute about 80.4 and 81.73% in Gasera and Ginnir, respectively of the total annual ME supply of roughage feeds in the districts with the inclusion of wheat straw (Table 8). However, in line with the utilization of wheat straw, farmers (FGD) underlined that, they used wheat straw as their animal feed when there are no other feed options in the area particularly during prolonged dry season. On the other hand the high proportion of crop residues utilization as animal feed in the study are indicate that high cropping intensity in the area. The overall ME supplied by crop residues of the total ME available in the study districts was higher than the contribution of crop residues (45-48%) observed in the highlands of Arsi (Abdinasir, 2000) and 55.75% reported by Solomon (2004) in Sinana-Dinsho district.

The annual mean DCP supply in Gasera and Ginnir was 5,248,782.00 and 5,157,519.08 gm, respectively (Table 8). The DCP supply was not statistically different in the study districts. This is probably due to relatively comparable feed resources availability in the area, which supplies CP to the equivalent extent either in content or bulkiness. It was estimated that crop residues supplied 70.97% of the available DCP including wheat straw in roughage feeds of both study districts where the remaining was from natural grasses of different grazing lands.

Nutrient requirement for maintenance by livestock types were estimated based on the recommendations of Kearl (1982) and McCarthy (1986) for tropical livestock unit (TLU). Annual DM, ME and DCP requirement for maintenance were 24,920.65 kg, 173,159.28 MJ and 1,697,268.80 gm, respectively in Gasera while it was 30,323.84 kg, 203,723.65 MJ and 1,980,976.54 gm, respectively in Ginnir area (Table 8). Dry matter available from major feed resources per household satisfied the maintenance DM requirement of livestock in the study districts according to the figure in Table 8. According to Ørskov and Ryle (1990), ME from roughage feed is less efficiently used due to low digestibility and voluntary intake than other feeds by the animals. Therefore, ME available is below the requirement for livestock production in the study districts, which implies an enormous requirement of fermentable feed supplementation for livestock, especially for production and other energy demanding activities accordingly.

Crude protein available from major feed resources was even deficient for maintenance in the study districts. Crude protein supply satisfied only the maintenance requirement of the animals in wet season when there is availability of green forages that are efficiently digestible by the animals. On the other side protein deficiency could also aggravate the problem of energy deficiency since both DCP and ME available in the feed are interrelated and affected each other through the activity of rumen microorganisms (Ørskov and Ryle, 1990). This suggests that, the need of supplementing animals with good quality feed even for maintenance, especially during the dry season (Table 4). In general, related to available feed resources, there is huge amount of poor quality roughage feed particularly crop residues that not permit the experts to say there is feed shortage in the study districts rather the bottleneck is feed quality which invites timely intervention of all the concerned bodies for the betterment of future livestock productivity.

5. CONCLUSION

Natural grass, crop residues, crop aftermath and fallow lands were the major feed resource in both study districts. Productivity of livestock in the area is poor because available feed resources can't met the nutritional requirement of livestock especially the content of protein is less than 7%, which did not meet, even the requirements of rumen microbes. Natural grass and fallow land grass contributed 18.94% and 29.03% ME and CP respectively and crop residues accounts 81.06% and 70.97% ME and CP respectively in both study districts. Season is a major factor that determines feed resource availability in the study area and during dry season animals face shortage of feed.

Therefore, based on the above conclusion the following recommendations are forwarded:

a. The chemical composition of crop residue should be studied further at different level of crop management practices and at different cereal crop residue varieties because type of cereal crop variety, different level of management like pesticide and herbicide application have effect on nutritional value and palatability of cereal crop residue.

b. In order to compensate poor protein content of major feeds and to increase palatability of crop residue, it is better to give supplementary feeds and practice crop residue treatment (chopping and urea treatment) respectively.

c. Awareness should be given for farmers how to conserve forage to overcome feed shortage during dry season.

ACKNOWLEDGEMENTS

The authors would like acknowledge Maddawalabu University for financial support and Holeta Agricultural Research Center's Nutrition Laboratory for their technical support to analyze the chemical composition of feed samples.

REFERENCES

1. Abdinasir I (2000). Smallholder Dairy Production and Dairy Technology Adoption in the Mixed Farming System in Arsi Highland, Ethiopia. Ph.D. Thesis, Humboldt University, Berlin, Germany. 146 p.
2. Alemayehu M, Sissay A (2003). Integrated Livestock Development Project (ILDP). Livestock Feed Resources Survey. North Gondar, Ethiopia. P 75.
3. Alemu T (1982). Grassland Composition and Current Livestock feeding System in Nekemte Awraja. M. Sc. Thesis. Alemaya College of Agriculture, Addis Ababa University. Alemaya. P 73.
4. AOAC (Association of Official Analytical Chemists) (1990). Official Method of Analysis. 15th ed. AOAC Inc. Anc. Arlington, Virginia, USA. 1298p.
5. ARC (Agricultural Research Council). (1980). The Nutrient Requirement of Ruminants Livestock. Commonwealth Agricultural Bureaux, Faraham Royal, London, UK.
6. Arsham H (2005). Questionnaire design and surveys sampling, 9th Ed.
7. Church DC, Pond WC (1982). Basic Animal Nutrition and Feeding Records. John Wiley and Sons, USA.
8. CSA 2012/13. Agricultural Sample Survey; Livestock and Livestock Characteristics. Volume II, 570 statistical bulletin, April 2013, Addis Ababa.
9. Daniel K (1988). Role of crop residues as livestock feed in Ethiopian Highlands. In: B. H. Dzowela (ed.). African Forage Plant Genetic Resources, Evaluation of Forage Germplasm and Extensive Livestock Production Systems. Proceedings of a Workshop Held in Arusha, Tanzania, 27-30 April.
10. Dawit A, Ajebu N, Sandip B (2013). Assessment of feed resource availability and livestock production constraints in selected Kebeles of Adami Tullu Jiddo Kombolcha District, Ethiopia.
11. Deressa T, Hassan R, Ringler C (2008). Measuring Ethiopian farmers' vulnerability to climate change across regional states, IFPRI discussion paper No. 806.
12. FAO (Food and Agriculture Organization of the United Nations) (1987). Master Land Use Plan, Ethiopia Range Land/Livestock Consultancy Report Prepared for the Government of the Peoples Democratic of Ethiopia. Technical Report. AG/ETH/82/020/FAO, Rome. P 94.
13. Gashaw G (1992). Assessment of Feed Resource Base and Performance of Crossbred Dairy Cows Distributed to Smallholder in the Selale Dairy Development Project Area. M. Sc. Thesis. Alemaya University of agriculture, Alemaya. 171 p.
14. Gryseels G (1988). Role of Livestock on a Mixed Smallholder Farms in Debre Berhan, PhD Dissertation, Agricultural University of Wageningen, the Netherlands. 249 p.
15. Holloway G, Nicholson C, Delgado C, Staal S, Ehui S (2000). Agro-industrialization through Institutional Innovation Transaction Costs, Cooperatives and Milk-market Development in the Eastern African Highlands. Ruminant Livestock Development Strategy, Addis Ababa. Agric. Econ. 23:279-288.
16. Kearl LC (1982). Nutrient Requirement of Ruminants in Developing Countries International Feed stuffs Institute, Utah Agricultural Experiment Station, Utah State University, Longman 84322. USA.
17. MAFF (Ministry of Agriculture Fisheries and Food) (1984). Energy allowances and feeding systems for ruminants, Reference Book 413 HMOs, London.
18. McCarthy G (1986). Donkey Nutrition. In: J. Reed, B. Capper and J. Neate (eds.) the professional Hand book of the Donkey (Compiled for the donkey sanctuary). Sid mouth (UK). P 248.
19. McDonald P, Edwards RA, Greenhalgh JFD, Morgan CA (1995). Animal Nutrition. Longman group, Harlow, UK. 607 p.
20. MoA (Ministry of Agriculture of Ethiopia) (1998). The Role of Village Dairying Co-operative in Small-holder Dairy Development: Prospects for Improving Dairy in Ethiopia.

21. MoARD (Ministry of Agriculture and Rural Development) (2007). Livestock development master plan study phase I report, Data collection and analysis volume V, Policy and Institutions; GRM International.
22. Mosi AK, Butterworth MH (1985). The voluntary intake and digestibility of diets containing different proportions of teff (Eragrotis tef) straw and Trifolium tembense hay when fed to sheep. Trop. Anim. Prod. 10:19-22.
23. Ørskov ER, Ryle M (1990). Energy Nutrition in Ruminants. ELSEVIER Science Publisher Co., Inc. 149 p.
24. Preston TR, Leng RA (1984). Supplementation of diets based on fibrous
25. Reed JD, Yilma K, Fossel LK (1988). Factors affecting the nutritional value of sorghum and millet crop residues. Plant Breeding and Nutritive Value of Crop Residues. Proceedings of a workshop held at ILCA, Addis Ababa, Ethiopia, 7-10 December 1987. Addis Ababa. pp. 233-253.
26. Seyoum B, Zinash S (1989). The Composition of Ethiopian feed stuffs. Institute of Agricultural Research. Addis Ababa, Ethiopia. Research Report 6:33.
27. Seyoum B, Getinet A, Abate T, Dereje F (2001). Present status and future direction in feed resources and nutrition research targeted for wheat based crop livestock production system in Ethiopia. In: P. C. Wall (Eds.). Wheat and Weed: Food and Feed. Proceedings of Two Stakeholder Workshops. CIMMYT, Mexico City. Improving the productivity of Crop Livestock Production in Wheat-based Farming Systems in Ethiopia, Addis Ababa, 10-11 October 2000. pp. 207-226.
28. Singh GP, Oosting SJ (1992). A model for describing the energy value of straws. Indian dairy man XLIV:322-327.
29. Solomon B (2004). Assessment of livestock production system and feed resource base in Sinana-dinsho district of bale highlands southeast Oromia. An M. Sc. Thesis. Alemaya University, Alemaya.
30. Steinfeld H, Hokkonen J (1996). A classification of livestock production systems. World Animal Review pp. 83-94.
31. Thornton P, Herrero M (2001). Integrated crop-livestock simulation models for scenario analysis and impact assessment. In: Proceedings of Third International Symposium on Systems Approaches for Agricultural Development, Lima, Peru, 8-10 November 1999.
32. Tolera A, Yami A, Alemu D (2012). Livestock feed resources in Ethiopia: Challenges, Opportunities and the need for transformation. Ethiopia Animal Feed Industry Association, Addis Ababa, Ethiopia.
33. UNESC (United Nation Economic and Social Council) 2012. Report on Livestock Value Chains In Eastern And Southern Africa: A Regional Perspective, Eighth Session of the Committee on Food Security and Sustainable Development and Regional Implementation Meeting for the Twentieth Session of the Commission on Sustainable Development. Addis Ababa, Ethiopia, 19 - 21 November 2012.
34. Van Soest PJ (1982). Nutritional Ecology of the Ruminants, O and B Books. Inc. Oregon, USA.
35. Van Soest PJ, Robertson JB (1985). Analysis of forages and fibrous foods. Cornell University.
36. Van Soest PJ (1985). Nutritional Ecology of the Ruminants, O and B Books. Inc. Oregon, USA.
37. World Bank (2008). World Development Report 2008: Agriculture for development. Washington DC.
38. World Bank (2011). Climate-Smart Agriculture: Increased Productivity and Food Security, Enhanced Resilience and Reduced Carbon Emissions for Sustainable Development. Opportunities and Challenges for a Converging Agenda: Country.
39. Yeshitila A (2007). Assessment of livestock feed resources utilization in Alaba woreda, southern Ethiopia.
40. Yihalem D (2004). Assessment of Botanical Composition and Stage of Harvesting of Selected Natural Pasture for Optimum Utilization as Hay at Andassa Livestock Research Center, Northwestern Ethiopia. M. Sc. Thesis. Alemaya University of Agriculture, Alemaya. 69 p.

41. Yitaye A (1999). A Study on Livestock Production System, Feed Resource and Feed Allocation Practices in three PAs' of the Awassa Woreda, Sidama Zone, Southern Ethiopia. M. Sc. Thesis. Alemaya University of Agriculture, Alemaya. 99 p.
42. Zewdie W (2010). livestock production systems in relation with feed availability in the highlands and central rift valley of Ethiopia MSc. thesis, Haramaya University.

CHAPTER 7

Livestock-Handling Related Injuries and Deaths

Kamil Hakan Dogan[1] and Serafettin Demirci[2]

[1] *Selcuk University, Turkey*

[2] *Necmettin Erbakan University, Turkey*

1. INTRODUCTION

Handling livestock is a dangerous activity. Few farm people look upon their livestock as a source of danger. However, a number of serious injuries and deaths occur every year as a result of animal-related accidents. Livestock handlers are involved in a variety of activities such as feeding, moving animals to different locations, loading animals on trucks/trailers, artificial insemination, shearing, grooming, basic animal care such as hoof care, dehorning, and cleaning animals, roping animals, applying topical insecticides, giving vaccinations, applying topical or administering oral medications, castrating, pulling teeth, ear tagging, milking, branding, shoeing, assisting with delivery of newborns, and assisting veterinarians with treatment or handling of injured animals. Other activities involving animals may include work tasks such as plowing fields; pulling equipment such as wagons to transport farm goods; riding animals, primarily horses, for farm or ranch activities such as corralling cattle; teaching others to ride; butchering animals for food; and, rarely, euthanizing or destroying sick or aggressive animals (Langley & Morrow, 2010).

The World Health Organization predicts that by the year 2020, injuries will be responsible for more death, morbidity, and disability than all communicable diseases combined (Murray & Lopez, 1998). Injuries account for 1 in 7 potential life-years lost worldwide, but by 2020 they will account for 1 in 5, with the developing countries bearing the brunt of this increase (Vilardo, 1988; Zwi et al., 1996; Murray & Lopez, 1998). Injury control has gained attention and enormous support with the infusion of funding for injury control in developed countries and particularly the creation of the National Center for Injury Control and Prevention within the Centers for Disease Control and Prevention in the United States (Myers, 1992; Myers, 1997). During the last decade of the 20th century, workers in the US agriculture industry received particular attention because of the high risk of fatal injuries and suspected risk for serious nonfatal injuries (Aherin et al., 1992; Myers, 1998).

Farming ranks among the highest of United States (US) industries for work-related fatal and non-fatal injuries. The lack of information regarding agricultural injuries has been recognized as an obstacle in the development of effective injury prevention measures (Zhou & Roseman, 1994). Within the past decades increased emphasis has been placed on quantifying and limiting farm-work injury hazards. Studies have consistently reported that farm machinery, livestock, and falls are major contributors to agricultural injuries (Brison & Pickett, 1992; Pratt et al., 1992; Zhou & Roseman, 1994; Nordstrom et al., 1995).

The weight of farm mammals varies from less than a few pounds in newborns to over 3000 pounds in adult bulls. Animals can cause serious injuries to animal handlers through various mechanisms. Farm animals can bite, kick, gore, trample, fall on, step on, knock down, crush or pin between other animals or farm structures, peck, scratch, throw or buck off, drag, and ram or butt (Langley & Morrow, 2010). Horses and cattle, rather than any single type of agricultural machinery, are reported the leading cause of injury (Cogbill & Busch, 1985; Shireley & Gilmore, 1995). Anyone who works with livestock knows each animal has its own personality. Animals sense their surroundings differently than humans. Their vision is in black and white, not in color. They also have difficulty judging distances. And differences exist between the vision of cattle, swine and horses. For example, cattle have close to 360-degree panoramic vision. A quick movement behind cattle may "spook" them. Animals have extremely sensitive hearing and can detect sounds that human ears cannot hear. Loud noises frighten animals, and research proves that high-frequency sounds actually hurt their ears. These factors explain why animals are often skittish and balky, particularly in unfamiliar surroundings (Baker & Lee, 1993).

2. EPIDEMIOLOGY

Agriculture ranks among the most hazardous occupations in the United States, with a fatality injury rate 8.5 times greater than for all other occupations combined (28.7 vs 3.4 per 100,000 workers) (National Safety Council, 2008). Shared working and living environments associated with agriculture place all residents at risk, including children (Bancej & Arbuckle, 2000; Gerberich et al., 2001).

There have been many local or regional surveys that have looked at risk to farm animal handlers (Stallones, 1990; Brison & Lawrence, 1992; Pratt et al., 1992; Waller, 1992; Zhou & Roseman, 1994; Layde et al., 1996; Boyle et al., 1997; Casey et al., 1997a; Casey et al., 1997b; Hwang et al., 2001; Sprince et al., 2003; Douphrate et al., 2009). Most of the studies show that livestock handling activities are the second or third leading cause of injuries on the farm, causing from 12% to 24% of farm injuries. It is estimated that about 30 farmers are killed each year from contact with farm animals, primarily horses and cattle (Austin, 1998). Dairy bulls may be more likely to injure humans than beef bulls. Dairy bulls generally have more frequent contact with humans than do beef cattle, and are known to be possessive of their herd and occasionally disrupt routine feeding, cleaning, and milking operations (Boyle et al., 1997).

Unfortunately, the number of animal handlers is not exactly known, but tens of thousands of people are exposed daily to farm animals. A recent survey estimated that there were at least 54,000 workers on swine and poultry establishments in the United States (Gray et al., 2007). Numerous hazards exist on poultry and livestock farms. However, most of the information in the recent medical literature has primarily focused on respiratory symptoms in pork production facilities, and injuries primarily involving cattle and horses, especially recreational activities with horses. There is very little published on injuries associated with handling swine or sheep/goats or other farm animals (Langley & Morrow, 2010).

Farming is one of the few industries in which families are at increased risk. In particular, farm surveys indicate that the injury rate is highest among children age 15 and under and adults more than 65 year of age. Unlike other occupations, farmers routinely work beyond the average retirement age. Data from the National Institute for Occupational Safety and Health (NIOSH) reveals that farmers aged 75 and older are more than twice as likely to die on the job than their younger counterparts. Age-related conditions, such as arthritis, vision or hearing problems make farming potentially more dangerous for senior farmers (Hernandez-Peck, 2001).

According to the Bureau of Labor Statistics, from 1992-1997, more than 75,000 workers received injuries and 375 workers were killed from animal-related injuries. Cattle are responsible for most injuries caused by farm animals. A 1997 study conducted by Oklahoma State University (OSU), Biosystems and Agricultural Engineering Department, found 150 cases of cattle handling-related injuries among 100 Oklahoma cow-calf operations. The study also showed that more than half of the injury cases resulted from human error (Hubert et al., 2003). In 2002 alone, 730 deaths and 150,000 disabling injuries occurred on U.S. farms. Each day, about 500 agricultural workers suffer lost-time injuries, 25 of which result in permanent impairment. Farm operators and their family members accounted for most of the injuries reported (Myers, 2003).

Previous studies of work-related injuries among farmers have described patterns of farmers' injuries and have evaluated a variety of potential risk factors (Aherin et al., 1992; Zhou & Roseman, 1994; Myers, 1997; Stallones et al., 1997; Browning et al., 1998; Crawford et al., 1998; Lewis et al., 1998; Myers, 1998; Norwood et al., 2000; Sheldon et al., 2009). In general, the risk factors have been categorized into 2 domains: physical characteristics of the farming environment and personal characteristics of the farmers. With respect to characteristics of the farming environment, the patterns of injury have been fairly consistently reported among these studies, with farm machinery, falls, and animal-related injuries being the 3 major external causes of injury (Zhou & Roseman, 1994; Myers, 1997; Stallones et al., 1997; Myers, 1998; Xiang et al., 1999). With respect to personal characteristics of the farmers, males were found to be at higher risk for injury than females, regardless of hours spent in farm activities (Stallones, 1990; Pratt et al., 1992; Myers, 1998). Although results of several studies indicated that younger farmers have the highest risk of nonfatal injuries (Stallones, 1990; Crawford et al., 1998; Lewis et al., 1998; Stallones, 1998; Xiang et al., 1999), older farmers tend to account for the greatest proportion of agricultural fatalities (May, 1990; Stallones, 1990).

In the United States, studies have consistently reported that the leading causes of agricultural injuries are machinery and other equipment, falls, and livestock (Myers, 1997; Myers, 1998; Lewis et al., 1998; Xiang et al., 1999). Because of the increasing mechanization of farming over the past half century, and the high fatality rate associated with injuries due to machinery and tractors (McFarland, 1968; Simpson, 1984; McKnight and Hetzel, 1985; Hopkins, 1989; Etherton et al., 1991; Lee et al., 1996; Bernhart and Langley, 1999; Carlson et al., 2005; Cole et al., 2006; Dogan et al., 2010), most studies of agricultural injuries have focused on issues related to interactions with machinery or tractors also. Animals may bite, kick, scratch, trample, crush, gore, buck or throw, or drag the livestock-handler (Langley, 1999). Studies demonstrated non-fatal injury rates are elevated on operations with livestock, especially beef and dairy cattle (Brison and Pickett, 1992; Pratt et al., 1992; Zhou & Roseman, 1994; Nordstrom et al., 1995). Researchers have reported between 12% and 33% of injuries on the farm are caused by animals (Cleary et al., 1961; Cogbill et al., 1985; Hoskin et al., 1988; Myers, 1990; Brison and Pickett, 1992; Pratt et al., 1992; Zhou & Roseman, 1994; Layde et al., 1995; Nordstrom et al., 1995; Pickett et al., 1995; Gerberich et al., 1998; Lewis et al., 1998; Sprince et al.,

2003) and livestock-related injuries account for the highest rate of lost work days (Thu et al., 1997).

The 2007 United States Department of Agriculture (USDA) Census of Agriculture estimated the number of farms in the United States at approximately 2.2 million. The Census found that the number of cattle operations totaled 956,400, including 757,000 beef cow and 67,000 milk cow operations. The number of hog operations totaled 73,150, sheep operations 82,330, goat operations 149,800, and farms with any poultry 187,000. There are 575,900 farms with horses and 99,000 farms with mules/donkeys/burros. As of July 1, 2009, there were approximately 102 million head of cattle and calves, 67 million hogs and pigs, 7 million sheep and lambs, and 3.7 million goats on farms in the United States. In 2007, there were 266 million turkeys and 9 billion broilers produced. The USDA estimated in 2007 that there were 4,028,000 horses and ponies and 283,000 mules/burrows/donkeys on US farms. The American Horse Council estimated there were 9.2 million horses involved in farming, sports, entertainment, and recreational activities in the United States in 2005 (Langley & Morrow, 2010).

In a survey of US farm operations 1993-1995, the National Institute for Occupational Safety and Health (NIOSH) found that livestock was the number two cause of nonfatal injuries (99,310 injuries), almost the same as machinery, which was the number one cause of injuries (99,402 injuries). The rate of nonfatal occupational injuries from 1993 to 1995 was 7 per 100 workers on livestock farms, and cattle/hog/sheep operations had the highest number of injuries by type of farm operation. Of the injuries, 37% were due to horses and 31% due to cattle (Myers, 2001). The major source of injuries on US farming operations were machinery excluding tractor (21.3%), livestock (20%), and slips, trips, and falls on working surfaces (8.5%). Beef, hog, or sheep operations were found to have the highest number of lost work-time injuries (84,736) and restricted workdays (1,869,561). When looking at activities in which workers were injured, 45.7% involved livestock handling. When looking at restricted workdays by source of injury, livestock were responsible for the largest percent (34.9%), followed by machinery. Austin (1998) studied nonvenomous animal related fatalities in the workplace 1992-1994. She found that about 40 deaths per year occur, and 27 of these occur in farmers. Of 144 deaths obtained using the US Department of Labor (US DOL) Census of Fatal Occupational Injuries (CFOI), there were 22 transportation fatalities involving animals. Of the 122 nontransportation deaths, 68 were due to cattle and 41 from horses, and 13 from other animals. Bulls caused 54% of the cattle related deaths. Of workers that were farmers, cattle caused 54 deaths and horses caused 27 deaths. Of the deaths from cattle, 40% were due to multiorgan trauma, 35% trauma to trunk and chest, and 18% from head trauma. Of the horse-related deaths, 46% were due to head trauma.

Among 20 reviewed studies of stress and occupational injuries, all found a statistically significant association between stress and injuries, and 12 of the 17 studies with quantitative measures had odds ratios greater than 1.0, indicating that stress increased the risk of injuries (Gadalla, 1962; Xiang et al., 2000). Other factors such as education (Johnston, 1995), preexisting diseases and use of medications (Browning et al., 1998; Crawford et al., 1998; Lewis et al., 1998; Xiang et al., 1999), alcohol consumption (Zhou & Roseman, 1994), family incomes (Aherin et al., 1992; Browning et al., 1998; Lewis et al., 1998; Xiang et al., 1999), and knowledge of safe practices and safety behaviors (Lewis et al., 1998) also have been evaluated, but the conclusions have been inconsistent. However, outside North America, western Europe, and Australia, information about injury problems and solutions is particularly sparse because the injury control efforts from communities and government in developing countries are well below the level of those directed at other health problems (Li & Baker, 1991; Gumber, 1994; Zwi et al., 1996; Li and Baker, 1997).

Drudi (2000) studied occupational injuries due to animals from 1992 to 1997 using the US DOL Census of Fatal Occupational Injuries and the Surveillance of Occupational Injuries and Illnesses (SOII) databases. He found there were 4600 bovine-related, 5100 equine-related, and 1900 porcine-related nonfatal occupational injuries reported over this time period. For these nonfatal injuries, it should be noted that self-employed and farms with less than 11 employees were excluded from the SOII. From CFOI information, of the 375 occupational deaths from animals, cattle caused 141 deaths and equine 104. Fewer than 5 deaths were reported from sheep and swine. On average about 30 deaths per year were from cattle and equine. Bulls, which account for only 2% of cattle, were responsible for 48% of the deaths from cattle. Five deaths were reported to farmers that were related to birthing or maternal defensiveness, such as cows attacking farmers trying to midwife them. Cattle attacks accounted for four fifths of the cattle-related deaths and about three fifths of cattle-related nonfatal injuries involved the worker being attacked by cattle. Likewise about half of the deaths from equine were due to attacks and more than half of the cases of nonfatal injuries from equine involved an attack.

In a follow-up study using CFOI data, Meyer (2005) evaluated fatal occupational injuries to farmers 55 years and older for the period 1995-2002. He found of 190 farm fatalities due to assaults from animals, 113 occurred in older workers (5% of the work-related deaths in older farmers). The median age of workers assaulted by an animal was 62 years, 4 to 6 years older than the average farmer age of 56 to 58 years. Fifty-nine percent of the deaths were from cattle.

A national study of youth on farms (younger than 20 years of age) was conducted in 1998. Approximately 1,260,000 youth under age 20 lived on farms. An estimated 32,808 injuries occurred with about 20% being due to animals. Of the animal-related injuries, 41% occurred in youth <10 years old, 29% in those 10-15 years old, and 30% in those 16-19 year olds (Hendricks & Adekoya, 2001). Seventy percent of the animal-related injuries occurred to farm residents and 69% were work related. In a survey of youth on racial minority farms and Hispanic farms, horses accounted for the second highest number and rate of nonfatal injuries, followed by all terrain vehicles (ATVs) and tractors (Myers et al., 2003).

NIOSH also conducted a national survey of injuries in farmers age 20 and older in 2001 and 2004. From the information collected, Myers et al. (2009) looked at injuries in a subgroup of farmers 55 years and older. Workers 20 to 54 years old had an injury rate of 5.8 per 100 compared to 5.3 per 100 in workers 55 and older, but the older farmers had more severe injuries and a fatality rate 2.6 times greater than farm workers <55 years old. They found that 10% of nonfatal injuries in older farmers were due to animals. Animals were the third leading mechanism of death in older farmers behind tractors (46%) and trucks (7%) with 5% of fatalities due to animals.

The most comprehensive evaluation of bull-related injuries on persons or property was recently completed, finding 287 cases using different databases (Sheldon et al., 2009). The majority of reports were from 1980 to 2008. Of the 261 cases involving attacks on people, 57% resulted in a fatality. Of the fatal injuries, 94% were males. The average age of the victims was 56 years with a range from 3 to 91 years. The nature of the injury was classified as follows: unspecified attack (110 cases); charged (49 cases); gored (21 cases); trampled (15 cases); mauled (11 cases); and other (56 cases). The type of injuries reported appeared to be related to the nature of the attack. When a person was charged by a bull, the injury usually involved broken ribs, puncture wounds due to goring, or blunt force chest trauma. When the victim was trampled, the injuries usually involved broken bones, crushing-type injuries, internal injuries, or head injuries. The authors concluded that a bull that shows aggressive behavior should be culled from the herd and sent to slaughter and not sold at a general livestock auction in order to prevent it from posing a risk to a new owner.

Fatal and non-fatal traumatic injuries associated with agricultural production are a major public health problem that needs to be addressed through comprehensive approaches that include further delineation of the extent of the problem, particularly in children and older adults, and identification of the specific risk factors through analytic efforts (Hard et al., 2002). While these activities are in progress, additional endeavors will also be essential because of the numerous exposures involved that will require more intensive and specific investigations. Only through such efforts can appropriate prevention efforts be developed (Gerberich et al., 1992; Gerberich et al., 1994). Integral to this process is the incorporation of comprehensive surveillance systems that can be used to monitor the magnitude of the problem, over time, and evaluate the efficacy of any intervention efforts that are implemented (Gerberich, 1995). While surveillance is a key element for assessing the magnitude of the traumatic agricultural injury problem and identifying appropriate intervention strategies, based on quality risk factor information, it will not reduce injuries on its own. It is apparent that effective interventions are imperative in the alleviation of this major public health problem. Continued development of relevant surveillance systems and implementation of appropriate interventions are the primary challenges for the current decade.

The major sources/vehicles of injury for the farming-related injuries were livestock (30%), machinery, other than tractors (20%), and tractors (9%), accounting for 59% of the events. Among all of the farming-related injury cases, only 6% resulted in hospitalization which has implications relevant to the limitations imposed if only hospital-based surveillance is used; however, 80% were treated by a health care professional. Furthermore, the fact that substantial proportions of cases were actually restricted for a week or more (37%, with 19% restricted for a month or more) and/or had some type of persistent problem, including some permanent disabilities (25%), is very important when looking at the overall impact (Gerberich et al., 1998). Farmers are exposed to a variety of hazards including tractors, machinery, enclosed structures such as grain bins and silos, overhead power lines, tools, ponds, and animals. In addition, they often work long hours under severe time constraints and many use older model farm equipment that lack safety features (Sterner, 1991). Farm machinery is involved in approximately 50% of farm work-related deaths (Murphy et al., 1990) and 18% of nonfatal injuries (Ehlers et al., 1993). When not fatal, these injuries can cause serious, permanent disability. For example, tractor roll-overs can cause crushing, evisceration, and amputation of limbs; and entanglement in rotating shafts or drivelines can result in limb amputation or scalping. Other types of farm injury include suffocation from silo-gas and engulfment in grain, physical trauma from working with livestock, drowning, and electrocution. The spouses and children of farmers are also at risk for injury because they come into contact with hazards on the farmstead, regardless of whether or not they are working (Cordes & Foster, 1988; Rivara, 1997; Dogan et al., 2010).

In a recent study, McCurdy et al. (2012) reported that injury risk was related in step-wise fashion to annual hours spent in farm work among rural California public high school students. After adjustment for farm work hours, ethnicity, and sex, we observed increased overall injury risk (i.e., not limited to injuries occurring in performance of the specified tasks) associated with each of eight selected farm tasks. Mixing chemicals, feeding large animals, and welding each showed a statistically significant approximate doubling of odds for injury. The number of these selected tasks performed showed a significant trend with injury risk.

3. LIVESTOCK BEHAVIOR

An understanding of livestock behavior will facilitate handling, reduce stress for the animal and handler, and improve animal welfare and handler safety. Handlers who understand livestock behavior can reduce animal stress. Research has proven that reducing stress during handling will improve productivity. "Reducing stress also should help improve weight gain, reproductive performance and animal health."

Cows, horses, sheep, goats, donkeys, and even chickens all have a herding instinct to protect survival of the group from predators. Many animals can tolerate sharing space with other species, but they are more at ease when sharing barn space with their own kind.

Understanding cattle behavior can help farm and ranch workers avoid dangerous situations. Temple Grandin, Colorado State University animal behavior specialist, states that "handling practices can be less stressful to the animals and safer for the handler if one understands the behavioral characteristics of livestock." An animal's physical structure, psychological makeup, environment, and individual personality can influence behavior.

An animal's senses function like those of a human; however, most animals detect and perceive their environments very differently as compared to the way humans detect and perceive the same surroundings. While cattle have poor color recognition and poor depth perception, their hearing is extremely sensitive relative to humans. Knowing these characteristics, one can better understand why cattle are often skitish or balky in unfamiliar surroundings (Hubert et al., 2012).

3.1. Cattle

Cattle handling skills are essential for managing cattle. Good cattle handlers learn these skills through observation and trial and error. Good cattle handling saves time and effort, and reduces stress for people and animals. Inefficient and rough handling causes financial losses because of bruising, poorer meat quality and lower milk production. Working with cattle may be dangerous, especially in yards, races and dairy sheds where people and cattle are close together. High risk activities include working with bulls and with cows and newborn calves.

Cattle live in social hierarchy with dominant and subordinate animals. People usually behave as the most dominant animal by forcing cattle to move, restricting movement and controlling access to feed. Handlers need to be confident and to establish authority from the start so that the cattle know who is boss.

The leader cattle are not always the dominant animals. Cows coming in for milking are led by middle-ranking cows, which are followed by the dominant animals and then the lowest-ranking animals. Forcing the hindmost animals along will not necessarily speed up the movement of cattle, as the dominant animals will not be forced by their subordinates.

Cattle behavior in yards is influenced by rank. Low-ranking animals may try to avoid dominant animals. Dominant cattle may turn and attack subordinate animals and as the defeated animal escapes it may run over the handler. The dominant cattle in a mob may stay as far away from people as possible.

Many farmers sustain minor injuries while working with cattle. Common injuries include cuts, bruises, fractures, sprains and strains. Serious injuries cost farmers in lost time, additional help and many other ways. Financial costs alone are large (Stafford, 1997).

Cattle have sensory functions similar to those in humans. However, cattle often detect and perceive their environments quite differently from humans. A better understanding of the sensory functions in animals is gained by observing cattle anatomical structure and comparing it to those in humans.

Vision - Livestock have their eyes located on the sides of their heads. As a result, they have wide-angle (panoramic) vision and can see all the way around them except for small blind spots at the nose and the rear. This is perhaps the most important factor involved in livestock handling. For example, cattle often "spook" at sudden movements behind them because they perceive such movements as threats. Handlers are often injured by being pinned or kicked by animals that are frightened in this manner. Approaching livestock from the side or front is less startling to these animals than approaching from behind. Proper approach angle will reduce the risk of injury.

The eye placement that gives livestock excellent wide-angle vision causes them to have relatively weak eye muscles. Therefore, livestock cannot focus quickly on nearby objects. This is another reason that causes livestock to "spook" or balk at sudden movements. One of the most common causes of balking in handling facilities are loose chain ends that makes a rapid movement. Fan blades or flapping cloth (a coat hung on a post) will also cause livestock to balk.

Livestock are either color-blind or have very limited color vision and often perceive shadows as holes. Therefore, cattle tend to balk if a shadow is across their path. A handler may be injured by being pinned between a balking animal and a fixed object or an animal moving from behind. Using diffuse, uniform lighting in areas where cattle are handled minimizes shadows and bright spots and can prevent balking. Cattle also have a tendency to move from dimly illuminated areas to brightly lit locations if the light is not glaring in their eyes. A spot light directed onto a ramp or other entry points will improve cattle movement provided the light does not shine directly into the eyes of approaching animals (Gay & Grisso, 2012).

Hearing - Cattle are more sensitive than humans to high frequency noises. Excessive yelling and hollering while handling and herding livestock causes stress and may make cattle more difficult to handle. Furthermore, unexpected loud or novel noises such as banging gates or load exhaust from air cylinders can startle cattle. Use of rubber bumpers on gates and squeeze chutes will prevent these loud noises. Cattle will readily adapt to reasonable levels of continuous noise such as a radio. Continuous playing of the radio may help reduce the reactions of some animals to sudden noises.

Flight zone - An animal's "personal space" is referred to as the flight zone. When a handler enters the flight zone the animal will move away. Understanding and respecting the flight zone can reduce animal stress and help prevent injuries to handlers. The flight zone is basically a circle and its size depends on the tameness of the animal. For example, wild cattle can have a flight zone with a 160 feet diameter. In contrast, a tame dairy cow has almost no flight zone and is difficult to drive. Understanding the flight zone is the key to easy, quiet handling of cattle. The size of an animal's flight zone depends on its fearful or docile behavior, the angle of handler's approach, and its state of excitement. Work at the edge of the of flight zone at a 45 to 60 degree angle behind the animal's shoulder (Fig. 1). Cattle will circle away. The flight zone radius can range from five to over 25 feet for feedlot cattle and as far as 300 feet for range cattle. If one is within their flight zone, the animal moves away or retreats. When moving cattle, avoid approaching them directly. Try to work them close to the point of balance, moving back and forth on a line parallel to the direction the animal is travelling (Hubert et al., 2012).

Isolation - Cattle are social animals used to living in herds. They are easier to move in groups. They do not like to be isolated. Cattle may become stressed and dangerous when they are separated from the herd. Cattle should always be able to see another member of the herd. They will follow a leader through yards, even in single file, without being stressed, as long as they can see the beast in front. Individuals that break away will usually rejoin the herd, given time and space.

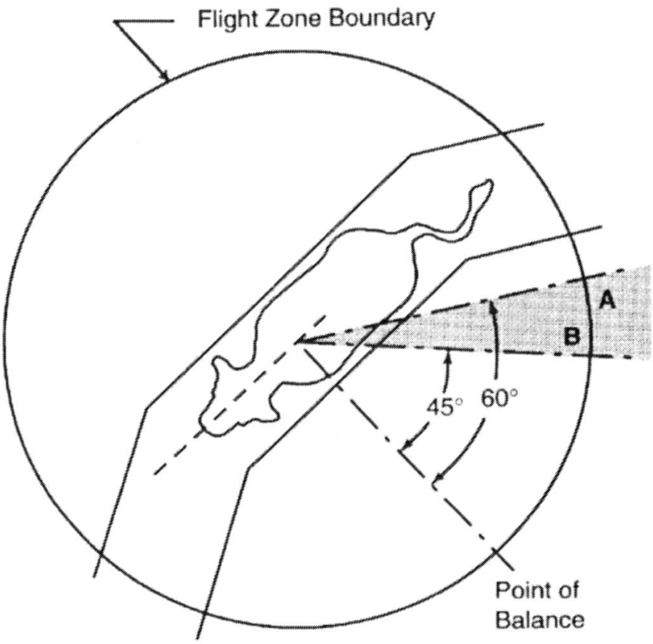

Figure 1. Cattle flight zone. Note: Animal movement stops if handler is in position "A". Handler moves to position "B" to start movement.

Memory - Cattle remember painful or frightening experiences and react accordingly. Therefore, an unpleasant experience during a given routing can affect subsequent routines. Unpleasant experiences due to rough handling, electric shock, pokes, slipping on smooth floors and standing on shaky uncomfortable grates will impact future behavior (Stafford, 1997; Gay & Grisso, 2012).

Wild cattle were prey for wolves. They reacted to wolves by running away, kicking as they ran if the wolves were close, or by turning and fi ghting by butting and goring. Under some circumstances they responded by remaining immobile until the danger disappeared. Domesticated cattle retain these characteristics and are capable of defending themselves by using their head to bunt, their horns to gore and their legs to kick.

Cattle regard people either as predators, as cattle, or as irrelevant features of the environment. People chasing cattle imitate the behavior of wolves and engender fear. Young cattle should not be chased but should be moved slowly, using their working distance (the distance at which cattle start to move away from people).

Stationary cattle can kick forward to their shoulder and outward with their hind legs. Moving cattle usually kick directly backwards. The safest place when handling cattle is either close up against the beast or at a distance greater than the animal can kick. It is important to identify an escape route when closing in on cattle.

The size and speed of these animals mean that getting in close contact is always potentially dangerous.

However, when you are close to a cattle beast the power of a kick is reduced and becomes more of a push. A gentle touch and soothing words may calm an animal down. Good control requires that cattle regard people as dominant members of the herd.

- Cattle are social animals
- Cattle are easier to work in groups
- Aggressive cows should be culled
- Gentle handling of calves will improve their temperament as adults
- Breed is only one factor influencing temperament
- Cattle remember bad experiences
- Bulls, and cows with newborn calves, may be dangerous
- Make sure the cattle know you are the boss
- Cattle don't see like we do
- Music may reduce the likelihood of sudden noises startling cattle
- Use your voice to let cattle know where you are
- Keep cattle calm. Overexcited cattle are difficult to handle
- Use your voice to soothe and calm cattle
- Keep your distance while moving cattle
- Beating cattle will overexcite them
- Give cattle time to settle when they come into yards
- Use the working distance like an accelerator pedal
- Keep your distance - getting too close to cattle may cause them to scatter
- Work at the edge of the working distance
- Learn to balance yourself on stock

- Use balance points by moving through them
- Keep an eye on what is going on around you
- Cattle learn from experience
- Gentle handling of calves should result in easy-to-handle cows
- Make yarding as pleasant as possible
- Cattle learn to avoid places where they have been hurt
- Never trust a bull - never turn your back on a bull
- Dairy bulls are particularly dangerous
- Bulls can kill people
- Keep the calf between you and the cow
- Keep the calf's mouth shut to stop it bawling
- Calving heifers are very sensitive to handling
- Remain calm and collected when milking heifers
- Cull nervous or aggressive heifers (Stafford, 1997).

Dairy producers recognize the ability of cows to communicate despite an inability to speak. Cattle display characteristic signs of fear and aggression. Good handlers are sensitive to these warnings, which include:

- Raised or pinned ears
- Raised tails
- Raised hair on back
- Bared teeth
- Pawing the ground
- Snorting

If an animal exhibits any of the above signs, practice extra caution around them (Gay & Grisso, 2012).

3.2. Horse

3.2.1. Basic horse behavior

- Horses detect danger through their vision, sense of smell and keen sense of hearing. They have wideangle vision, but they also have blind spots directly behind and in front of themselves. For example, when a horse lifts its head and pricks its ears, it is focusing on something far away. The horse lowers its head when focusing on low, close objects. Keep these blind spots in mind and know where your horse's attention is focused so you do not scare it.

- Your horse's ears will give you clues; they will point in the direction in which its attention is focused. Ears that are "laid back," or flattened backward, warn you that the horse is getting ready to kick or bite.
- Always work with calm but deliberate movements around horses. Nervous handlers can make horses nervous, creating unsafe situations.

3.2.2. Approaching the horse

- When catching a horse, approach from its left shoulder. Move slowly but confidently, speaking to the horse as you approach. Read the horse's intention by watching its body language.
- Be careful when approaching a horse that is preoccupied, such as when its head is in a hay manger.
- When approaching a horse in a stall, speak to the horse to get its attention and wait until it turns and faces you before entering. Make sure the horse moves over before you walk in beside it.
- Speak to your horse and keep your hands on it when moving around. Even if a horse is aware of your presence, it can be startled by quick movements.
- When approaching from the rear, advance at an angle speaking to the horse, making sure you have its attention. Touch the horse gently as you pass by its hindquarters.

3.2.3. Leading the horse

- Hold the lead line with your right hand, 8 to 10 inches away from the horse's head, while holding the end, or bight, of the line with your left hand. Always use a lead line so you have this "safety zone" and to prevent getting a hand caught in the halter.
- Teach your horse to walk beside you so that you are walking at its left shoulder, with your right elbow near the horse's shoulder so you can anticipate its actions. Do not let the horse "walk" you. Do not allow it to get behind you either, as it could jump into you if spooked.
- To lead a horse through a doorway, you should step through first and then quickly step to the side out of the horse's way. Keep an eye on it, as some horses try to rush through narrow spaces.

- Never wrap any piece of equipment attached to a horse around your hand, even with small loops, as it could wrap around the hand and cause serious injury.
- After you remove the halter, make the horse stand quietly for several seconds before letting it go completely. This will help prevent the horse from developing a habit of bolting away and kicking at you in the process.
- Some horses can become sour and begin nipping at you if they anticipate discomfort during grooming. Do not hurry the grooming procedure, especially with a young or spooky horse. Stay near the horse and keep a hand on it at all times so you can anticipate its movements.
- Do not climb over or under the lead line of a tied horse. The horse may pull back and cause you to trip over the line, and you will have no quick escape should the horse lunge forward, paw or try to bite. Never walk under the belly of any horse (Gay & Grisso, 2012).

3.3. Swine

- Your best safety aid for the jobs is a lightweight hurdle or solid panel with a handle attached. The panel should be slightly narrower than the alleys through which the animals are being driven.
- As with most animals, make yourself known quietly and gently to avoid startling your hogs. A knock on the door or rattling the door handle will usually suffice.
- Don't let small children reach through pens or fences to pet or feed hogs. Keep unauthorized people out of pens or away from the facility. Bio security can be an important issue.
- Keep boars separate at all times.
- Use a drafting board when moving boars.
- Use nose ropes and crushes to restrain pigs when necessary.
- Lifting a pig should be avoidable, but if you must lift a pig, sit it down facing away from you, draw it close to your body and pick it up by the back legs, making sure to lift with your thigh muscles (NCFM, 2012).

3.4. Sheep

- A common accident involving sheep is being butted by a ram. Ewes will also protect their young and should be handled carefully. A sheep can be immobilized for safe handling by sitting it up on its rump on the ground.

- Plan musters in advance.
- Assume that rams will act unpredictably.
- Use suitably trained sheep dogs to control the mob.
- Lifting a sheep should be avoidable, but if you must lift a sheep, sit it down facing away from you, draw it close to your body and pick it up by the back legs, making sure to lift with your thigh muscles.
- When shearing, use a harness to support your back (NCFM, 2012).

Watching animals for signs of aggressiveness or fear alerts you to possible danger. Warning signs may include raised or pinned ears, raised tail or hair on the back, bared teeth, pawing the ground or snorting. Although handling methods may vary greatly for different types of livestock, there are some generally accepted rules for working with any animal:

- Most animals will respond to routine; be calm and deliberate.
- Avoid quick movements or loud noises.
- Be patient; never prod an animal when it has nowhere to go.
- Respect livestock - don't fear it!
- Move slowly and deliberately around livestock; gently touch animals rather than shoving or bumping them.
- Always have an escape route when working with an animal in close quarters (Baker & Lee, 1993).

4. TYPES OF LIVESTOCK-HANDLING RELATED INJURIES AND DEATHS

The injuries caused by animals are related to the ecological structure of the environment as well as the sociocultural and socioeconomic properties of the society. People who work with animals including farmers, veterinarians, butchers, and workers in zoos and circuses are all at risk (Wiggins et al., 1989). A high-energy trauma is applied to the body as a result of attack generally by bull, horse, pig, or large dogs, possibly resulting in serious injury or death. People at greatest risk for these injuries are those whose occupation or livelihood involves large animals (Nogalski et al., 2007). It has been reported that animals are one of the main causes of injury in the farming industry in the United States of America (Purschwitz, 1997).

Although the animals are implicated in such injuries, in most cases the incidents occur due to inappropriate behaviors of people or a lack of control of the animals (Nogalski et al., 2007). Langley et al. (2001) reported that large animals (cows and horses) caused the majority of deaths (67%) among workers in farms. Farmers and farm workers can easily be injured by livestock. Cattle, pigs, horses, sheep, dogs and other farm animals can be unpredictable and should be treated with caution at all times. Attempting to lift or push animals can cause injury and animals are capable of transmitting certain diseases (NCFM, 2012). The majority of studies find that activities involved in livestock handling are a leading cause of injury on the farm and that contact with animals causes 12% to 25% of injuries. Various researchers have found factors that may increase the risk of incurring an injury in livestock handlers. Among the risk factors found are presence of hearing loss or joint problems (Stallones, 1990; Sprince et al., 2003), working more hours per day or week (Layde et al., 1996; Stallones, 1990), younger age (Stallones, 1990; Zhou & Roseman, 1994; Sprince et al., 2003), older age (Brison & Lawrence, 1992; Pratt et al., 1992), alcohol consumption (Zhou & Roseman, 1994), male sex (Brison & Lawrence, 1992; Pratt et al., 1992), and prescription drug use (Brison & Lawrence, 1992).

In animal attacks, health and life can be threatened by the direct effect of trauma (hitting, kicking, biting, etc.) or by subsequent wound infections or contagious illnesses transmitted by the animal. It has been reported that most serious animal-related injuries are caused by large animals such as bulls, horses, or pigs (Busch et al., 1986; Conrad, 1994).

The size and speed of these animals create forces similar to those which occur in motor vehicle accidents (Norwood et al., 2000). Nogalski et al. (2007) reported that most of the injuries that occurred as a result of bull and horse attacks necessitated hospitalization of the victim.

Interactions with animals of all kinds may result in human fatalities from a wide variety of causes ranging from blunt force trauma to envenomation. In the United States between 1991 and 2001, a total of 1943 people died after encounters with venomous and nonvenomous animals, representing an average of 177 deaths per year. Of these fatalities, 61% were caused by nonvenomous animals, with dog attacks resulting in 17.6% of these deaths and "other specified animals" responsible for 71.5% (Langley, 2005).

Occupation may have an effect on the occurrence of injury and death, with the highest rates occurring in situations of greatest exposure, such as farm workers, veterinarians, cowboys and rodeo riders, animal caretakers, and hunters (Langley, 2001). Animal-related deaths were cited as the fourth most common cause of death in those aged over 55 years in the agricultural industry in Australia from 2001 to 2004, causing 7.1% of deaths. In those aged 15 years and older, horses were the main culprit causing 3.5% of deaths, followed by cattle at 1.9% (Bury et al., 2012). Those most at risk in the United States are men over the age of 65 years (Langley, 2001).

4.1. Blunt force injuries

Animals may be responsible for an array of potentially lethal injuries. Blunt force injuries characteristically involve larger animals such as cattle or horses that may kick, crush, or trample a victim causing head and facial injuries. Farm workers in particular are at high risk of lethal injuries involving the head and torso (Bury et al., 2012). There are many reported cases of large farm animals butting or kicking individuals, resulting in death, with horses, bulls and cows being the most common offenders (Karkola et al., 1973; Rabl & Auer, 1992; Langley, 1994; Holland et al., 2001; Dogan et al., 2008). Occupation and sex bias are demonstrated in the report of nine cattle-related deaths in which eight of the victims were men with the deaths occurring on farms. Predictably, kicks to the head and chest result in more immediately fatal outcomes, most likely due to the effects of direct trauma to the brain, heart, or lungs (Langley, 1994). Because of their small size and physical vulnerability, children are at high risk of being kicked by farm animals such as horses. In addition, riding accidents are not uncommon, when children either fall from or are thrown off a horse. Deaths in such cases are usually caused by severe head trauma and may be related to failure to wear protective helmets (Bury et al., 2012).

4.2. Penetrating injuries

Bulls and buffalos, along with other horned animals, gore their victims as may sometimes be seen with matadors in Spanish bull fighting (Langley, 1994; Bakkannavar et al., 2010). In Pamplona, Spain, there is a yearly tradition of running with the bulls with 15 people having died since 1922 (BBC, 2009). Both domestic and wild pigs may fatally attack humans. Wild boars incapacitate prey by repeatedly charging with their tusks causing multiple penetrating injuries, sometimes resulting in intestinal evisceration (Manipady et al., 2006). Two deaths caused by domestic pigs were reported by Langley (1994) where both victims sustained significant tearing and shredding injuries of the skin because of biting, with limb amputation in one case. There were also multiple fractures and extensive crush injuries. Exsanguination can result from vascular injury inflicted by the tusk of wild boar (Prahlow et al., 2001). Although the agricultural environment poses a number of dangers to people in all age groups, there are also notable hazards for children that are important sources of childhood mortality and morbidity (Marlenga et al., 2001). Because of the increasing mechanization of agricultural operations over the past half-century, and the high fatality rate associated with injuries due to machinery and tractors, much of the literature has focused on injuries due to these sources. However, animal-related injuries are also an important source of operation-related injuries (Layde et al., 1996) and have frequently been reported as a major source of nonfatal agricultural injury (Cogbill et al., 1985; Boyle et al., 1997; Gerberich et al., 2001; Gerberich et al., 2003; Gerberich et al., 2004). Several studies have identified animals as a leading source of injury to children on agricultural operations (Gerberich et al., 2001; Gerberich et al., 2003; Cogbill et al., 1985; Pickett et al., 2001). Jones & Field (2002) found that 33% of all fatalities to Amish adults and children were directly or indirectly related to animals. A major barrier to progress in the prevention of agricultural injuries has been a lack of knowledge about the magnitude of the problem and of specific risk factors associated with these injuries (Gerberich et al., 2001). Despite the large number of documented animal/livestock injuries, little is known about the consequences of this problem or factors that increase or decrease the risk of injury (Boyle et al., 1997).

4.3. Needle stick injuries

Needle stick injuries are not uncommon in animal handlers (Berkelman, 2003; O'Neill et al., 2005; Weese & Jack, 2008). Very little research has been done on needle sticks in farmers, but there are several studies that have looked at needle sticks in veterinarians. Some vaccines contain live bacteria and infection has occurred in the person who was accidentally injected. Some cases of sterile abscess formation and even tissue necrosis have developed after unintentional vaccine injections, especially ones with oil-based adjuvants. Injuries from needles sticks can be prevented by following the recommendations of the NASPHV and others (Weese & Jack, 2008; Centers for Disease Control and Prevention, 2009).

- Read the package insert, label, and material safety data sheet (MSDS) for any medications administered, and use the product only as directed on the package or as directed by your veterinarian
- Never remove needle caps by using your mouth
- Do not recap needles except in rare instances when required as part of a medical procedure or protocol
- Dispose of all sharps in designated puncture-proof sharps containers
- Dispose of the used syringe with attached needle in the sharps container when injecting live vaccines or aspirating body fluids
- For most other veterinary procedures, use the needle removal device on the sharps container and dispose of the syringe in the regular trash

- Sharps containers should be located in every area of the workplace where sharps are used
- Do not transfer sharps from one container to another
- Devices that cut needles prior to disposal should not be used because they increase the potential for aerosolization of the contents
- Never dispose of sharps in the regular trash
- Work with your veterinarian to determine how often you should change needles
- Don't place your hands between the animal and the side of the stall when injecting drugs
- Don't carry a syringe and needle in your pocket
- Don't use damaged or bent needles
- Develop, communicate, and enforce standard operating procedures for safe sharps handling
- Make sure employees are trained on hazards of handling drugs, hormones, and vaccines

4.4. Other health and safety hazards

There are numerous hazards besides animal contact that present a risk to livestock handlers on the farm or ranch. As previously noted, most fatal injuries are due to machinery and equipment. Tractors, augers, and other feeding equipment are frequent sources of injuries. Trucks and trailers used to move livestock may be involved in traffic accidents. Injuries from corral gates, either from the animal kicking the gate that strikes the worker or from other contact with gates, caused 4% of compensation claims in livestock handlers in Colorado. Injuries were also associated with contact with milking units, milking rail, walls, squeeze chute, and even a hose. Falls and slips on farms and ranches from slippery surfaces or stepping in holes or uneven surfaces are often recorded as being in the top three leading causes of farm injuries. Strains from heavy lifting caused 13% to 21% of the claims among dairy farm workers, cattle raisers and dealers in Colorado (Douphrate et al., 2009).

Waller (1992) evaluated dairy farmers and the various sources/activities that led to their injuries. Of 147 injuries, 29% were due to cattle but other causes included equipment repair/use 7.5%, chemicals/biologics 7%, hay/haying 6%, tractors 6%, slip/trip/fall same level 5%, foreign body in eye 5%, bumped fixed object 4%, pitchfork 3%, fall from height 3%, and miscellaneous 5%.

Asphyxiation from oxygen deficient environments may occur when farmers enter manure pits or silos. Meyer (2005) found 73 fatalities among farm workers due to oxygen-deficient exposures from 1995 to 2002. He also detected 157 electrocutions among farm workers.

Although injuries from chemical use on farms are not uncommonly reported, thousands of tons of various chemicals and fertilizers are used on farms, creating risk not only to the handler but also to farm family members, farm animals, and the environment. There are thousands of pounds of herbicides also used on farms that may potentially cause adverse effects if not handled properly (Langley & Morrow, 2010).

4.5. Risk factors associated with farm injuries

Risk factors associated with farm injuries can be categorized into two levels: (1) characteristics of the farm environment, and (2) characteristics of the farmer. Studies examining farm environment factors have reported that larger farms, farms with more workers, and farms with higher annual production were associated with higher injury risks for the farmer (Zhou & Roseman, 1994; Pickett et al., 1995). Studies have demonstrated that injury risk is elevated on farms with animals, especially on beef and dairy farms (Brison & Pickett, 1992; Pratt et al., 1992; Zhou & Roseman, 1994; Nordstrom et al., 1995). The patterns of injury have been fairly consistently reported across these studies, with farm machinery, accidental falls, and animal related injuries being the three major external causes of injury (Brison & Pickett, 1992; Zhou & Roseman, 1994; Nordstrom et al., 1995).

With respect to individual risk factors for the farmer, greater number of hours spent farming, full-time farm work, greater cumulative years of farm work experience, and male gender have shown positive associations with higher injury rates (Stallones, 1990; Brison & Pickett, 1992; Zhou & Roseman, 1994; Layde et al., 1996). Whereas several studies have suggested that farmers holding off-the-farm employment may be at greater risk for injury (Zhou & Roseman, 1994; Nordstrom et al., 1995; Zhou & Roseman, 1995), other studies have not confirmed the finding (Brison & Pickett, 1992). Although other personal risk factors for injuries, including alcohol consumption, prior traumatic injury, lower levels of education and training, and the use of prescription medications have been investigated, the evidence associating these risk factors with farm injuries remains inconclusive (Elkington, 1990; Zwerling et al., 1995; Zhou & Roseman, 1995).

A substantial portion of the risk for injury associated with working on the farm can be attributable to the farm environment itself (Browning et al., 1998). There may be several reasons for the elevated risk of injury on farms with beef cattle. Research by Elkington (1990) suggests that farms with livestock tended to be the most time-consuming enterprises and that injury risk increased with increasing number of hours worked on the farm. Brison & Pickett (1992) reported a harvest-related peak for injuries on beef farms. They argued that the excessive fatigue, missed meals, and stress associated with harvest time is particularly felt by beef farmers. They indicated that beef farmers worked larger amounts of land with fewer manpower resources and used older equipment. Proper animal handling and transportation practices and the exercise of caution around bulls, steers, and pregnant cows could reduce some of these injuries (Layde et al., 1996; Browning et al., 1998).

Potential risks can be assessed in many ways:

- Walk through all animal-handling areas and look for hazards, such as broken gate latches, broken posts, restraining equipment not working.
- Consult with farm safety advisers from the Victorian WorkCover Authority - they may provide free consultations.
- Reflect on injury records to pinpoint recurring dangers, including less obvious ones like lacerations and sprains.
- Talk over safety issues with family members, workers and other animal handlers.
- Make sure at least one person on the farm is trained in first aid.
- Remember that inexperienced workers and bystanders are more likely to be injured (NCFM, 2012).

5. PREVENTIVE MEASURES FOR LIVESTOCK-HANDLING RELATED INJURIES AND DEATHS

Efforts to reduce or eliminate on-farm animal-related injuries should be directed principally to farm residents. Efforts should be made to educate farm residents about the many dangers presented by farm animals. In addition to educational efforts, structural modification of barns to limit animal interaction, isolation of dangerous animals, and wearing protective gear in recreational and work activities are a few of the prevention strategies which can be employed to reduce animal-related farm injuries.

Animals may not purposely hurt a worker, but their size and bulk make them potentially dangerous, especially when in close contact. Farmers, veterinarians and other animal handling businesses should implement the following measures to prevent injuries (Hendricks & Adekoya, 2001).

5.1. Facilities

Many livestock handling injuries are directly related to equipment or building structures. Poor facilities and equipment can also cause injuries to animals. This can mean considerable economic loss at market time. Tripping hazards such as high door sills, cluttered alleyways and uneven walking surfaces can cause serious injury and a considerable amount of lost work time. Concrete floors are best for livestock. The finish on concrete floors should be roughened to prevent slips under wet conditions. High traffic areas, such as alleyways, should be grooved. Floors should allow water to drain easily. Slatted floors often are used to keep animals dry in a confinement system. Fencing and gates should be strong enough to contain crowded livestock. A variety of materials are available, but the key is strength and durability. A protruding piece of lumber, a nail or a bolt can cause painful and infectious injuries. If backed or pushed into, one of these objects can cause a serious back injury. Alleys and chutes should be wide enough to allow animals to pass, but not wide enough to allow them to turn around. A width of 30 inches is recommended for a cow-calf operation. For cattle in the range of 800 to 1,200 pounds, a 26-inch width is recommended. Solid wall chutes, instead of fencing, will lower the number of animals that balk in the chute. Lighting should be even and diffused. Bright spots and shadows tend to make animals more skittish, especially near crowding or loading areas. Animals move more readily from dark areas into light, but avoid layouts that make them look directly into the sun. Handling equipment can speed up livestock confinement work operations, reduce time and labor requirements, cut costs, and decrease the risk of injury (Baker & Lee, 1993).

5.1.1. Corral systems design & pens

In crowding pens, consider handling cattle in small groups up to ten head. The cattle need room to turn. Use their instinctive following behavior to fill the chute. Wait until the single file chute is almost empty before refilling. A crowding gate is used to follow the cattle, not to shove against them. If a lone animal refuses to move, release it and bring it back with another group. An animal left alone in a crowding pen may become agitated and attempt to jump the fence to rejoin the herd.

Corral and working facilities are constructed to confine cattle safely and efficiently for close observation and to perform routine health and management procedures. Good cattle handling systems make working livestock easier with limited manpower. The operation of any cattle facility depends on cattle behavior, corral design, and the skill and technique of the handler.

Pens serve several purposes, including catching and holding cattle being worked, sorting cattle into groups, and holding cattle being quarantined. When designing and constructing pens for working facilities, consider the following:

- Provide at least 20 square feet per head for mature cattle.
- Size pens for a maximum of about 50 head of mature cattle.
- Larger, wider pens can make effective sorting difficult for a single worker.
- Pens too small or narrow can result in workers entering the animal's flight zone. The smallest pen dimensions should be no less than 16 feet.
- Too few pens can make separating animals difficult. This can also put workers at risk, as they must physically enter pens with large numbers of agitated animals. Consider adding in a 14-inch wide pass-through for worker escape in pen corners.
- Use proper gate placement to facilitate animal movement from pen to pen and to other areas. Poor animal movement puts workers at risk by having to force the movement. If there are too few gates, some animals can become separated. Thus, when animals enter the alley, separated herdmates will follow along the inside of the pen. This is often referred to as "backwash." There may be problems guiding these pen-bound animals back to the exit gate as their herdmates move away from them down the alley. Separated animals can become confused or agitated, putting workers at further risk.

The proper design, construction and operation of a cattle handling facility is important to ensure safe working conditions for animals and humans. Understanding the inherent behavior of cattle, plus working them slowly and quietly, will reduce injuries and help make an operation run more smoothly and efficiently (Hubert et al., 2012).

5.2. The human element

Human error is the primary cause of many types of accidents. These errors in judgement and action are due to a variety of reasons, but occur most when when people are tired, hurried, upset, preoccupied, or careless.

Remember that human physical, psychological, and physiological factors greatly affect the occurrence of life threatening accidents. Using this information in combination with proper cattle handling techniques can reduce the risk of injury.

Stress - When the handlers are under physical or psychological stress, avoid major animal handling activities. Fatigue and stress can predispose handlers to serious farm accidents.

Training - Teach new and young workers how to work safely with farm animals. During training sessions, emphasize all known animal behavior problems. Also demonstrate the use and effectiveness of all animal restraining equipment and facilities available.

Sure Footing - Reduce the hazards of falls, provide slip-resistant footing for both handlers and livestock with a roughened surface on concrete ramps and floors in animal facilities.

Lifting - Use a hip lifter to lift or assist a downed cow, and have additional handlers help with the handling to prevent any strain or back injuries (Gay & Grisso, 2012; Hubert et al., 2012).

5.3. Additional handling tips

In addition to the flight zone, an understanding of the "herd instinct" is important. Cattle follow the leader and are motivated to follow each other. Each animal should be able to see others ahead of it. Make single file chutes at least 20 feet long, or 30 to 50 feet for larger facilities. Don't force an animal in a single file chute unless it has a place to go. If the cow balks, it will continue balking (Hubert et al., 2012).

5.3.1. Specific handling methods

Specific handling methods vary with species. However, some general handling rules for all animals include the following:

- Most animals respond favorably to routines having calm, deliberate responses.
- Avoid loud noises and quick movements.
- Be patient; never prod an animal when it has no place to go.
- Move slowly and deliberately.
- Touching animals gently can be more effective than shoving and/or bumping them.

- Respect rather than fear livestock. Breeding stock are highly protective and often irritable. Disposition deteriorates with age and parturition. Old breeding stock can be cantankerous, deceptive, unpredictable, and large enough to be dangerous.
- Special facilities should be provided for breeding stock, especially for large males. Most animals are highly protective of their young. Be especially careful around newborn animals.
- Male animals should be considered potentially dangerous at all times. Proper equipment and facilities are necessary to assure safety. Extreme caution should be practiced when handling male animals.
- The size, mass, strength, and speed of an individual and herd's of animals should never be taken lightly. Animals will defend their territory and should be worked around keeping in mind that there is always the potential for harm (Gay & Grisso, 2012).

5.3.2. Reducing animal risks

Management of a dairy farm involves animal handling activities such as milking, feeding, and providing health care. This constant close contact with animals increases your chance of having an animal-related accident. How can the accident risk be reduced?

Gates - The gate is the simplest and often most versatile item for handling animals. When improperly placed or hinged in the wrong direction, it can also be the most obtrusive, resulting in blocked freestalls, inability to move cows in the desired direction, unnecessary injury to cattle and people and damage to the gate.

Properly located gates can block off a travel lane and direct the cow into a desire area. Gates may also be used within pens to form a funnel to direct a reluctant cow into a stanchion or other lockup. They can also be part of a smaller confinement area for breeding or rectal examination.

Gates should be high enough to discourage jumping (6-7 feet). The bottom rail should be low enough to discourage animals from turning around or crawling under it, and high enough to allow a trapped person to roll underneath (16-20 inches). Cross bars or rails should be spaced no farther than 10 inches apart to prevent animals from getting their heads trapped between them.

For young animals, cross bars or rails should be 4-6 inches apart. In open buildings, gates may be covered with plywood to minimize drafts on animals during cold weather conditions.

Stanchions - A stanchion or yoke is a fixed or swiveled device that consists of two vertical bars or slates that can be closed against the animal's neck. A swivel stanchion pivots around its vertical axis, making it easier for the animal to turn its head. Self-closing stanchions are used for post calving examinations, pregnancy checks, vaccinations, tail head chalking, heat detection, and artificial insemination. They may be used in large groups along a feed line in a freestall barn, in small groups in a treatment holding area, or individually in a maternity pen. Satisfactory fenceline stanchions self-lock upon cow entry, provide individual or group release, have a method to lock out animals when the system is not in use, and allow for easy removal of "downed" animals.

When evaluating a self-locking stanchion be sure the top opening is wide enough for the cow to easily insert and extract her head without twisting, turning or banging. The height of the pivot point is also critical to assure the unit will lock closed as the cow reaches down to eat, but also will swing open as her head reaches a normal position. Stanchions that require the cow to raise her head too far upward will result in more banging and possible injury to the animals. If the pivot point is too low the cow may not trip the stanchion. Recommended mounting heights will vary among manufacturers. The bottom rail is also critical for unobstructed access by the animal to feed. Very young calves often have trouble using self locking stanchions if the pivot point is too high. In general, a fixed vertical or slant bar divider works better until calves learn to reach through a hole for eating.

Headgates - Self-locking headgates have many features not found in a typical stanchion. The force of the cow trying to walk through opens, closes, and latches the headgate. Headgates should be adjustable for size and be on hinges to allow them to open completely from top to bottom, to avoid injury to the hips and legs of the animals. The restricted opening should extend to the ground to prevent choking a "downed" animal and to permit easy removal. Some operators feel that self-locking headgates result in excessive bruising to an animal's shoulders. Manually controlled mechanical, hydraulic or air operated closing mechanisms are also used.

Chutes - Chutes, usually used in conjunction with headgates, help direct the animal into a headgate and prevent side to side cow movement during examination and treatment. It is necessary that the sides of the chute be open, and removable, to permit access to the sides of the cow for examining the udder, body cavity and feet. For rectal examinations, a gate or pass through is required behind the animal. A head grate and working chute can usually be incorporated in lanes where cows walk single file (Gay & Grisso, 2012).

5.3.3. Restraining equipment and facilities

Over 75% of the animal-related human injuries are due to *insufficient restraining equipment and facilities on most dairy farms.* Proper application and the right choice of restraining equipment and facilities are very important consideration for reducing potential injuries to the dairy farmer.

Before selecting the animal restraining equipment and/or the facility, ask: Will it be safe for the animal handler? Will it be safe for the animal? Will it accomplish the intended purpose?

Equipment for various activities - When milking in the barn or cleaning or examining the udder use anti-kicking devices on cows that are chronic kickers. Use a rope halter, squeeze chute, and headgate when you engage in major animal handling activities such as hoof trimming, breeding, and applying medication. Use a squeeze chute with a headgate to protect yourself from the animal's violent movements. Use a tail holder to prevent eye injuries when milking or examining the animal.

Mangates - Many veterinarians recommend mangates in cattleyard pens. Mangates are small passages between two posts about 14 inches apart in the fence around a yard pen, through which a person can easily escape from unexpected, dangerous situations without having to open the animal gate or climb the fence.

Dairy bulls - If you keep a dairy bull on your farm for breeding purposes, have all the necessary restraining equipment and facilities. An ideal confinement unit for a dairy bull should be designed so you never come in direct contact with the bull for feeding or breeding. Dairy bulls are much more aggressive by nature than cows. Although some dairy bulls appear gentle and calm, they may react to unexpected movements, inflicting serious injuries or death onto the bull handler. Never consider a bull safe, and do not let your children play with a bull even if you have raised him.

Treatment stall - Maintain a treatment stall on your farm to reduce the risk of injuries to yourself as well as the veterinarian during activities such as pregnancy examination, vaccination, medication, deworming, and artificial insemination.

Cows with new calves - A cow with her new calf is usually more defensive and more difficult to handle. Let her calf stay as close to her as possible.

Horned animals - Horned cows or bulls are more prone to attacking handlers. Make sure all cattle on the farm are dehorned.

Kicking - Cows commonly kick forward and out to the side. They also have a tendency to kick toward the side where they have pain from inflammation or injuries. Therefore, if a cow is suffering from mastitis of only one quarter, you may want to consider approaching her from the side of the non-affected udder when examining or milking.

Dry cows - Dry cows usually exhibit more aggressive behavior after coming back from the pasture. It may take them a week or so before they get used to barn life again (Gay & Grisso, 2012).

5.3.4. Handling of escaped animals

If horses, cattle, or other large animals escape from an auction ring, show ring, or slaughter plant, they must never be chased. Chasing escaped cattle often cause them to run wildly through crowds of people, injuring people and damaging property. If an escaped cow or horse is located and they are not an immediate threat to people, allow them to be alone for 30 minutes so they will calm down. Twenty minutes is required for the animal's heart rate to return to normal. When the escaped animal has calmed down, it can be quietly moved. Interestingly, a lone animal often returns of its own volition to other horses and cattle. A panicked cow or horse may crash through a chain link fence, because they can not see the thin mesh. Escaped cattle have knocked over a chain link fence when being chased. Chain link fences hold a calm cow, but a frightened bovine may either knock it over or go under the bottom edge of the mesh. Fences that present a visual barrier, such as a board fence, are less likely to be broken down by excited animals.

Allow only experienced livestock handlers to approach an escaped animal. A panicked horse, steer, or bull can be calmed when it hears its owner's familiar voice. Some of the most dangerous incidents with escaped animals have occurred when security guards or the police became involved (Gay & Grisso, 2012).

5.3.5. Personal protective equipment

Farm safety experts estimate more than half of farm injuries can be prevented by using some type of personal protective equipment, depending on the work activity. Use safety glasses, gloves, long trousers, steel-toed work shoes, and a bump hat for activities such as handling bulls, hoof trimming, and shipping the animals. This type of equipment will reduce the injury potential to the head, feet, hands, and other parts of the body.

Use a chemical respirator, eye goggles, hard shell hat, rubber gloves, trousers, and a longsleeved shirt when preparing and applying pesticides. Exposure to pesticide chemicals through breathing, swallowing, or skin contact is a significant health hazard and may lead to poisoning or serious skin problems.

Use a dust respirator when working in areas that generate dust. Avoid breathing any dust generated by moldy forage or grain because it may cause farmer's lung disease (Gay & Grisso, 2012).

5.3.6. Controlling diseases

Handlers should also be concerned with zoonotic diseases, which are illnesses that can be transmitted between humans and animals. Leptospirosis, rabies, brucellosis, salmonellosis and ringworm are especially important.

To reduce exposure to disease, use basic hygiene and sanitation practices, which include prompt treating or disposal of infected animals, adequate disposal of infected tissues, proper cleaning of contaminated sites, and proper use of personal protective equipment.

6. CONCLUSION & RECOMMENDATIONS

Education is a key component to prevent injuries and illnesses in the livestock industry. Frequent staff turnover means that training new employees may be needed several times a year. Employees in livestock operations need to be trained on such issues as proper animal handling, including animal welfare concerns, correct methods to employ during animal loading and transportation, proper use of personal protective equipment, and proper use of antibiotics, pesticides, and other chemicals. Employees also need to be trained on hazards of confined spaces, weather extremes, and proper cleaning and first-aid of any injury that occurs. Workers should be instructed to report to their supervisor any on-the-job injury or illness that may affect their health or the health of the animal herd. Use of vaccines, such as influenza vaccine, to prevent spread of disease to coworkers or to animals should also be a component of a worker health and safety program (Langley & Morrow, 2010).

Proper design of animal facilities may relieve animal stress and make movement of animals easier. Properly designed facilities should also reduce injuries from slippery uneven surfaces. Dr. Temple Grandin has developed plans for animal facility construction that incorporates basic animal behavior issues (Grandin & Deesing, 2008).

More research is needed to develop programs that lead to a decrease in injuries, that is acceptable and easy to use, and that maintains animal welfare. The following are some suggestions for future research. It is recommended that NIOSH update national surveys of injuries in livestock handlers to look at the types of injuries that occur and the circumstances behind the injuries. Studies should be conducted to determine if there is a decrease in animal handler injuries following species-specific training programs for the handlers. These studies should include pig, sheep, and goat farms as well as the traditional cattle and horse farms. Studies should be conducted to determine the incidence of needle stick injuries by species and agent injected and if redesign of equipment is needed. An evaluation of needleless injection systems should be included to see if they can be successfully used on animals. Studies to determine the incidence of all-cause/all-species injuries and death on small farms compared to larger farms are needed to see if there are different risk factors that may play a role. Insurance industries' databases for all deaths/injuries on livestock and poultry farms should be evaluated to look for trends in types of injuries (Langley & Morrow, 2010).

Safety Reminders for Livestock Handling

- Good housekeeping is essential, not only for your personal safety, but also for the health and well being of your stock.
- Keep children away from animals, particularly in livestock handling areas.
- Most male animals are dangerous. Use special facilities for these animals and practice extreme caution when handling them.
- Be calm and deliberate when working with animals.
- Always leave yourself an "out" when working in close quarters.
- Respect all animals. They may not purposely hurt you, but their size and bulk make them potentially dangerous.
- Most animals tend to be aggressive when protecting their young; be extra careful around newborn animals.
- Stay clear of animals that are frightened or "spooked." Be extra careful around strange animals (Baker & Lee, 1993).

Provide workers with the knowledge of animal behaviors and habits

A good understanding of animal behaviors and habits can help handlers maintain control of routine handling as well as emergency situations.

- Most animals respond favorably to calm and deliberate movement and responses from a handler.
- Animals have a personal space or "flight zone": if a handler gets too close, they will move away.
- Animals have difficulty judging distances and cannot see directly behind them. Quick movement behind them may frighten them.

- Animal vision is in black and white, not in color. Animals move more readily from dark areas into light. Bright lights and shadows tend to make animals skittish.
- Animals have sensitive hearing and can detect sounds that human ears cannot. Loud noises frighten animals. High frequency sounds actually hurt their ears, causing animals to become skittish and balky.
- Animals do not like surprises. Cattle become uneasy or skittish when their routines or surroundings change. Cattle can also be easily frightened by strangers or around small children who tend to make sudden movements.
- Stressed animals that are sick, injured, in heat or just mated can be easily agitated and highly unpredictable. Males are generally more aggressive by nature. Female animals tend to be more aggressive when with their young.
- Animals that are stressed show signs of fear or aggressiveness. Warning signs may include raised or pinned ears, raised tail or hair on the back, bared teeth, pawing the ground or snorting.

Provide employee training in the hazards associated with animal handling and in safe handling techniques

- Approach and handle animals in a calm, steady and consistent manner; don't shove or bump them.
- Approach animals from the front or side. Move slowly and deliberately. Avoid startling animals with quick movements or loud noises.
- Wait until an agitated animal calms down before resuming working with the animal.
- Stop and seek additional help when handling animals that are unusually aggressive or in a situation that is not safe to perform the work. Use proper animal restraints and take adequate safety precaution.
- Promptly remove dangerous animals from farms or facilities to prevent worker injury.
- Plan ahead to allow plenty of time when moving animals. Be patient. Never prod an animal when it has nowhere to go.
- Whenever possible, move and isolate animals from livestock areas prior to performing work in those areas.
- Use a soft light that does not cast shadows that could spook the animals when trying to move cattle at night.

- Exercise extra caution when handling animals that are sick, hurt, new mothers, in heat or just mated.
- Position yourself so that you are not between a mother and her young.
- Use special facilities to separate male animals.
- Keep strangers and children out of animal handling areas.
- ALWAYS plan an escape route when working with animals in close quarters.
- Require workers to wear proper and necessary personal protective equipment such as protective safety shoes or boots with non-slip soles, sturdy clothing, gloves and helmets.

Design, construct and maintain a safe animal handling facility

Many animal handler injuries are directly related to unsafe features or design of the animal handling facilities. A safe animal handing facility can both protect workers and prevent injuries to animals that often cause considerable economic loss.

- Eliminate tripping hazards such as high door sills, cluttered alleyways and uneven walking surfaces that can contribute to falls.
- Build concrete floors with a roughened finish.
- Groove high traffic areas such as alleyways.
- Keep floors dry in working and walking areas. These areas should drain easily.
- Provide non-slip surfaces when possible.
- Construct "man passes" or narrow escape routes in barns where a person can safely get away, but an animal cannot follow.
- Plan and design escape routes from open animal yards.
- Build alleys and chutes wide enough for animals to move through but not wide enough to allow them to turn around.
- Build strong fences and gates to contain crowded livestock using strong and durable materials.
- Build chutes with solid walls instead of wires, fences, or open sides. Solid walls can shield the animals from outside distractions.
- Eliminate or remove any potentially hazardous protrusions and sharp objects in the livestock area, such as nails, bolts and broken boards. These may startle or distract animals and create a dangerous situation for workers.

- Provide lighting that is even and diffused to avoid casting shadows. Avoid layouts that force animals to look directly into the sun.
- Use handling equipment in livestock confi nement work operations to reduce the risk of injury. These include hydraulic chutes, portable alleys and headbenders (FACE, 2012).

REFERENCES

1. Aherin, R.A., Murphy, D.J. & Westby, J.D. (1992). *Reducing Farm Injuries: Issues and Methods.* American Society of Agricultural Engineers, St Joseph, MI
2. Austin, C.C. (1998). Nonvenomous animal-related fatalities in the United States workplace, 1992-1994. *J Agromed,* Vol. 5, pp. 5-16
3. Baker, D.E. & Lee, R. (1993). *Animal Handling Safety Considerations.* Missouri University Extension, the University of Missouri-Columbia, Columbia, USA
4. Bakkannavar, S.M., Monteiro, F.N., Bhagavath, P. & Pradeep Kumar, G. (2010). Death by attack from a domestic buffalo. *J Forensic Leg Med,* Vol. 17, pp. 102-104
5. Bancej, C. & Arbuckle, T. (2000). Injuries in Ontario farm children: population based study. *Inj Prev,* Vol. 6, pp.135-140
6. BBC. (2009). Bull gores man to death in Spain, Available from: http:// http://news.bbc.co.uk/2/hi/8143744.stm?ls
7. Berkelman, R.L. (2003). Human illness associated with use of veterinary vaccines. *Clin Infect Dis,* Vol. 37, pp. 407-414
8. Bernhart, J. & Langley, R. (1999). Analysis of tractor-related deaths in North Carolina from 1979 to 1988. *J Rural Health,* Vol. 15, pp. 285-295
9. Boyle, D., Gerberich, S.G., Gibson, R.W., Maldonado, G., Robinson, R.A., Martin, F., Renier, C. & Amandus, H. (1997). Injury from dairy cattle activities. *Epidemiology,* Vol. 8, pp. 37-41.
10. Brison, R.J. & Pickett, C.W. (1992). Nonfatal injuries in 117 eastern Ontario beef and dairy farms: A one-year study. *Am J Ind Med,* Vol. 21, pp. 623-636
11. Browning, S.R., Truszczynska, H., Reed, D. & McKnight, R.H. (1998). Agricultural injuries among older Kentucky farmers: the Farm Family Health and Hazard Surveillance Study. *Am J Ind Med,* Vol. 33, pp. 341-353
12. Bury, D., Langlois, N, & Byard, R.W. (2012). Animal-related fatalities--part I: characteristic autopsy findings and variable causes of death associated with blunt and sharp trauma. *J Forensic Sci,* Vol. 57, pp. 370-374
13. Busch, H.M. Jr., Cogbill, T.H., Landercasper, J. & Landercasper, B.O. (1986). Blunt bovine and equine trauma. *J Trauma,* Vol. 26, pp. 559-560
14. Carlson, K., Goodwin, S., Gerberich, S., Church, T., Ryan, A., Alexander, B., Mongin, S., Renier, C., Zhang, X., French, L. & Masten, A. (2005). Tractor-related injuries: A population-based study of a five-state region in the Midwest. *Am J Ind Med,* Vol. 47, pp. 254-264
15. Casey, G.M., Grant, A.M., Roerig, A.D., Boyd, J., Hill, M., London, M., Gelberg, K.H., Hallman, E. & Pollock, J. (1997a). Farm worker injuries associated with bulls. *AAOHN J,* Vol. 45, pp. 393-396
16. Casey, G.M., Grant, A.M., Roerig, A.D., Boyd, J., Hill, M., London, M., Gelberg, K.H., Hallman, E. & Pollock, J. (1997b). Farm worker injuries associated with cows. *AAOHN J,* Vol. 45, pp. 446-450
17. Centers for Disease Control and Prevention. (2009). Fatalities caused by cattle-four states, 2003-2008. *MMWR Morb Mort Wkly Rep,* Vol. 58, pp. 800-804
18. Cleary, J., Benzimiller, J., Kloppedal, E., Gallagher, D. & Evans, A. (1961). Farm injuries in Dane County, Wisconsin. *Arch Environ Health,* Vol. 3, pp. 83-90

19. Cogbill, T.H. & Busch, H.M. (1985). The spectrum of agricultural trauma. *J Emerg Med,* Vol. 3, pp. 205-210
20. Cogbill, T.H., Busch, H.M. & Stiers, G. (1985). Farm accidents in children. *Pediatrics,* Vol. 76, pp. 562-566
21. Cole, H., Myers, M. & Westneat, S. (2006). Frequency and severity of injuries to operators during overturns of farm tractors. *J Agric Saf Health,* Vol. 12, pp. 127-138
22. Conrad, L. (1994). The maul of the wild. Animal attacks can produce significant trauma. Emerg Med Serv, Vol. 23, pp. 71-72, 76.
23. Cordes, D.H. & Foster, D. (1988). Health hazards of farming. *Am Fam Physician,* Vol. 38, pp. 233- 244
24. Crawford, J.M., Wilkins, J.R., Mitchell, G.L., Moeschberger, M.L., Bean, T.L. & Jones, L.A. (1998). A cross-sectional case control study of work-related injuries among Ohio farmers. *Am J Ind Med,* Vol. 34, pp. 588-599
25. Dogan, K.H., Demirci, S., Erkol, Z., Sunam, G.S. & Kucukkartallar, T. (2008). Injuries and deaths occurring as a result of bull attack. *J Agromedicine,* Vol. 13, pp. 191-196
26. Dogan, K.H., Demirci, S., Sunam, G.S., Deniz, I. & Gunaydin, G. (2010). Evaluation of farm tractor-related fatalities. *Am J Forensic Med Pathol,* Vol. 31, pp. 64-68
27. Douphrate, D.I., Rosecrance, J.C., Stallones, L., Reynolds, S.J. & Gilkey, A.P. (2009). Livestock-handling injuries in agriculture: an analysis of Colorado workers' compensation data. *Am J Ind Med,* Vol. 52, pp. 391-407
28. Drudi, D. (2000). Are animals occupational hazards? *Compensation and Working Conditions,* Vol. Fall, pp. 15-22
29. Ehlers, J., Connon, C., Themann, C., Myers, R. & Ballard, T. (1993). Health and safety hazards associated with farming. *Am Assoc Occup Health Nurs,* Vol. 41, pp. 414-421
30. Elkington, J.M. (1990), *"A case-control study of farmwork-related injuries in Olmstead County, Minnesota"* [Dissertation], The University of Minnesota, Minneapolis
31. Erkal, S., Gerberich, S.G., Ryan, A.D., Renier, C.M. & Alexander, B.H. (2008). Animalrelated injuries: a population-based study of a five-state region in the upper Midwest: Regional Rural Injury Study II. *J Safety Res,* Vol. 39, pp. 351-363
32. Etherton, J., Myers, J., Jensen, R., Russell, J. & Braddee R. (1991). Agriculture machinerelated deaths. *Am J Public Health,* Vol. 81, pp. 766-768
33. FACE (The New York State Fatality Assessment and Control Evaluation). (2012). Fatal injuries among animal handlers in New York State. Available from: http://www.health.ny.gov/publications/6021.pdf
34. Forjuoh, S.N. & Mock, C.N. (1998). Getting together for injury control. *World Health Forum,* Vol. 19, pp. 39-41
35. Gadalla, S.M. (1962). *Selected Environmental Factors As Associated With Farm and Farm Home Accidents in Missouri.* University of Missouri College of Agriculture, Rural Health Series Publication 16, Columbia
36. Gay, S.W. & Grisso, R. (2012). Safe handling of livestock. Available from: http://bsesrv214.bse.vt.edu/Grisso/Ext/AnimalHandling.pdf
37. Gerberich, S.G. (1995). Prevention of death and disability in farming. In: McDuffie, H.H., Dosman, J.A., Semchuk, K.M., Olenchock, S.A., Senthilselvan, A., editors, *Human Sustainability in Agriculture: Health, Safety, Environment, Proceedings,* Third International Symposium: Issues in Health, Safety and Agriculture, Saskatoon, Canada
38. Gerberich, S.G., Gibson, R.W., Gunderson, P.D., Melton, III L.J., French, L.R., Renier, C.M., True, J.A. & Carr, W.P. (1992). Surveillance of injuries in agriculture. In: *Proceedings: Surgeon General's Conference on Agricultural Health and Safety.* DHHS (NIOSH) Publication Number 92-105
39. Gerberich, S.G., Gibson, R.W., Gunderson, P.D., French, L.R., True, J.A., Renier, C. & Carr, P. (1994). The regional rural injury study-I (RRIS-I): A Model for national surveillance of injuries in agriculture. In: *Proceedings of the Ninth International Symposium on Epidemiology in Occupational Health,* National Institute for Occupational Safety and Health

40. Gerberich, S.G., Gibson, R.W., French, L.R., Lee, T.Y., Carr, W.P., Kochevar, L., Renier, C.M. & Shutske, J. (1998). A population-based study of machinery-related injuries: Regional rural injury study - I (RRIS - I). *Accid Anal Prev,* Vol. 30, pp. 793-804
41. Grandin, T. & Deesing, M. (2008). *Humane Livestock Handling.* Storey Publishing, North Adams, MA
42. Gray, G.C., Trampel, D.W. & Roth, J.A. (2007). Pandemic influenza planning: shouldn't swine and poultry workers be included? *Vaccine,* Vol. 26, pp. 4376-4381
43. Gumber, A. (1994). *Burden of Injury in India: Utilization and Expenditure Pattern.* Takemi Program in International Health, Harvard School of Public Health, Research Paper No. 88, Boston, MA
44. Hard, D.L., Myers, J.R. & Gerberich, S.G. (2002). Traumatic injuries in agriculture. *J Agric Saf Health,* Vol. 8, pp. 51-65
45. Hendricks, K.J. & Adekoya, N. (2001). Non-fatal animal related injuries to youth occurring on farms in the United States, 1998. *Inj Prev,* Vol. 7, pp. 307-311
46. Hernandez-Peck, M. (2001). Older Farmers: Factors Affecting their Health and Safety. Available from: http://www.cdc.gov/nasd/docs/d001701-d001800/d001760/d001760.html
47. Holland, A.J., Roy, G.T., Goh, V., Ross, F.I., Keneally, J.P. & Cass, D.T. (2001). Horse-related injuries in children. *Med J Aust,* Vol. 12, pp. 609-612
48. Hopkins, R. (1989). Farm equipment injuries in a rural county, 1980 through 1985: The emergency department as a source of data for prevention. *Ann Emerg Med,* Vol. 18, pp. 758-762
49. Hoskin, A., Miller, T., Hanford, W. & Landes, S. (1988). *Occupational injuries in agriculture: A 35-state summary.* National Safety Council, Chicago, IL
50. Hubert, D.J., Huhnke, R.L. & Harp, S.L. (2003). Cattle handling safety in working facilities. Available from: http://pods.dasnr.okstate.edu/docushare/dsweb/Get/ Document-4821/BAE-1738web.pdf
51. Hwang, S., Gomez, A.I., Stark, A., St. John, T.L., May, J.J. & Hallman, E.M. (2001). Severe farm injuries among New York farmers. *Am J Ind Med,* Vol. 40, pp. 32-41
52. Johnston, J.J. (1995). Occupational injury and stress. *J Occup Environ Med,* Vol. 37, pp. 1199-1230
53. Jones, P. J., & Field, W. E. (2002). Farm safety issues in older Anabaptist Communities: unique aspects and innovative intervention strategies. *J Agric Saf Health,* Vol. 8, pp. 67-81
54. Karkola, K., Mottonen, M. & Raekallio, J. (1973). Deaths caused by animals in Finland. *Med Sci Law,* Vol. 13, pp. 95-97
55. Langley, R.L. (1994). Fatal animal attacks in North Carolina over an 18-year period. *Am J Forensic Med Pathol,* Vol. 15, pp. 160-167
56. Langley, R.L. (1999). Physical hazards of animal handlers. *Occup Med,* Vol. 14, pp. 181-192
57. Langley, R.L. (2005). Animal-related fatalities in the United States—an update. *Wilderness Environ Med,* Vol. 16, pp. 67-74
58. Langley, R.L. & Hunter, J.L. (2001). Occupational fatalities due to animal-related events. *Wilderness Environ Med,* Vol. 12, pp. 168-174
59. Langley, R.L. & Morrow, W.E. (2010). Livestock handling--minimizing worker injuries. *J Agromedicine,* Vol. 15, pp. 226-235
60. Layde, P.M., Stueland, D., Nordstrom, D., Olson, K., Follen, M. & Brand. L. (1995). Machine-related occupational injuries in farm residents. *Ann Epidemiol,* Vol. 5, pp. 419-426
61. Layde, P. M., Nordstrom, D. L., Stueland, D., Wittman, L. B., Follen, M. A. & Olson, K. A. (1996). Animal-related occupational injuries in farm residents. *J Agric Saf Health,* Vol. 2, pp. 27-37
62. Lee, T., Gerberich, S., Gibson, R., Carr, W., Shutske, J. & Renier, C. (1996). A populationbased study of tractor-related injuries: Regional Rural Injury Study-I (RRIS-I). *J Occup Environ Med,* Vol. 38, pp. 782-793
63. Lewis, M.Q., Sprince, N.L., Burmeister, L.F., Whitten, P.S., Torner, J.C. & Zwerling, C.Z. (1998). Work-related injuries among Iowa farm operators: an analysis of the Iowa Farm Family Health and Hazard Surveillance Project. *Am J Ind Med,* Vol. 33, pp. 510-517

64. Li, G.H. & Baker, S.P. (1991). A comparison of injury death rates in China and the United States, 1986. *Am J Public Health,* Vol. 81, pp. 605-609
65. Li, G.H. & Baker, S.P. (1997). Injuries to bicyclists in Wuhan, People's Republic of China. *Am J Public Health,* Vol. 87, pp. 1049-1052
66. Manipady, S., Menezes, R.G. & Bastia, B.K. (2006). Death by attack from a wild boar. *J Clin Forensic Med,* Vol. 13, pp. 89-91
67. Marlenga, B., Pickett, W., & Berg, R. (2001). Agricultural work activities reported for children and youth on 498 North American Farms. *J Agric Saf Health,* Vol. 7, pp. 241-252
68. May, J. (1990). Issues in agriculture health and safety. *Am J Ind Med,* Vol. 18, pp. 121-131
69. McCurdy, S.A., Xiao, H. & Kwan, J.A. (2012). Agricultural injury among rural California public high school students. *Am J Ind Med,* Vol. 55, pp. 63-75
70. McFarland, T. (1968). On-farm accidents: Howto make the farm machineman-environment system function as it should. *Agric Eng,* Vol. 49, pp. 581-614
71. McKnight, R. & Hetzel, G. (1985). Trends in farm machinery fatalities. *Agric Eng,* Vol. 66, pp. 15-17
72. Meyer, S. (2005). Fatal occupational injuries to older workers in farming, 1995-2002. *Monthly Labor Review,* Vol. 128, pp. 38-48
73. Murray, C.J. & Lopez, A.D. (1998). *The Global Burden of Disease: A Comprehensive Assessment of Mortality and Disability From Diseases, Injuries, and Risk Factors in 1990 and Projected to 2020.* Harvard University Press, Boston, MA
74. Myers, J.R. (1990). National surveillance of occupational fatalities in agriculture. *Am J Ind Med,* Vol. 18, pp. 163-168
75. Myers, J.R. (1997). *Injuries among Farm Workers in the United States, 1993.* National Institute for Occupational Safety and Health, DHHS (NIOSH) publication 97-115, Washington, DC
76. Myers, J.R. (1998). *Injuries among Farm Workers in the United States, 1994.* National Institute for Occupational Safety and Health, DHHS (NIOSH) publication 98-153, Washington, DC
77. Myers, J.R. (2001). *Analysis of the Traumatic Injury Surveillance of Farmers (TSIF) Survey, 1993-1995.* National Institute for Occupational Safety and Health, Morgantown, WV
78. Myers, J.R., Hendricks, K., Goldcamp, E. & Layne, L. (2003). *Analysis of the Minority Farm Operator Childhood Agricultural Injury Study (M-CAIS).* National Institute for Occupational Safety and Health, Morgantown, WV
79. Myers, J.R., Layne, L.A. & Marsh, S.M. (2009). Injuries and fatalities to U.S. farmers and farm workers 55 years and older. *Am J Ind Med,* Vol. 52, pp. 185-194
80. Murphy, D.J., Seltzer, B.L. & Yesalis, C.E. (1990). Comparison of two methodologies to measure agricultural occupational fatalities. *Am J Public Health,* Vol. 80, pp. 198-200
81. Murphy, D.J., Kiernan, N.E. & Chapman, L.J. (1996). An occupational health and safety intervention research agenda for production agriculture: does safety education work? *Am J Ind Med,* Vol. 29, pp. 392-396
82. National Centre for Farmer Health (NCFM). (2012). *Farm safety - handling animals.* Available from: http://www.betterhealth.vic.gov.au/bhcv2/bhcpdf.nsf/ByPDF/Farm_safety_handling_animals/$File/Farm_safety_handling_animals.pdf
83. National Safety Council. (2008). *Injury Facts,* National Safety Council, Itasca, IL
84. Nogalski, A., Jankiewicz, L., Cwik, G., Karski, J. & Matuszewski, L. (2007). Animal related injuries treated at the Department of Trauma and Emergency Medicine, Medical University of Lublin. *Ann Agric Environ Med,* Vol. 14, pp. 57-61
85. Nordstrom, D.L., Layde, P.M., Olson, K.A., Stueland, D.T., Brand, L. & Follen, M.A. (1995). Incidence of farm-work-related acute injury in a defined population. *Am J Ind Med,* Vol.28, pp. 551-554
86. Norwood, S., McAuley, C., Vallina, V.L., Fernandez, L.G., McLarty, J.W. & Goodfried, G. (2000). Mechanisms and patterns of injuries related to large animals. *J Trauma,* Vol. 48, pp. 740–744
87. O'Neill, J.K., Richards, S.W., Ricketts, D.M. & Patterson, M.H. (2005). The effects of injection of bovine vaccine into a human digit: a case report. *Environ Health,* Vol. 11, pp. 21-23

88. Pickett, W., Brison, R.J., Niezgoda, H. & Chipman, M.L. (1995). Nonfatal farm injuries in Ontario: A population-based survey. *Accid Anal Prev,* Vol. 27, pp. 425-433
89. Prahlow, J.A., Ross, K.F., Lene, W.J.W. & Kirby, D.B. (2001). Accidental sharp force injury fatalities. *Am J Forensic Med Pathol,* Vol. 22, pp. 358-366
90. Pratt, D.S., Marvel, L., Darrow, D., Stallones, L., May, J. & Jenkin, P. (1992). The dangers of dairy farming: The experience of 600 workers followed for two years. *Am J Ind Med,* Vol. 21, pp. 637-650
91. Purschwitz, M.A. (1997). Epidemiology of agricultural injuries and illnesses. In: Langley, R., McLymore, R., Meggs, W. & Roberson, G., editors. *Safety and Health in Agriculture, Forestry and Fisheries.* Government Institute Press, Rockville, MD
92. Rabl, W. & Auer, M. (1992). Unusual death of a farmer. *Am J Forensic Med Pathol,* Vol. 13, pp. 238-242
93. Reynolds, S.J. & Groves, W. (2000). Effectiveness of roll-over protective structures in reducing farm tractor fatalities. *Am J Prev Med,* Vol. 18, pp. 63-69
94. Rivara, F.P. (1997). Fatal and non-fatal farm injuries to children and adolescents in the United States, 1990-1993. Inj Prev, Vol. 3, pp. 190-194
95. Sheldon, K.J., Deboy, G., Field, W.E. & Albright, J.L. (2009). Bull-related incidents: their prevalence and nature. *J Agromed,* Vol. 14, pp. 357-369
96. Shireley, L.A. & Gilmore, R.A. (1995). A statewide surveillance program of agricultural injuries and illnesses - North Dakota. In: McDuffie, H.H., Dosman, J.A., Semchuk, K.M., Olenchock, S.A. & Senthilselvan, A., editors. *Agricultural Health and Safety. Workplace, Environment, Sustainability.* Lewis Publishers, Boca Raton, FL
97. Simpson, S. (1984). Farm machinery injuries. *J Trauma,* Vol. 24, pp. 150-152
98. Sprince, N.L., Park, H., Zwerling, C., Lynch, S.F., Whitten, P.S., Thu, K., Burmeister, L.F., Gillette, P.P. & Alavanja, M.C. (2003). Risk factors for animal-related injury among Iowa large-livestock farmers: a case-control study nested in the Agricultural Health Study. *J Rural Health,* Vol. 19, pp. 165-173
99. Stafford, K.J. (1997). *Cattle Handling Skills.* ACC Wellington, New Zealand
100. Stallones, L. (1990). Surveillance of fatal and non-fatal farm injuries in Kentucky. *Am J Ind Med,* Vol. 18, pp. 223-234
101. Stallones, L., Keefe, T.J. & Xiang, H.Y. (1997). Characteristics associated with increased farm work-related injuries among male resident farm operators in Colorado, 1993. *J Agric Saf Health,* Vol. 3, pp. 195-201
102. Sterner, S. (1991). Farm injuries. How can the family farm be made a safer place? *Postgrad Med,* Vol. 90, pp. 141-142, 147, 150
103. Thu, K., Zwerling, C. & Donham, K. (1997). Health problems and disease patterns. Livestock rearing. In: Stellman, J., editor. *International Labor Encyclopedia of Occupational Health and Safety,* (4th ed), International Labor Organization, Geneva, Switzerland
104. Vilardo, F.J. (1988). The role of the epidemiological model in injury control. *J Safety Res,* Vol. 19, pp. 1-4
105. Waller, J.A. (1992). Injuries to farmers and farm families in a dairy state. *J Occup Med,* Vol. 34, pp. 414-421
106. Weese, J.S. & Jack, D.C. (2008). Needlestick injuries in veterinary medicine. *Can Vet J,* Vol. 49, pp. 780-784
107. Wiggins, P., Schenker, M.B., Green, R. & Samuels, S. (1989). Prevalence of hazardous exposures in veterinary practice. *Am J Ind Med,* Vol. 16, pp. 55-66
108. Xiang, H., Stallones, L. & Chiu, Y. (1999). Nonfatal agricultural injuries among Colorado older male farmers. *J Aging Health,* Vol. 11, pp. 65-78
109. Xiang, H., Wang, Z., Stallones, L., Keefe, T.J., Huang, X. & Fu, X. (2000). Agricultural workrelated injuries among farmers in Hubei, People's Republic of China. *Am J Public Health,* Vol. 90, pp. 1269-1276
110. Zhou, C. & Roseman, J.M. (1994). Agricultural injuries among a population- based sample of farm operators in Alabama. *Am J Ind Med,* Vol. 25, pp. 385-402
111. Zhou, C. & Roseman, J.M. (1995). Agriculture-related residual injuries: prevalence, type, and associated factors among Alabama farm operators—1990. *J Rural Health,* Vol. 11, pp. 251-258

112. Zwerling, C., Sprince, N., Wallace, R., Davis, C., Whitten, P. & Heeringa, S. (1995). Occupational injuries among agricultural workers 51-61 years old: A national study. *J Agric Saf Health,* Vol. 1, pp. 273-281
113. Zwi, A.B., Forjuoh, S., Murugusampillay, S., Odero, W. & Watts, C. (1996). Injuries in developing countries: policy response needed now. *Trans R Soc Trop Med Hyg,* Vol. 90, pp. 593-595

CHAPTER 8

Influence of breed, season and age on quality bovine semen used for artificial insemination

Hirwa Claire D'Andre1[1*], Kugonza Donald Rugira[1,2], Amahoro Elyse[1], Ingabire Claire[1], Niyiragira Vincent[1], Myambi Celestin[3], Manzi Maximillian[3], Murekezi Tiba1, Nyabinwa Pascal1, Nshimiyimana Alphonse Marie[3], Kanyandekwe Christine[3] and Gahakwa Daphrose[3]

[1]*Large Stock Program, Rwanda Agriculture Board (RAB), P. O. Box 5016, Kigali, Rwanda*

[2]*Department of Agricultural Production, College of Agricultural and Environmental Sciences, Makerere University, P. O. Box 7062, Kampala, Uganda*

[3]*Rwanda Agriculture Board (RAB), P. O. Box 5016, Kigali, Rwanda*

ABSTRACT

The success or failure of artificial insemination starts with the quality status of semen used, hence, this study aimed to investigate the effects of breed, season and year on bovine semen quality of the National Bull Stud of Rwanda kept at Masaka bull station, Rwanda. A total of 1475 semen samples were collected bi-weekly from nine bulls of Holstein Friesian (n = 3), Inyambo (n = 3) and Jersey (n = 3) breed using an artificial vagina. Semen volume, colour, concentration, mass motility, live sperm percentage and post-freeze motility were evaluated. Libido of the bulls at the time of semen collection was scored. Ejaculate volume, mass motility, individual motility, density and post freeze motility significantly differed ($p < 0.05$) among seasons of the year, bull breeds and age/year of collection. Friesian bulls had superior ($p < 0.05$) semen volume (5.76±0.08 ml) to that of Jersey (4.29±0.09 ml) and Inyambo (3.37±0.1 ml). However, Friesians had an inferior ($p < 0.05$) lighter coloured semen, with only 52.7% of Friesian samples having the preferred cream colour, as compared to 65.2% of Jersey and 64.9% of the Inyambo semen samples. Year of collection (2011 or 2012) and in essence age of the bull negatively affected ($p < 0.05$) all the parameters studied, with semen volume dropping from 5.16 to 5.10 ml, colour lightened from 62.8 to 54% of the samples being cream; mass motility fell from 2.98 to 2.65, while live sperm percentage in ejaculates dropped from 65 to 63%. Of the eight parameters studied, only post-freeze motility was not affected by passage of time. Semen collected during the October to December period had the best quality characteristics, though collections in the long rains (March to May) had comparable mass motility and post-freeze motility. Semen volume (4.61 ml per bull ejaculate) and post-freeze motility (39%) were poorest in the long dry season (January to February). In conclusion, the Friesian breed should be promoted at the bull station. Most semen should be collected during the rain season, particularly the short rains (October to November). Bulls below three years of age should be of focus.

Keywords: Cattle, mortality, seasonality, semen evaluation, pedigree bulls.

1. INTRODUCTION

Artificial insemination (AI) is the most valuable breeding management tool available to cattle breeders to improve the genetic potential of their herds. The optimal use of genetically superior bulls through artificial insemination is highly dependent on precise seminal quality which allows for reasonable estimations of field fertility with normal or low-dose inseminations (Fuerst-Waltl et al., 2006; Christensen et al., 2011; Ahmed et al., 2016). There are factors determining the success of AI such as environmental seasonal variations and temperature (McDowell, 1972). In areas where seasonal variation in temperature occurs, sperm morphological characteristics were variable (Curtis, 1983; Mathevon et al., 1998; Brito, 2010). Mathevon et al. (1998) reported that the interaction between age and season could have a significant effect on the semen characteristics of Friesian bulls. seasonal factors, especially temperature, have an important effect on the spermatic production in most bulls, but the individual response to thermal stress is different. Detrimental effect on the spermatic parameters has been registered mostly in the warm seasons, during months with average temperatures of about 20°C. Numerous studies were dedicated to identification of the factors which affect the spermatic production and the quality of the semen. To date, some differences and even contradictions regarding the effect of the season on spermatic production were identified. Consequently, some studies (Mathevon et al., 1998 and Brito, 2010) demonstrated significant influence of season on the spermatic production, while some other studies did not detect any effect of season on the spermatic production (Brito et al., 2002; Girdhar, 2003; Vilakazi and Webb, 2004). Meanwhile, other researchers did observe the highest concentration and total number of spermatozoa occurs during the summer season (Stalhammar et al., 1989; Mathevon et al., 1998; Ghasemi and Ghorban, 2014).

Season of the year influences the secretion of luteinizing hormone (LH), the average concentration of testosterone in bulls, and the number of spermatozoa per ejaculate is consequently affected (Jimenez-Severiano et al., 2003). The study of Jimenez-Severiano (2003) also revealed that highest LH average values in young bulls were registered in spring and the lowest in the winter period and that average testosterone concentrations were also higher in spring and summer than in the cold season. The same effect was revealed in bulls of a beef breed from the tropical region, in which there was observed a depreciation of semen quality and a diminution of testicular dimensions in winter (Nichi et al., 2006). The optimal environmental temperature for spermatic production is estimated to range between 15 and 20°C (Parkinson, 1987). It is considered that not only the temperature registered in the day of semen collection affects the production, but also the temperature registered during the entire spermatogenesis period, until 70 days before collection. Furthermore, it was indicated that for young bulls, superior morphological characteristics were observed during the winter and spring as compared to summer and autumn. Dombo (2002) reported that the semen quality does not remain the same throughout the year, especially as the season changes and as the bull advances in age.

Many efforts have been made to improve milk production in Rwanda and a crossbreeding programme is being implemented on a large scale (Girdhar, 2003; MINAGRI, 2011). This programme is aimed to maximally depend on the use of artificial insemination both for small-scale/smallholder/zero grazer units, and for large scale dairy farms. The artificial insemination programme in Rwanda relies on both imported and in-country produced bovine semen, with a plan to produce most of the semen domestically. In tropical regions, semen output, quality and mating behaviour of bulls can vary from season to season, depending mainly on availability and quality of feed and climatic conditions. Therefore, this study aimed to study the influence of breed of the bull, season of semen collection and age of the bull on bovine semen quality of Friesian, Jersey and Inyambo bulls kept at Masaka Bull Station, Rwanda.

2. MATERIALS AND METHODS

2.1 Experimental animals and housing

A total of 1475 semen samples were collected from the National Bull Stud of Rwanda from three Holstein Friesian bulls (946 semen samples), three Jersey bulls (371 semen samples) and three Inyambo bulls (174 semen samples) maintained at the semen production facility at Masaka Bull Station. At the onset of the study, the breeder bulls were in good health status and were aged between three and seven years. Data recorded over a period of two years were used for semen quality evaluation. Bulls were housed individually in pens, which were constituted into two large sheds/barns. The sheds were designed conventionally, to provide sufficient natural cross-ventilation and to minimize heat stress. Each bull pen was facilitated with a water point and a feeding manger. The pens were provided with floor litter in form of dry grass to increase the comfort of the bulls.

2.2 Feeding and health management

The feeding requirements of animals were calculated according to their body weight, with daily feed Although the production system and marketing are offering on dry weight basis of 3% of body weight. The feeding regime per bull consisted of Napier grass and 5 to 6 kg of protein-energy concentrate mixture.

Water was provided *ad libitum* to the bulls via a water point. Regular vaccination was carried out; twice a year in June and December against hemorrhagic septicemia; and foot and mouth disease in February and September; and once a year in April against black quarter. Preventive measures against external and internal parasitic infestation were also undertaken three times a year. Multivitamins were given once a month by intramuscular injection.

2.3 Climate

The season condition of the study area was considered as indicated in Table 1.

Table 1. Mean temperature recorded by Rwanda meteorological service.

Seasons	Maximum temperature (°C)	Average temperature (°C)	Minimum temperature (°C)
Long rain (March-May)	25	20	15
Long dry (June-September)	27	22	15
Short rain (October-November	26	21	15
Short dry (December-February)	26	21	15

Source: Rwanda meteorological service.

2.4 Semen collection and processing

Semen collection was undertaken twice a week. Bulls to be used on a particular day went through preparatory measures including showering, drying and cleaning, brushing, grooming and half an hour exercise. The bulls were then allowed to mount a teaser bull and the semen was collected with the help of a pre-warmed (42 to 45°C) artificial vagina. Ejaculate volume was recorded directly from the graduated semen collection tube. Each ejaculate was examined for mass motility, individual motility and sperm concentration and ejaculates having less than 50% sperm motility were discarded. The fresh semen was then diluted depending upon microscopic evaluation of sperm concentration at 32°C with tris, citrate, fructose, egg yolk. All collections were performed early in the morning between 6.00 and 8.00 h. After collection, the tubes containing semen were immediately transferred from the collection room to the laboratory and placed in a water bath at 36°C to maintain the temperature of sperm cells. The characterizations of semen require laboratory tests and depend on norms and standards fixed by the management of the station. After the initial observations, the semen was diluted using andromed extender (AndroMed®, Tiefenbach, Germany).

2.5 Characteristics of semen after collection

The collected semen was firstly examined with naked eyes to observe the colour and was then confirmed from a published colour chart; and then the volume was read off from a graduated semen collecting tube. The semen was then transferred to a water bath at 33-36°C. Microscopic evaluation was done using a microscope (Minitube, Germany), for mass activity. Individual motility and determination of concentration was then done with a calibrated photometer (Minitube, Germany). Mass motility was then tested by placing a small drop of semen (approximately 5 μl) on a slide using a microscope at 10x magnification. Another small drop was deposited on the slide and carefully placing a cover slip over it and then was visualized at 40x to observe individual motility. A score was given on a hedonic scale of 0 to 4 for the mass motility and percentage for individual motility by each of three independent observer technicians. The scores were not significantly different and were then summed up and mean computed.

To continue the process, the minimum mass motility score needed was 3, and 60% for individual motility, also the colour was scored on a scale of 1 to 3 (Javed et al., 2000). Score 3 was given to semen with a high concentration of sperm cells, and this was characterized by a creamy consistence, and also marked as good semen. Score 2 was given to semen with tolerable concentration of sperm cell, and was characterized by a milky consistence. Score 1 was given to semen with a very low concentration of sperm cells; and characterized by a watery consistence. Score 0 was given to ejaculate without sperm cells and characterized by a clear watery consistence. The semen with score 0 and 1 was discarded.

2.6 Data analysis

The data were typed into spreadsheets of the MS-Excel software. Preliminary analysis was conducted using Chi-square procedure of statistical analysis system (SAS, 2004), to determine whether there were significant differences between the levels of the class variables. This was because the number of animals used as the samples were few and possibly, the small sub-class numbers could hinder the effects of the independent variables to be quantified. On finding significant differences (Table 2), further analyses particularly, least square means were then computed using the generalized linear models of statistical analysis system (SAS, 2004) and were separated using Duncan's multiple range test. The results were presented as means ± standard error of the mean (S.E.M.). Data on scrotal circumference was not collected hence correlation with semen volume could not be done.

3. RESULTS

The results showed that ejaculate volume, mass motility, individual motility and post freezing motility significantly differed ($p < 0.05$) among the three cattle breeds studied (Table 2). Friesian bulls had superior ($p = 0.001$) semen volume to that of Jerseys and Inyambo. However, Friesian bulls had an inferior ($p = 0.003$) semen colour score (3.12) to that of Jersey (3.31) and Inyambo (3.29), but the latter two did not differ (Table 2). Only post freeze motility was not affected by time factor. The short rain season (October to November) gave the best semen quality characteristics but was not significantly ($p=0.838$) better than the long rains except in mass motility and post-freeze motility. Semen volume (4.61) and post-freeze motility (38.51%) were poorest in the long rainy season (March to May) (Table 3).

The results of the study indicated a significant interaction between breed and season in the effect on ejaculate volume ($p = 0.003$), libido score ($p = 0.0014$),semen concentration, mass motility ($p = 0.0014$) and post freeze motility ($p = 0.0009$). Age of the bulls negatively affected ($p=0.025$) all the parameters studied, with semen collected during 2011 possessing better attributes than that collected in 2012.

Table 2. Least square means of various semen quality parameters as influenced by breed.

Parameter	n	Bull breed Friesian (n = 946)	Jersey (n = 371)	Inyambo (n = 174)	GLM p-Value	χ^2 p-Value*
Volume (ml)	1468	5.76±0.08[a]	4.29±0.09[b]	3.37±0.1[c]	***	<0.001
Libido score	1463	3.14±0.01[a]	3.97±0.01[b]	2.81±0.03[c]	***	<0.001
Semen color (%)					*	<0.001
Creamy	854	52.7	65.2	64.9		
Milky	456	31.9	25.3	34.5		
Watery	62	4.3	5.4	0.6		
Yellowish	99	11.1	4.1	0		
Semen concentration (10^9 spz/ml)	1471	3.12±0.03[b]	3.31±0.05[a]	3.29±0.07[a]	***	<0.001
Mass motility (%)	1360	2.67±0.03[c]	3.05±0.04[a]	2.90±0.06[b]	***	<0.001
Individual motility (%)	1400	62.04±0.6[b]	68.14±0.9[a]	66.01±1.04[a]	***	<0.001
Live-dead Percentage (%)	1375	35.96±0.5[a]	29.37±0.7[b]	34.12±1.05[a]	***	<0.001
Post freezing motility (%)	762	40.43±0.5[b]	43.45±0.9[a]	42.20±1.1[a]	**	0.0012

[abcd] Values within the same column with different superscripts differed significantly { ***(p<0.001); **(p<0.01); * p<05)}. The Chi-square analysis was done at preliminary stage, to justify further ANOVA analysis.

Table 3. Influence of season on semen quality.

Parameters	Season$^{\Phi}$				P. value
	Long rain season (n = 274)	Long dry season (n = 380)	Short rain season (n = 401)	Short dry season (n = 389)	
Volume (ml)	4.61 ± 0.15^{b}	5.31 ± 0.12^{a}	5.01 ± 0.09^{a}	5.28 ± 0.13^{a}	***
Libido score	3.05 ± 0.04^{b}	2.97 ± 0.02^{c}	3.0 ± 0.09^{c}	3.22 ± 0.02^{a}	***
Semen color (%)					
Cream (%)	59.5	53.5	67.5	49.7	
Milky (%)	35.6	34.4	21.2	36.6	
Watery (%)	4.9	3.8	5.0	3.1	
Yellowish (%)	-	7.0	6.3	9.8	
Semen concentration (10^{9}spz/ml)	3.14 ± 0.08^{b}	3.11 ± 0.04^{b}	3.36 ± 0.04^{a}	3.07 ± 0.05^{b}	***
Mass motility (%)	2.71 ± 0.06^{b}	2.58 ± 0.04^{c}	2.86 ± 0.04^{a}	2.94 ± 0.04^{a}	***
Individual motility (%)	65.29 ± 1.17^{a}	63.79 ± 0.78^{b}	62.09 ± 0.98^{a}	65.85 ± 0.78^{a}	**
Live-dead Percentage (%)	34.70 ± 1.17^{ab}	36.06 ± 0.79^{a}	32.7 ± 0.74^{c}	33.97 ± 0.78^{ab}	*
Post freezing motility (%)	38.51 ± 1.41^{b}	42.45 ± 0.76^{a}	42.85 ± 0.78^{a}	39.68 ± 0.96^{b}	**

^{abcd} Values within the same column with different superscripts differed significantly { ***p<0.001); **(p<0.01); * p<05)}. $^{\Phi}$A: Long rain season (March-May); B: Long dry season (June-September); C: Short rain season (October–November); D: Short dry season (December- February).

Table 4. Influence of year on semen quality.

Parameter	Years		
	2011 (n = 568)	**2012 (n = 823)**	**P-Value**
Volume (ml)	5.16±0.09	5.10±0.08	
Libido Score	$3.27±0.02^{a}$	$2.95±0.01^{b}$	***
Semen colour (%)			*
Cream (%)	62.8	54.0	
Milky (%)	21.5	37.5	
Watery (%)	6.5	2.4	
Yellowish (%)	8.2	5.7	
Semen concentration (10^{9} spz/ml)	$3.28±0.04^{a}$	$3.11±0.03^{b}$	***
Mass motility (%)	$2.98±0.03^{a}$	$2.65±0.03^{b}$	***
Individual motility (%)	$65.08±0.8^{a}$	$63.12±0.5^{b}$	**
Live-dead %	$30.44±0.6^{b}$	$36.76±0.5^{a}$	**
Post freezer motility (%)	$40.50±0.8^{a}$	$41.93±0.5^{a}$	*

[abcd] Values within the same column with different superscripts differed significantly { ***(p<0.001); **(p<0.01); * p<05)}.

While the semen volume dropped with age between 2011 and 2012, from 5.16 to 5.10 ml, semen colour also lightened from a 3.28 hedonic scale score to a 3.11 score. Over the test period, semen mass motility score reduced from 2.98 to 2.65, while the proportion of live sperm in ejaculates dropped from 65 to 63% (Table 4).

4. DISCUSSION

Normal reproduction in male livestock is measured by the production of semen with normal and adequate spermatozoa as well as the desired ability to mate. These sexual functions (sexual development, production of spermatozoa and desired ability to mate) are under the control of gonadotrophin hormones such as testosterone, LH and FSH. These gonadotrophins are influenced to a larger extent by a combination of environmental fctors such as temperature, nutrition and animal management practices (Vilakazi and Webb, 2004). Temperature is one of the important factors affecting reproduction. Periods of high temperature damage the spermatogenic cells and lead to testicular degeneration, reduction in the efficiency of spermatogenesis and hence poor semen quality (Vilakazi and Webb, 2004). Season factors include temperature, photoperiod, humidity and feed quality (Andrabi et al., 2002; Barth and Waldner, 2002; Menon et al., 2011). Differences in the quantity of feed or in feed composition, environmental temperature, humidity and seasonal variation do affect semen output (McDowell, 1972; Castillo et al., 1987; Soderquist, 1996; Koivisto et al., 2009).

In the area where there is marked seasonal variation in environmental temperature, bull semen quality tends to be lower during summer or in tropical terms, warm months (Curtis, 1983; Vilakazi et al., 2004) as this results in thermal stress. Thermal stresses cause testicular degeneration, abdominal scrotal thermogram and hence lower the semen output (McDowell, 1972; Curtis, 1983). Significant seasonal variations are also observed in the incidence of sperm head abnormalities and total sperm abnomalities; and appear to be strongly associated with dry seasons (spring and summer) as compared to rain seasons. The results of this study indicated that a higher percentage of normal sperm occured during short rainy season and this was in agreement with Mathevon et al. (1998) where short rain (spring) season was also associated with higher percentage of normal sperm regardless of the age and breed of the bull.

In this study, age significantly affected semen characteristics. Previous studies also reported that the bulls aged 36 to 48 months were found to produce sperm of better morphology than bulls of 72 months age and older (Vilakazi and Webb, 2004). This succinctly shows the importance of bull age on influencing semen morphological traits. Several researchers have attributed similar observations to the scrotal circumference, the regulation balance mechanism, fat deposition in the brain and reproductive tissues which affect semen production and quality (McDowell, 1972; Salisbury et al., 1978; King, 1993; Coe, 1999; Mamabolo, 1999; Brito et al., 2002). Dairy bulls reach puberty at the age of 12 months (Bearden and Fuquay, 1997; Vilakazi and Webb, 2004) and attain maturity at the age of 3 to 4 years (Almquist, 1982). These studies found that bulls recorded higher sperm defects prior to maturity. This is probably due to the fact that in younger bulls, the testicles are still developing and hence the semen in the ejaculate is of low quality (Coulter and Foote, 1979; Vilakazi and Webb, 2004). Lower semen outputs in older bulls may be associated with the degenerative changes in seminiferous tubule (Coe, 1999), fat deposition which may take place in scrotum (Salisbury et al., 1978; King 1993) and the break down of body tissues, particularly, testicular tissues (King, 1993) with advancement in age. Fat deposition as a bull progresses in age may take place around the scrotum. This may affect semen quality by reducing the heat radiation capacity from the scrotal neck (Brito et al., 2002). Indeed, supportive findings by Vilakazi and Webb (2004) reported that the semen quality starts to deteriorate in bulls older than 72 months.

This deterioration may be associated with the accumulative hazards of life including non-specific infections, nutritional stress, diseases and accidents, which all combine to cause the direct positive relationship between semen quality and age to disappear (Salisbury et al., 1978; Vilakazi and Webb, 2004; Christensen et al., 2011). Its therefore very interesting for us to report that age as a factor is a necessary evil and paradoxical, in that at a young age, low age is disadvantageous and leads to poor quality semen, in mid-age, the age effect becomes positive, with bulls producing high grade/quality semen, before the bulls age effect becomes negative, causing the production of abnormal and sub-standard semen. Studies that have combined the effects of age and season have found that younger bulls recorded poor semen morphology during winter, while old bulls showed poor morphology during long dry season (Vilakazi and Webb, 2004; Koivisto et al., 2009; Ghasemi and Ghorban, 2014).

The current study hence recognizes that breed, age and season and their interactions are important sources of variation in semen quality. This implies that for a successful artificial insemination programme, semen collection should be done at the younger age, from two years to a maximum of 5 years for all breeds. It is therefore recommended that age, breed and season should be given urgent attention in any bull management system employed in Rwanda in order to obtain the best semen quality.

Indeed, when breed effects are kept out of the picture, past studies (Vilakazi and Webb, 2004) showed that the summer season is associated with very poor semen quality and is not ideal for semen collection from bulls that are five years or older. Manipulation of temperature in dairy cattle through the provision of management practices to reduce the effect of heat stress on bull reproduction is hereby suggested as a primary tool in optimising the quality of semen harvested in AI dairy bulls in Rwanda.

5. CONCLUSION AND RECOMMENDATIONS

In conclusion, Friesian bulls had better semen quality, the short rainy season had the best semen quality, as was the semen collected earlier. Hence, the Friesian breed should be promoted at the bull station ahead of the Jersey. Most semen should be collected during the rainy season, particularly the short rains. Particularly, bulls below three years of age should be of focus. It can be recommended that the relationship between the nutrition, reproduction physiology and management should be given a high priority in dairy production systems and further studies in this regard should be undertaken.

REFERENCES

1. Ahmed N, Kathiresan D, Ahmed FA, Lalrintluanga K, Mayengbam P, Gali JM (2016). Pattern of induced estrus and conception rate following Ovsynch and Ovsynch based gonadotropin-releasing hormone treatments initiated on day 6 of estrous cycle in repeat breeding crossbred cows. Vet. World 9(4): 342-345.
2. Andrabi SMH, Naheed S, Khan LA, Ullah N (2002). Semen characteristics of crossbred (Friesian × Sahiwal) and Sahiwal young bulls maintained under sub-tropical conditions of Punjab. Pak. Vet. J. 23:100-102.
3. Almquist JO (1982). Effect of long term ejaculation at higher frequency on output of sperm, sexual behaviour and fertility of Holstein bulls relation of reproduction capacity to higher nutrient allowance. J. Dairy Sci. 65:815-820.
4. Barth AD, Waldner CL (2002). Factors affecting breeding soundness classification of beef bulls examined at the Western College of Veterinary Medicine. Can. Vet. J. 43:274-284.
5. Bearden HJ, Fuquary JW (1997). Applied Animal Reproduction (5th ed.) Upper Saddle, New Jersey. Prentice Hall Inc.
6. Brito LFC (2010). A multi-stud study on variation of semen evaluation. In: Proceedings of the National Association of Animal Breeders and 23rd Technical Conference on Artificial Insemination and Reproduction, Columbia, MO, pp. 112-115.
7. Brito LFC, Silva ADF, Rodrigues LH, Vieira FV, Deragon LAG, Kastelic JP (2002). The effect of environmental factors, age and genotype on sperm production and semen quality of *B. indicus* and *B. taurus* AI bulls in Brazil. Theriogenol. 70:181-190.
8. Castillo E, Tizol G, Alverez JL, Perez M, Baez R (1987). Reduction of protein concentrate level in ration for Holstein sires. 1. Effect on semen quality. Cub. J. Agric. Sci. 21:247-251.
9. Christensen P, Labouriau R, Birck A, Boe-Hansen GB, Pedersen J, Borchersen S (2011). Relationship among seminal quality measures and field fertility of young dairy bulls using low-dose inseminations. J. Dairy Sci. 94(4):1744-1754.

10. Coe PH (1999). Association among age, scrotal circumference and proportion of morphological normal spermatozoa in young beef bulls during an initial breeding soundness examination. Vet. Med. Assoc. 214:1664-1667.
11. Coulter GH, Foote RH (1979). Bovine testicular measurement as indicators of reproductive performance and the relationship to reproductive traits in cattle: a review. Theriogenol. 11:297-311.
12. Curtis SE (1983). Environmental management in animal agriculture. Iowa State University Press.
13. Dombo HM (2002). Seasonal effect on semen quality of Gorno Altai and South African indigenous goats. M. Inst. Agrar. Thesis, University of Pretoria, South Africa.
14. Fuerst-Waltl B, Schwarzenbacher H, Perner C, Sölkner J (2006). Effects of age and environmental factors on semen production and semen quality of Austrian Simmental bulls. Anim. Reprod. Sci. 95(1-2):27-37.
15. Ghasemi MV, Ghorban AI (2014). Environmental and genetic factors affecting semen quality in Iranian Holstein bulls. Iranian J. Appl. Sci. 4:33-37.
16. Girdhar N (2003). Influence of nutrition and season on breeding values of bull. Agric. Rev. 24:75-78.
17. Javed MT, Khan A, Kausar R (2000). Effect of age and season on some semen parameters of Nili-Ravi buffalo (*Bubalus bubalis*) bulls. Veterinarski Arhiv 70(2):83-94.
18. Jimenez-Severiano H, Quintal-Franco J, Vega-Murillo V, Zanella E, Wehman ME, Lindsey R, Melvin EJ, Kinder JE (2003). Season of the year influences testosterone secretion in bulls administered luteinizing hormone. J. Anim. Sci. 81:1023-1029.
19. King GJ (1993). Reproduction in Domesticated animals. Elsevier Science Publisher B.V. London, New York.
20. Koivisto MB, Costa MT, Perri SH, Vicente WR (2009). The effect of season on semen characteristics and freezability in *Bos indicus* and *Bos taurus* bulls in the South Eastern region of Brazil. Reprod. Domest. Anim. 10: 1439-1531.
21. Mamabolo MJ (1999). Dietary, seasonally and environmental influences on semen quality and fertility status indigenous goats in Mpumalanga province, South Africa. M. Inst. Agrar. (Thesis). University of Pretoria, South Africa.
22. Mathevon M, Bhyr RM, Dekkers JC (1998). Environmental, management, and genetic factors affecting semen production in Holstein bulls. J. Dairy Sci. 81:3321-3330
23. McDowell RE (1972). Improvements of livestock production in warm climates. Cornell University. Wlt, Freeman and Company, NY.
24. Menon AG, Barkema HW, Wilde R, Kastelic JP, Thundathil JC (2011). Associations between sperm abnormalities, breed, age, and scrotal circumference in beef bulls. Can. J. Vet. Res. 75(4):241-247.
25. MINAGRI (Ministry of Agriculture and Animal Resources), (2011) Annual report, FY 2010/2011. Available at: www.minagri.gov.rw
26. Nichi M, Bols PE, Zuge RM, Barnabe VH (2006). Seasonal variation in semen quality in *Bos indicus* and *Bos taurus* bulls raised under tropical conditions. Theriogenology 66(4):822-1828.
27. Parkinson TJ (1987). Seasonal variations in semen quality of bulls: correlations with environmental temperature. The Vet. Record 120(20):479-482.
28. Salisbury GW, Van Denmark NL, Lodge JR (1978). Physiology of Reproduction and AI of Cattle. W.N. Freeman and Co. San Francisco.
29. SAS (Statistical Analysis Systems) (2004). SAS User's Guide: Statistics. SAS Institute Inc., Cary, NC, USA.
30. Soderquist L, Janson L, Haard M, Einarsson S (1996). Influence of season, age and breed and some other factor on the variation in sperm morphological abnormalities in Swedish dairy AI bulls. Anim. Reprod. Sci. 44:91-98.
31. Stalhammar EM, Janson L, Philipsson J (1989). Genetic studies on fertility in AI. bulls. I. Age, season and genetic effects on semen characteristics in young bulls. Anim. Reprod. Sci. 19:1-17.
32. Vilakazi DM, Webb EC (2004). Effect of age and season on sperm morphology of Friesland bulls at an artificial insemination centre in South Africa. S. Afr. J. Anim. Sci. 34:62-69.

CHAPTER 9

Lactation Responses toward Milk Indigenous Enzymes

Aneela Hameed, Majid Hussain* and Saeed Akhtar

Institute of Food Science & Nutrition, Faculty of Agricultural Sciences, Bahauddin Zakariya, University, Multan, Pakistan

ABSTRACT

Milk being a highly nutritious food in its natural form provides energy. There are various factors influencing the composition of milk: breed, stage of lactation, nutritional status, health, and milking intervals. A number of indigenous enzymes present in milk are being affected by stages of lactation period. Their concentration varies during early, mid and late lactation periods. This varied behavior ultimately affects the quality of dairy products. In this chapter, the level of milk enzymes: lipases and esterases, plasmin (PL), plasminogen (PLG) phosphatases (alkaline phosphatase ALP; acid phosphatase (ACP), lysozyme (LZ), lactoperoxidase (LP), xanthine oxidoreductase (XOR), and catalase (CAT) will be reviewed with respect to the stages of lactation periods.

Keywords: milk, indigenous enzymes, lactation stages, parity, season

1. INTRODUCTION

Milk is one of the perfect, complete, and primitive dairy food known by mankind. It is white and nutritious physiological secretion from the mammary glands of mammals, serves as nourishment for their neonates [1,

2]. It is a major product obtained from healthy and highly productive dairy animals. Physiological and nutritional provisions of each species are more or less distinctive. The breed, health, nutritional status, stage of lactation period, and milking intervals are some of the factors that affect the milk composition [3, 4]. The variation in constituents occurs entire lactation period. Lactation stage is the prime factors that affect the milk properties and some of the enzyme activities [5, 6]. Solids-non-fat (SNF) content is frequently highest throughout first 2–3 weeks of lactation.

2. MILK INDIGENOUS ENZYMES

Numerous enzymes have been indigenously identified in milk from 1924 to 1970 [7]. A large number of enzymes with multiple functionalities are present in milk. Additional enzymes contribute in quality of milk products and also perform an antibacterial action (LP). In bovine milk more than 70 enzymes are detected [8, 9]. A 50–60 substantial number of milk enzymes with multiple functions are present in abundance in milk and are concerned with processing stability and general customer safety [10] and additionally processing suitability (ALP). Some enzymes (LP) having antibacterial characteristics are with significant importance in preservation of milk and milk products and some, e.g., plasmin (PL) and lipoprotein lipase (LPL) connected with the serum, plasma, fat globules, casein, or leukocytes are important in maintaining of quality of milk and milk products. More than 40 enzymes have been recognized in cow milk [9, 11].

Lactation period in animals involves colostrum, developed milk, peak, and production with compositional variations. Numerous indigenous enzymes present in milk are secreted by epithelial cells and their composition changes with the lactation stages.

In already recognized indigenous enzymes in milk [7], almost 20 enzymes have been well characterized and the rest of the 40 enzymes are of little significance but can be identified through their activity. These enzymes indicate the efficient process of milk pasteurization (ALP, γ-glutamyl transferase GGT) or of mastitis (phosphatases, CAT). Additional enzymes can be of significance in processing and ultimately providing safety to human beings. They play an antibacterial activity (LP) and contribute quality to milk products (e.g., LPL, PL) associated with the serum, plasma, fat globule, casein, or leukocytes.

2.1. Lipases and esterases

Lipolytic enzymes have capability to hydrolyze triacylglycerols are considered as carboxylesterases [12, 13]. Those enzymes that can hydrolyze acyl glycerol having <10 carbon atom fatty acids are known as esterases or carboxylases (Enzyme Commission, EC 3.1.1.1) while those can hydrolyze ≥10 carbon atom fatty acids are considered as lipases, or triacylglycerol acyl hydrolases (EC 3.1.1.3) [14, 15].

Esterases are different from lipases due to their functions for being relatively soluble compared to emulsified ester substrates. Several esterases are present in milk [15, 16], the most prominent are carboxylesterase (EC 3.1.1.1), acetylcholinesterase (EC 3.1.1.7), and cholinesterase (EC 3.1.1.8). In bovine colostrum, lipase is not connected with casein and not activated by blood serum, therefore exhibited low lipase activity and showed slight lipolysis in early lactation. However, after few days of calving, normal milk from early lactation exhibited higher lipase activity [17, 18].

Lipases are naturally a critical group of enzymes since they are connected with the fat digestion system. Lipases are more dynamic at pH 8–9 and catalyze the advancement of hydrolytic rancidity in milk. Investigation of lipases is more alluring in the light of the fact that it would add to our comprehension about the properties and modes of these enzymes [19, 20].

The phenomenon of lipolysis is correlated with the lactation days. Higher activity is associated with its presence in fat fraction of milk. Activity of lipase in milk fat increases with the advancement in lactation stages [21]. The lipolysis process is of major apprehension in the dairy industry, as rancid off flavors are produced in milk and milk products during this phenomenon [22].

Earlier research has well established that milk lipase is sensitive to heavy metals. Copper, cobalt, and nickel have been shown to be more powerful inhibitors of lipase than iron, chromium, manganese, and silver. Enzyme activity is stimulated by blood serum albumin, ammonium, calcium ions, and mercaptoethanol. The buffer solutions, citrate, acetate, and phosphate buffers damage the enzyme activity, whereas borate and barbiturate buffers do not [23, 24].

LPL in cow milk is altered due to the breed, lactation phase, feed and fodder, season, and milk yield [22, 25]. Lipase activity increased from 0.32 to 2.98 U/mL of milk. At the point when milk fat globule membrane (MFGM) is damaged, lipolysis takes place rapidly and leads to hydrolytic rancidity and ultimately may cause variations in functionality and flavor of dairy products throughout storage period [15]. LPL found in goat milk is of low concentration in the early and late lactation stage [26].

The membrane lipase is available in higher concentration in milk from dairy animals in late lactation [27]. They additionally reported that lipase action in milk showed inclined pattern with reference to lactation stages. Hameed et al. [28] reported the expanding pattern of lipase activity with lactation stages in bovine milk. Lipase action (1.55 U/mL) was recorded higher ($p < 0.01$) in milk, examined at the last of lactation, followed by other lactations (1.29 and 1.16 U/mL, respectively).

2.2. Plasmin

Plasmin (PL; EC 3.4.21.7) is an alkaline serine proteinase enzyme that proteolytically cleaves the blood clots [29]. This enzyme has affinity toward arginine (Arg) and lysine (Lys) residues, specifically breaks the Arg-X and Lys-X bonds [30, 31]. The activity of enzyme is increased with the multiple factors that include lactation stage, lactation number and severity of mastitis infection [32–34].

On the basis of origin, the PL and plasminogen (PLG) are considered to be migrated from blood to milk, and higher activity of PL in peak lactation designating more conversion of PLG into PL in bovine milk [35–37].

PL is basically released in the form of PLG in normal milk. The concentration of PLG (0.8–2.8 μg/mL) in fresh milk is varied and its concentration is 2–30 times higher than that of PL (0.1–0.7 μg/mL). It is activated by storage or when milk is stayed in the lumen of mammary glands before milking. A considerable interest has been involved in the activation of PLG, upon which activity of PL depends [38–41]. It promptly hydrolyzes the bonding of β-casein, α_{s2}-casein, and α_{s1}-casein and affects the quality of dairy products [32, 42].

Advancing lactation stage is an essential factor that influences PL activity and percentage, suggesting that more PL activity in milk from goat and older cows is a result of increased PLG activation [43–45]. However, the relevant information about the varied concentration and activity of PL during lactation stages is controversial. Leitner et al. [46] declared significantly higher activities of PL in infected glands of sheep.

Caroprese et al. [47] and Albenzio et al. [33] found that there was decrease in PL activity in ewe's milk from the early to the late lactation stage whereas Koutsouli et al. [48] and Bianchi et al. [49] announced that PL activity significantly affected by udder health status and found an increased level of PL activity due to more somatic cell counts (SCCs) during the late lactation period.

The variation in PLG-derived activity and total PL plus PLG-derived activity is greatly influenced by lactation stage and seasonal changes. It is linked with reduction of milk yield and advancement in lactation stage [45, 50, 51].

Due to increased activity of plasminogen, more entry of PL occurred from blood to milk inside the mammary glands [52]. The PL and PLG activities were significantly increased in the advancement of lactation and a nonsignificant decrease in their ratio (PL:PLG) was observed as compared to camel milk [53, 54].

2.3. Phosphatases

2.3.1. Alkaline phosphatase

In 1925, for the first time, phosphatase enzyme in milk is documented by Demuth and then considered as an alkaline phosphatase (ALP; EC 3.1.3.1) indigenous to milk by Graham and Kay [55]. It became recognizable when it was confirmed that the requirement for time-temperature relationships to inactivate the ALP required slightly higher as compared to kill *Mycobacterium tuberculosis* [56, 57]. Almost 40% activity of ALP in raw cow milk is declared to be linked with the milk fat globule membrane (MFGM) in the cream phase, though the rest is soluble or dispersed in whey membrane particles (WMP) in skim milk [58]. Between individuals and herds, higher ALP levels vary significantly and its concerned activity is correlated with lactation stages and mastitis [59, 60]. Magnesium and zinc ions are promoter of ALP while tin, copper, cobalt, and ethylenediaminetetraacetic acid (EDTA) have inhibitory action and iron has no effect on activities of ALP [61].

ALP activity is in inverse relationship with yield but the other factors, e.g., fat content, breed, and feed, have no effect. For ovine milk, the ALP content is contrarily linked with milk production and directly to the milk fat substance, while infected milk (mastitis) has higher ALP activity [62, 63]. It is reported that ALP activity is low at the mature milk production stage, increased to maximum activity during the peak production stage and again decreased at the end production stage [28]. ALP activity in cow increases as lactation stage proceeds. Immediately after parturition, there is a decrease in ALP activity with a further sharp decrease after the first milking period. ALP activity then continues to decrease and noted minimum at the first week. Then increased slowly and found maximum by the 28th week of lactation [64]. In another study, ALP activity in milk was found lowest in the early lactation stages and progressed along with advancement of lactation stages and milk yields decreased. These ALP activities were also noticed greater in milk samples from evening milk as compared with morning milk [65].

2.3.2. Acid phosphatase

Acid phosphomonoesterase (ACP; EC 3.1.3.2) in milk was initially identified by Huggins and Talalay [66] and affirmed by Mullen [67], declared that ACP was ideally in the active form at 4.0 pH. It was thermally stable and for complete inactivation it required 88°C for 10 min. ACP in bovine milk hydrolyzes the phosphate group of casein particles [68]. There are some components that act as inhibitor and activator. Fluoride acts as an inhibitor for ACP activity but slightly activated by Mn^{2+}. In milk, the ACP level is just ~2% that of the ALP level. Approximately 75% of ACP was found generally in the skim milk phase and 20% of ACP in the MFGM [68, 69]. Reducing agents, ascorbic acid and 2-mercaptoethanol increases the ACP activity by 100% in skim milk, whereas the ACP activity in MFGM is unaffected by these agents. Casein acts as a substrate for the activity of ACP and major casein fractions αs (αs1 + αs2) > β > κ also serve as competitive inhibitors as the ACP enzyme binds with the phosphate group of casein. The ability to bind calcium with κ-casein to form micelles is reduced by dephosphorylation of casein [61].

ACP in milk might be of innovative significance due to three reasons. First, ACP exhibits thermal stability and because of this property it may be used as an indicator for severe heat treatment rather than normal. Second, numerous milk items may have a pH near to that of its optimum. Third, phosphoproteins such as caseins might be dephosphorylated readily. Technological milk properties and development of dairy products depend on the integrity of casein micelles. The enhanced activity of ACP may create problem in the inactivation of ACP without affecting nutritional qualities as it is linked with gelation of ultra-high temperature (UHT) and development of cheese flavor [70, 71].

Specific activity of ACP is greater in cream; however, about 80% ACP of milk is present in skimmed milk [60]. ACP levels in milk of Sahiwal dairy animals showed a declined pattern alongside lactation stages [28]. Shakeel-ur-Rehman and Farkye [72] observed the higher activity of ACP at 5–6 days postpartum, and afterward observed declined trend up to the end of lactation stage. Nevertheless, the range of ACP levels in their study was presented from 2.6 ± 10^{-4} to 2.6 ± 10^{-3} U/mL in normal cow milk. The ACP level is 4–10 times more in mastitis milk than normal cow milk [73, 74].

2.4. Lysozyme

Lysozyme (LZ; EC 3.2.1.17; muramidase) is a single polypeptide chain (14.3 KDa M.W.), cross-linked by four disulfide bonds [75, 76]. It is an important bacteriolytic protein in milk, component of the antibacterial system, that kills bacteria by cleaving the β-1,4-glycosidic bond between N-acetyl muramic acid and N-acetyl glucosamine residues in peptidoglycan of the bacterial cell wall [77, 78].

It helps in improving the human health status, especially neonate, to protect them from infections of invading pathogens with the promotion of gut microbiota until their own immune system is developed [79–81].

Basically, there are two types of LZ: hen egg-white (C-LZ) and goose egg-white (G-LZ). However, both C-LZ and G-LZ forms may be present in cow milk as these forms are present in other body fluids and in stomach tissue of the cow [82].

LZ is available at higher concentration (0.420 g/L) in human milk as compared to buffalos (3.85 μg/mL), cow (0.0013 g/L), and goat (0.0025 g/L) milk [83–86].

The activity of LZ was in greater extent and more stable in buffalo milk as compared to cow milk. However, colostrum possessed 5 times higher activity as compared to mature milk. It was also observed that various factors: parity of animal and lactation stage not influences the activity of LZ but it was increased during the peak summer and winter seasons [86–88]. A substantial increase of milk LZ in mastitis among different bovine species suggested that the neutrophils are the most probable source of LZ due to inflammation of mammary gland [89–91]

2.5. Lactoperoxidases

Lactoperoxidase (LP; EC 1.11.1.7) is the second most abundant enzyme after xanthine oxidase in bovine milk. The most generally prescribed industrial utilization of LP systems is the preservation of raw fresh milk during transportation and storage in dairy plants [92, 93]. It received a considerable attention as an optimum indicator of super-pasteurized milk [94]. Its level in bovine milk is about 30 mg/L constituting approximately 1% whey protein [95]. The LP system (LP-thiocyanate-H_2O_2) is a natural preservation system and has antimicrobial characteristics. Oxidation of thiocyanate in the presence of H_2O_2 is catalyzed by activated LP and produce hypothiocyanate (OSCN) or higher oxides (antimicrobial compounds). These compounds exhibited their antimicrobial properties by oxidizing the sulfhydryl groups of proteins to disulfides [96].

LP enzyme activities are affected by various factors, i.e., sexual cycle, season, lactation, diet, and breed [95, 97]. LP activity in bovine milk (1.2–19.4 U/mL) is about 20 times higher in peroxidase action than human milk [98]. LP levels in dairy animals milk is ranged from 1.5 to 2.7 U/mL with a general mean of 2.3 U/mL [99]. The LP level is low in colostrum of dairy animals, after that adopted inclined trend rapidly after 3–5 days postpartum [95]. LP enzyme activity is a precursor to diagnose the mastitis disease in dairy animals. The activity of LP increases as the somatic cell count (SCC) increases [100, 101].

The LP activity of cows adopted declined trend along with lactation stages. The activity of LP decreases with the advancement in lactation stages (9.64–6.66 U/mL) [28]. The decreasing trend along with lactation stages was also observed by Althaus et al. [102] who reported significant reduction in LP activity from the early stage of lactation toward the end of lactation. Reiter [95] observed a significant increase in LP activity between 4 and 5 days after calving of the lactation period, followed by a gradual decrease toward close of lactation. The reduction in action of LP activity in cow milk could be due to increase in the thiocyanate content as Fonteh et al. [99] described that the LP level promoted with an increase in 2,2'- azino-bis (3-ethylbenzothiazoline-6-sulphonic acid) or ABTS contents but reduced with an increase in thiocyanate contents. They also reported that LP activity is enhanced with whey protein, lactose, magnesium, sodium, and calcium chlorides, and reduced in occurrence of casein.

2.6. Catalases

Catalase (CAT; H_2O_2:H_2O_2 oxidoreductase; EC 1.11.1.6) dismutates hydrogen peroxide (H_2O_2) into water (H_2O) and free oxygen (O_2) [8, 103]. CAT was among the first enzymes present in milk. Babcock and Russell [104] portrayed that an extract of separator slime can break down H_2O_2. The CAT activity in milk fluctuates with feedstuff and lactation phase, level expanded particularly during mastitis [103, 105]. CAT has the ability to degrade the surplus hydrogen peroxide and reduce oxidative infection caused by reactive oxygen species (ROS) [106].

CAT and SCC contributed in the mastitis risk markers. Risk level of mastitis and losses in milk production increase with the advancement in parity, phase of lactation, and also in spring and winter seasons [107]. Measurement of CAT activity plays a distinct role in monitoring the health status of udders in cow. The antioxidant activity of enzyme CAT increases when SCC increases [100, 108, 109]. CAT antioxidant activity is higher in colostrum, then reductions occur as the lactation stage proceeds and again high in the late lactation [110, 111]. Its absence in milk is an indication of an efficient pasteurization process [7].

2.7. Xanthine oxidoreductase

Xanthine oxidoreductase (XOR; EC 1.13.22; 1.1.1.204) is a milk indigenous enzyme having capability of oxidizing hypoxanthine to xanthine and xanthine to uric acid with the reduction of O_2 to H_2O_2 [7, 112]. This protein is initially presented in milk; in 1902, Schardinger reported that this compound is competent for oxidizing aldehydes to acids by the lessening the methylene blue and after that generally called this chemical as

"Schardinger enzyme." XOR has been established to require FAD^{+} and $Mo^{++}$for its optimum catalytic action [103, 113, 114].

XOR is concentrated in MFGM, which is the second most abundant protein constituting, 20% of the MFGM protein. Milk is a good source of XOR, some of its portion is shifted to mammary glands by means of the blood circulation system. The XOR level in milk differs recognizably during lactation. However, bovine milk contains significant levels of XOR (1.4–1.8 U/mg) as compared to goat (0.27 U/mg) and sheep (0.69 U/mg) milk and camel (nd) milk because enzyme molecules lack molybdenum (Mo^{++}) [115–118]. This level can be amplified by complementing the diet with Mo^{++} [7].

In buffalo milk, XOR (0.75 U/mg) exists in the catalytically inactive form because of higher concentration of demolybdo and desulfo forms. Structural factors and lower contents of Fe/S might be the possible reason of lowering enzymatic activity of XOR in buffalo [119]. Surprisingly, camel milk exhibited no detectable XOR activity and its Mo^{++} contents were comparable to human and goat milk [120].

Being significant part of lactating cells, the levels of XOR mRNA began to increase during mid-pregnancy, turned upward at the onset of lactation and diminished quickly in constrained involution [121]. XOR expression remained constant, while specific activity enhanced at the initial lactation phase that facilitates in milk synthesis [122]. Physiologically, XOR contains hydrogen peroxide, nitric oxide, and superoxide ion, mainly functions as in the activation of various metabolic pathways [123]. XOR contributes to an antimicrobial defense mechanism in the gastrointestinal tract (GIT) tract and plays a significant role in the immune system of mammary glands [111, 124, 125]. XOR activity increases during infectious diseases and its cytotoxic action is useful for the defenses against bacteria [123].

3. CONCLUSIONS

Conclusively, intensive review of enzyme activities has revealed the significance of indigenous milk enzymes with varied concentration behavior during lactation stages. Lactation stage has a prominent effect on enzymes activities and ultimately it may affect the technological behavior of milk composition.

Generally, colostrum formation contains higher enzyme activities than during the established lactation period. Mastitis or several other progressions that increase leukocytes in milk increase enzyme activities such as CAT. LP, ACP, and LP decrease while lipase activity increases with progress of lactation. ALP activity first increases then decreases at the end of lactation. PL activity increases in the late lactation because of that it makes milk less suitable for cheese making.

This varied behavior of enzyme activities at early, mid, and late lactation stages can be a troubling problem for manufacturing of milk and milk products in various regions of the world. As enormous animals in late lactation periods and considerable seasonal variations affect the ultimate quality of milk and have a better increased choice to process the specified valued dairy product. Furthermore, milk from mid lactation would be a balanced source of energy to maintain the health status of the individuals.

REFERENCES

1. Pehrsson P, Haytowitz D, Holden J, Perry C, Beckler D. USDA's national food and nutrient analysis program: food sampling. J. Food Compos. Anal. 2000;13:379–389. doi:10.1006/jfca.1999.0867
2. Nickerson S. Milk production: Factors affecting milk composition. In: Harding F, editor.Milk Quality: Springer, USA; 1995. p. 3–24. doi: 10.1007/978-1-4615-2195-2_2
3. Misra S, Sharma A, Bhattacharya T, Kumar P, Roy S. Association of breed and polymorphism of αs1- and αs2-casein genes with milk quality and daily milk and constituent yield traits of buffaloes (*Bubalus bubalis*). Buffalo Bulletin. 2008;27:294–301.
4. Fox P, Guinee T, Cogan T, McSweeney P. Fundamentals of Cheese Science. AspenPublishers, Inc., Gaithersburg, MD; 2000.
5. Looper ML. Factors Affecting Milk Composition of Lactating Cows. Division of Agriculture, University of Arkansas System, USA, 2012.
6. Schutz M, Hansen L, Steuernagel G, Kuck A. Variation of milk, fat, protein, and somaticcells for dairy cattle. J. Dairy Sci. 1990;73:484–493. doi: 10.3168/jds.S0022-0302(90)78696-1
7. Fox P, Kelly A. Indigenous enzymes in milk: overview and historical aspects—Part 1.Int. Dairy J. 2006;16:500–516. doi: 10.1016/j.idairyj.2005.09.013
8. Fox PF. Significance of indigenous enzymes in milk and dairy products. Food Scienceand Technology, Marcel Dekker, New York; 2003. pp.255–278.
9. Kelly A, Fox P. Indigenous enzymes in milk: a synopsis of future research requirements.Int. Dairy J. 2006;16:707–715. doi: 10.1016/j.idairyj.2005.10.018
10. Moatsou G. Indigenous enzymatic activities in ovine and caprine milks. Int. J. DairyTechnol. 2010;63:16–31. doi: 10.1111/j.1471-0307.2009.00552.x
11. Walstra P, Jenness R. Dairy Chemistry & Physics. John Wiley & Sons, New York, USA;1984.
12. Zhao W-S, Hu S-L, Yu K, Wang H, Wang W, Loor J, Luo J. Lipoprotein lipase, tissueexpression and effects on genes related to fatty acid synthesis in goat mammary epithelial cells. Int. J. Mol. Sci. 2014;15:22757–22771. doi: 10.3390/ijms151222757
13. Beisson F, Tiss A, Rivière C, Verger R. Methods for lipase detection and assay: a criticalreview. Eur. J. Lipid Sci. Technol. 2000;102:133–153. doi: 10.1002/(SICI)1438-9312(200002)102:2<133::AID-EJLT133>3.0.CO;2-X
14. Chahiniana H, Sarda L. Distinction between esterases and lipases: comparative-biochemical properties of sequence-related carboxylesterases. Protein Pept. Lett.2009;16:1149–1161. doi: 10.2174/092986609789071333
15. Chen L, Daniel RM, Coolbear T. Detection and impact of protease and lipase activitiesin milk and milk powders. Int. Dairy J. 2003;13:255–275. doi: 10.1016/S0958-6946(02)00171-1
16. Acton QA. Esterases—Advances in Research and Application: 2013 edition. ScholarlyEditions, Atlanta, Gorgia; 2013.

17. Fuquay JW, Fox PF, McSweeney PL. Encyclopedia of Dairy Sciences. Academic Press,San Diego; 2011. doi: 10.1016/B978-0-12-374407-4.09006-3
18. Saito Z, Kim G. Effects of lactation stage on temperature-activated lipolysis and lipaseactivity in cow's milk. Japanese J. Dairy Food Sci. (Japan). 1994; 44:A139–145.
19. Deeth H, Fitz-Gerald C. Lipolytic enzymes and hydrolytic rancidity. Advanced DairyChemistry Volume 2 Lipids. Springer, New York; 2006. pp. 481–556. doi:10.1007/0-387-28813-9_15
20. Palmquist D. Milk fat: origin of fatty acids and influence of nutritional factors thereon.Advanced Dairy Chemistry Volume 2 Lipids. Springer, New York; 2006. pp. 43–92. doi:10.1007/0-387-28813-9_2
21. Ahrné L, Björck L. Lipolysis and the distribution of lipase activity in bovine milk inrelation to stage of lactation and time of milking. J. Dairy Res. 1985;52:55–64. doi:10. 1017/S002202990002389X
22. Deeth HC. Lipoprotein lipase and lipolysis in milk. Int. Dairy J. 2006;16:555–562. doi:10.1016/j.idairyj.2005.08.011
23. Chandan R, Shahani K. Milk lipases. A review. J. Dairy Sci. 1964;47:471–480. doi:10.3168/jds.S0022-0302(64)88694-X
24. Egelrud T, Olivecrona T. The purification of a lipoprotein lipase from bovine skim milk.J. Biol. Chem. 1972;247:6212–6217.
25. Toušová R, Stádník L, Ducháček J. Effect of season and the time of milking on spontaneous and induced lipolysis in bovine milk fat. Czech. J. Food Sci. 2013;31:20–26.
26. Chilliard Y, Ferlay A, Rouel J, Lamberet G. A review of nutritional and physiologicalfactors affecting goat milk lipid synthesis and lipolysis. J. Dairy Sci. 2003;86:1751–1770.doi: 10.3168/jds.S0022-0302(03)73761-8
27. Tarassuk N, Frankel E. The specificity of milk lipase. IV. Partition of the lipase systemin milk. J. Dairy Sci. 1957;40:418–430. doi: 10.3168/jds.S0022-0302(57)94496-X
28. Hameed A, Hussain R, Zahoor T, Akhtar S, Riaz M, Ismail T. Effect of oxytocin onenzyme activities in bovine milk. Int. Dairy J. 2014;39:229–231. doi: 10.1016/ j.idairyj. 2014.06.013
29. Turner RB, Liu L, Sazonova IY, Reed GL. Structural elements that govern the substratespecificity of the clot-dissolving enzyme plasmin. J. Biol. Chem. 2002;277:3306833074. doi: 10.1074/jbc.M203782200
30. Fox P, McSweeney P. Proteolysis in cheese during ripening. Food Rev. Int. 1996;12:457–509. doi: 10.1080/87559129609541091
31. Kitchen BJ. Indigenous milk enzymes. In: Fox P, editor. Developments in DairyChemistry—3: Springer, Netherlands; 1985. p. 239–279. doi:10.1007/978-94-009-4950-8_9
32. Ismail B, Nielsen S. Invited review: plasmin protease in milk: current knowledge andrelevance to dairy industry. J. Dairy Sci. 2010;93:4999–5009. doi: 10.3168/jds.2010-3122
33. Albenzio M, Caroprese M, Santillo A, Marino R, Muscio A, Sevi A. Proteolytic patterns and plasmin activity in ewes' milk as affected by somatic cell count and stage of lactation. J. Dairy Res. 2005;72:86–92. doi: 10.1017/S0022029904000676
34. Bastian ED, Brown RJ. Plasmin in milk and dairy products: an update. Int. Dairy J.1996;6:435–457. doi: 10.1016/0958-6946(95)00021-6
35. Di Luccia A, Picariello G, Trani A, Alviti G, Loizzo P, Faccia M, Addeo F. Occurrenceof β-casein fragments in cold-stored and curdled river buffalo (Bubalus bubalis L.) milk.J. Dairy Sci. 2009;92:1319–1329. doi: 10.3168/jds.2008-1220
36. Farrell H, Jimenez-Flores R, Bleck G, Brown E, Butler J, Creamer L, Hicks C, Hollar C,Ng-Kwai-Hang K, Swaisgood H. Nomenclature of the proteins of cows' milk—Sixthrevision. J. Dairy Sci. 2004;87:1641–1674. doi: 10.3168/jds.S0022-0302(04)73319-6
37. Kaminogawa S, Mizobuchi H, Yamauchi K. Comparison of bovine milk protease with plasmin. Agric. Biol. Chem. 1972;36:2163–2167. doi:/10.1271/bbb1961.36.2163
38. Crudden A, Kelly AL. Studies of plasmin activity in whey. Int. Dairy J. 2003;13:987–993.doi: 10.1016/S0958-6946(03)00140-7

39. Nielsen S. Plasmin system in milk. In: Roginski H, Fuquay J, Fox P, editors. Encyclopediaof dairy sciences. II: Bodmin, Cornwall, UK; 2003. p. 929–934. doi: 10.1016/B0-12-227235-8/00153-X
40. Ozen BF, Hayes KD, Mauer LJ. Measurement of plasminogen concentration anddifferentiation of plasmin and plasminogen using Fourier-transform infrared spectroscopy.Int. Dairy J. 2003;13:441–446. doi: 10.1016/S0958-6946(03)00055-4
41. Nielsen SS. Plasmin system and microbial proteases in milk: characteristics, roles, andrelationship. J. Agric. Food Chem. 2002;50:6628–6634. doi: 10.1021/jf0201881
42. Crudden A, Fox F, Kelly AL. Factors affecting the hydrolytic action of plasmin in milk.Int. Dairy J. 2005;15:305–313. doi: 10.1016/j.idairyj.2004.08.008
43. Cortellino G, Locci F, Rampilli M. An investigation of the plasmin–plasminogen systemin caprine milk and cheese. Int. Dairy J. 2006;16:619–622. doi: 10.1016/j. idairyj. 2005.11.012
44. Politis I. The role of plasminogen activator in the bovine mammary gland. In: Mol J,Clegg R, editors. Biology of the Mammary Gland. 480: Springer, USA; 2002. p. 203–207.doi: 10.1007/0-306-46832-8_25
45. Bastian ED, Brown RJ, Ernstrom CA. Plasmin activity and milk coagulation. J. DairySci. 1991;74:3677–3685. doi: 10.3168/jds.S0022-0302(91)78557-3
46. Leitner G, Chaffer M, Shamay A, Shapiro F, Merin U, Ezra E, Saran A, Silanikove N.Changes in milk composition as affected by subclinical mastitis in sheep. J. Dairy Sci.2004;87:46–52. doi: 10.3168/jds.S0022-0302(04)73140-9
47. Caroprese M, Schena L, Marzano A, Marino R, Sevi A, Albenzio M. Contribution ofmacrophages to plasmin activity in ewe bulk milk. Ital. J. Anim. Sci. 2007;6:545–547.doi: 10.4081/ijas.2007.1s.545
48. Koutsouli P, Patounas G, Massouras T, Bizelis I, Politis I. Plasmin–plasminogen systemand milk coagulation properties of two Greek dairy sheep breeds. Small Ruminant Res.2015;124:89–94. doi: 0.1016/j.smallrumres.2015.01.015
49. Bianchi L, Bolla A, Budelli E, Caroli A, Casoli C, Pauselli M, Duranti E. Effect of udderhealth status and lactation phase on the characteristics of Sardinian ewe milk. J. DairySci. 2004;87:2401–2408. doi: 10.3168/jds.S0022-0302(04)73362-7
50. Nicholas GD, Auldist MJ, Molan PC, Stelwagen K, Prosser CG. Effects of stage oflactation and time of year on plasmin-derived proteolytic activity in bovine milk inNew Zealand. J. Dairy Res. 2002;69:533–540. doi: 10.1017/S0022029902005745
51. Politis I, Hang KNK, Giroux R. Environmental factors affecting plasmin activity in milk.J. Dairy Sci. 1989;72:1713–1718. doi: 10.3168/jds.S0022-0302(89)79286-9
52. Richardson B, Pearce K. Determination of plasmin in dairy products. N.Z. J. Dairy Sci.Technol. 1981;16:209–220.
53. Hamed H, Trujillo A-J, Juan B, Guamis B, ElFeki A, Gargouri A. Interrelationshipsbetween somatic cell counts, lactation stage and lactation number and their influenceon plasmin activity and protein fraction distribution in dromedary (Camelus dromedaries)and cow milks. Small Ruminant Res. 2012;105:300–307. doi: 10.1016/j.smallrumres.2012.01.002
54. Chaiyabutr N, Thammacharoen S, Komolvanich S, Chanpongsang S. Effects of longtermadministration of recombinant bovine somatotropin on the plasminogen–plasminsystem and milk composition of crossbred Holstein cattle. Anim. Sci. J. 2007;78:251–258. doi: 10.1111/j.1740-0929.2007.00432.x
55. Graham W, Kay H. Phosphorus compounds of milk. V. The phosphorus partition inmilk, with preliminary observations on milk phosphatase. J. Dairy Res. 1933;5:54–62.doi: 10.1017/S0022029900000923
56. Rankin S, Christiansen A, Lee W, Banavara D, Lopez-Hernandez A. Invited review: theapplication of alkaline phosphatase assays for the validation of milk product pasteurization.J. Dairy Sci. 2010;93:5538–5551. doi: 10.3168/jds.2010-3400
57. Rosenthal I, Bernstein S, Rosen B. Alkaline phosphatase activity in *Penicillium roquefortiand* in blue-veined cheeses. J. Dairy Sci. 1996;79:16–19. doi: 10.3168/jds.S0022-0302(96)76328-2

58. Silanikove N, Shapiro F. Distribution of xanthine oxidase and xanthine dehydrogenaseactivity in bovine milk: physiological and technological implications. Int. Dairy J.2007;17:1188–1194. doi: 10.1016/j.idairyj.2007.03.003
59. Babaei H, Mansouri-Najand L, Molaei M, Kheradmand A, Sharifan M. Assessment oflactate dehydrogenase, alkaline phosphatase and aspartate aminotransferase activitiesin cow's milk as an indicator of subclinical mastitis. Vet. Res. Commun. 2007;31:419–425. doi: 10.1007/s11259-007-3539-x
60. ndrews A. Indigenous enzymes in milk: General introduction. In: Fox P, editor.Advanced dairy chemistry-1: Proteins. II ed: Elsevier Science Publishers; 1992. p. 286–292.
61. Fox PF, McSweeney PL. Dairy Chemistry and Biochemistry. Blackie Academic &Professional, an imprint of Chapman & Hall, London; 1998.
62. Anifantakis E, Rosakis P. Alkaline phosphatase activity of sheep's milk and some factorsaffecting it. Egypt. J. Dairy Sci. 1983;11:173–182.
63. Patil M, Nagvekar A, Ingole S, Bharucha S, Palve V. Somatic cell count and alkalinephosphatase activity in milk for evaluation of mastitis in buffalo. Vet. World. 2015;8:363.doi: 10.14202/vetworld.2015.363-366
64. Haab W, Smith L. Variations in alkaline phosphatase activity of milk. J. Dairy Sci.1956;39:1644–1650. doi: 10.3168/jds.S0022-0302(56)94905-0
65. Banks J, Muir D. Variability of alkaline phosphatase in goat milk in relation to its useas an effective index of pasteurisation. Project S01003. Hannah Research Institute, Ayr,Scotland. 2004.
66. Huggins C, Talalay P. Sodium phenolphthalein phosphate as a substrate for phosphatasetests. J. Biol. Chem. 1945;159:399–410.
67. Mullen J. 427. The acid phosphatase of cows' milk: I. Some properties of the enzyme. J.Dairy Res. 1950;17:288–305. doi: 10.1017/S0022029900005847
68. Akuzawa R, Fox PF. Acid phosphatase in cheese. Anim. Sci. J. 2004;75:385–391. doi:10.1111/j.1740-0929.2004.00202.x
69. Flynn N, Akuzawa R, Fox P, editors. Characterisation of isozymes of acid phosphataseisolated from bovine milk. Abstracts 28th Food Science and Technology ResearchConference. University College, Cork; 1998. p. 165.
70. Fox P, Uniacke-Lowe T, McSweeney P, O'Mahony J. Heat-induced changes in milk.Dairy Chemistry and Biochemistry: Springer International Publishing, Switzerland;2015. p. 345–375. doi: 10.1007/978-3-319-14892-2_9
71. Fox PF, Morrissey PA. Indigenous Enzymes of Bovine Milk. In: Birch GG, BlakebroughN, Parker KJ, editors. Enzymes and Food Processing. 3rd ed: Springer Netherlands;2012. p. 213–238. doi: 10.1007/978-94-011-6740-6
72. Shakeel-ur-Rehman, Farkye NY. Enzymes indigenous to milk: phosphatase. In:Roginski H, Fuquay J, Fox P, editors. Encyclopedia of Dairy Sciences: Amsterdam Academic Press; 2002. p. 934–938. doi: 10.1016/B0-12-227235-8/00154-1
73. Fox P, Stepaniak L. Enzymes in cheese technology. Int. Dairy J. 1993;3:509–530. doi:10.1016/0958-6946(93)90029-Y
74. Andrews A. Further studies on acid phosphatase of leucocyte origin in normal andmastitic bovine milks. J. Dairy Res. 1976;43:127–131. doi: 10.1017/S0022029900015673
75. Cegielska-Radziejewska R, Lesnierowski G, Kijowski J. Properties and application ofegg white lysozyme and its modified preparations—a review. Pol. J. Food Nutr. Sci.2008;58. doi: 10.1007/s00217-010-1347-y
76. Masschalck B, Michiels CW. Antimicrobial properties of lysozyme in relation tofoodborne vegetative bacteria. Crit. Rev. Microbiol. 2003;29:191–214. doi:10.1080/713610448
77. Wiesner J, Vilcinskas A. Antimicrobial peptides: the ancient arm of the human immunesystem. Virulence. 2010;1:440–464. doi: 10.4161/viru.1.5.12983 Park YW, Haenlein GF. Handbook of Milk of Non-Bovine Mammals. John Wiley & Sons,New York; 2008. doi: 10.1002/9780470999738
78. Al-Jabri AA. Honey, milk and antibiotics. Afr. J. Biotechnol. 2005;4:1580–1587.
79. Lönnerdal B. Nutritional and physiologic significance of human milk proteins. Am. J.Clin. Nutr. 2003;77:1537S-1543S.

80. Maga EA, Sargent RG, Zeng H, Pati S, Zarling DA, Oppenheim SM, Collette NM, MoyerAL, Conrad-Brink JS, Rowe JD. Increased efficiency of transgenic livestock production. Transgenic Res. 2003;12:485–496. doi: 10.1023/A:1024257906647
81. Losnedahl K, Wang H, Aslam M. Antimicrobial proteins in milk. J. Diary Sci.1996;84:1123–1127.
82. Król J, Litwinczuk Z, Brodziak A, Barlowska J. Lactoferrin, lysozyme and immunoglobulinG content in milk of four breeds of cows managed under intensive productionsystem. Polish J. Vet. Sci. 2010;13:357–361.
83. Benkerroum N. Antimicrobial activity of lysozyme with special relevance to milk. Afr.J. Biotechnol. 2008;7:4856–4867.
84. Abd El-Aziz M. Study on lysozyme level, distribution and effect of heat treatment inbuffalo and cow milk. Ann. Agric. Sci. (Cairo). 2006;51:439–446.
85. Montagne P, Cuilliere M, Mole C, Bene M, Faure G. Changes in lactoferrin and lysozymelevels in human milk during the first twelve weeks of lactation. In: Newburg D, editor.Bioactive Components of Human Milk: Springer, USA; 2001. p. 241–247. doi:10.1007/978-1-4615-1371-1_30
86. Hussain M, Bakalis S, Gouseti O, Zahoor T, Anjum FM, Shahid M. Dynamic and shearstress rheological properties of guar galactomannans and its hydrolyzed derivatives.Int. J. Biol. Macromol. 2015;72:687–691. doi: 10.1016/j.ijbiomac.2014.09.019
87. Priyadarshini S, Kansal VK. Purification, characterization, antibacterial activity and Nterminalsequencing of buffalo-milk lysozyme. J. Dairy Res. 2002;69:419–431. doi:10.1017/S002202990200554X
88. Persson K, Carlsson Å, Hambleton C, Guidry A. Immunoglobulins, lysozyme andlactoferrin in the teat and udder of the dry cow during endotoxin-induced inflammation.J. Vet. Med. B. 1992;39:165–174. doi: 10.1111/j.1439-0450.1992.tb01154.x
89. Radwan Y, Elmarimi A. The use of lysoplate test in the diagnosis of subclinical mastitis.Bull. Anim. Health. Prod. Africa. 1987;35:100–135.
90. Ismail M, Selim S, Soliman R. Changes in lysozyme activity in milk and its significancein the diagnosis of subclinical mastitis in goats. Egypt. Vet. Med. J. 1984;32:59–65.
91. De Wit J, Van Hooydonk A. Structure, functions and applications of lactoperoxidase innatural antimicrobial systems. Nederlands melk en Zuiveltijdschrift. 1996;50:227–244.doi: 10.1201/9781420039368.ch3
92. Pruitt KM, Kamau DN. The lactoperoxidase system of bovine and human milk. In: DSRobinson, NAM Eskin, Editors. Oxidative Enzymes in Foods. London: ElsevierApplied Science; 1991. 133–174p
93. Lorenzen PC, Martin D, Clawin-Rädecker I, Barth K, Knappstein K. Activities ofalkaline phosphatase, γ-glutamyltransferase and lactoperoxidase in cow, sheep andgoat's milk in relation to heat treatment. Small Ruminant Res. 2010;89:18–23. doi:10.1016/j.smallrumres.2009.11.013
94. Reiter B. The lactoperoxidase system of bovine milk. In: Pruitt K, Tenovou J, editorsThe lactoperoxidase system: chemistry and biological significance: Marcel Dekker,New York, USA; 1985. p. 123–141. doi: 10.1016/B978-0-08-031739-7.50030-6
95. Martin-Hernandez M, van Markwijk BW, Vreeman HJ. Isolation and properties oflactoperoxidase from bovine milk. Neth. Milk Dairy J. 1990;44:213–231.Lactation Responses toward Milk Indigenous Enzymeshttp://dx.doi.org/10.5772/66392
96. Kussendrager KD, Van Hooijdonk A. Lactoperoxidase: physico-chemical properties, occurrence, mechanism of action and applications. Br. J. Nutr. 2000;84:19–25. doi:10.1017/S0007114500002208
97. Gothefors L, Marklund S. Lactoperoxidase activity in human milk and in saliva ofnewborn infants. Infect. Immun. 1975;11:1210–1215.
98. Fonteh FA, Grandison AS, Lewis MJ. Factors affecting lactoperoxidase activity. Int. J.Dairy Technol. 2005;58:233–236. doi: 10.1111/j.1471-0307.2005.00227.x
99. Andrei S, Pintea A, Groza I, Bogdan L, Ciupe S, Matei S. Milk antioxidant enzymesactivity in cows with sub clinical mastitis. Scientific Papers – Veterinary Medicine,University of Agricultural Sciences and Veterinary Medicine "Ion Ionescu de la Brad"Iaşi. 2009;52:1–6.

100. Asadpour R, Tayefi-Nasrabadi H, Moghadam G, Nofouzi K. Correlation betweenlactoperoxidase activity and somatic cell count for diagnosis subclinical mastitis inearly lactation of dairy cows. J. Anim. Vet. Adv. 2008;7:777–779.
101. Althaus R, Molina M, Rodriguez M, Fernandez N. Analysis time and lactation stageinfluence on lactoperoxidase system components in dairy ewe milk. J. Dairy Sci.2001;84:1829–1835. doi: 10.3168/jds.S0022-0302(01)74622-X
102. McSweeney P, Fox P. Advanced Dairy Chemistry: Basic Aspects: Volume 1A, Proteins.4th ed. Springer, New York, USA; 2013. doi: 10.1007/978-1-4614-4714-6
103. Babcock SM, Russell HL. Unorganized ferments of milk: A new factor in the ripeningof cheese. Fourteenth annual report of the Wisconsin agricultural experiment station.1897;161–193.
104. Johnson A. The composition of milk. In: Webb B, Johnson A, Alford J, editors. Fundamentalsof dairy chemistry: AVI Publishing Co. Inc, Westport, CT; 1974. p. 1–57.
105. He Z, Yu S, Mei G, Zheng M, Wang M, Dai Y, Tang B, Li N. Maternally transmitted milkcontaining recombinant human catalase provides protection against oxidation formouse offspring during lactation. Free Radic. Biol. Med. 2008;45:1135–1142. doi:10.1016/j.freeradbiomed.2008.07.019
106. Atasever S, Erdem H. An investigation on the determination of mastitis risk levels andmilk production traits in Holstein cows. J. Appl. Anim. Res. 2008;34:13–16. doi:10.1080/09712119.2008.9706932
107. Atasever S, Erdem H. Association of somatic cell count with catalase enzyme activityin bovine milk. Trends Anim. Vet. Sci. 2010;1:24–27.
108. Lindmark-Månsson H, Åkesson B. Antioxidative factors in milk. Br. J. Nutr.2000;84:103–110. doi: 10.1017/S0007114500002324
109. Yuksel S, Yigit AA, Cinar M, Atmaca N, Onaran Y. Oxidant and antioxidant status ofhuman breast milk during lactation period. Dairy Sci. Technol. 2015;95:295–302. doi:10.1007/s13594-015-0211-z
110. Atasever S, Erdem H, Kul E. Relationship between somatic cell count and catalaseactivity in raw milk of Anatolian buffaloes. Sci. Res. Essays. 2011;6:4109–4112. doi:10. 5897/SRE11.609
111. Martin HM, Hancock JT, Salisbury V, Harrison R. Role of xanthine oxidoreductase asan antimicrobial agent. Infect. Immun. 2004;72:4933–4939. doi: 10.1128/IAI.72.9.4933-4939.2004
112. Harrison R. Physiological roles of xanthine oxidoreductase. Drug Metab. Rev.2004;36:363–375. doi: 10.1081/DMR-120037569
113. Massey V, Harris C. Milk xanthine oxidoreductase: the first one hundred years.Biochem. Soc. Trans. 1997;25:750–755. doi: 10.1042/bst0250750
114. Pandya AJ, Khan MMH. Buffalo milk. Handbook of milk of non-bovine mammals.2006:195–273. doi: 10.1002/9780470999738.ch8
115. Godber BL, Schwarz G, Mendel RR, Eisenthal R, Harrison R. Molecular characterizationof human xanthine oxidoreductase: the enzyme is grossly deficient in molybdenumand substantially deficient in iron-sulphur centres. Biochem. J. 2005;388:501–508. doi:10.1042/BJ20041984
116. Atmani D, Benboubetra M, Harrison R. Goats' milk xanthine oxidoreductase is grosslydeficient in molybdenum. J. Dairy Res. 2004;71:7–13. doi: 10.1017/ S0022029903006514
117. Benboubetra M, Baghiani A, Atmani D, Harrison R. Physicochemical and kineticproperties of purified sheep's milk xanthine oxidoreductase. J. Dairy Sci. 2004;87:1580–1584. doi: 10.3168/jds.S0022-0302(04)73311-1
118. Gadave K, Panda S, Singh S, Kalra S, Malakar D, Perugini MA. Structural and functionalinsights into the catalytic inactivity of the major fraction of buffalo milk xanthineoxidoreductase. PLoS One. 2014;9:1–11. doi: 10.1371/journal.pone.0087618
119. Baghiani A, Harrison R, Benboubetra M. Purification and partial characterisation ofcamel milk xanthine oxidoreductase. Arch. Physiol. Biochem. 2003;111:407–414. doi:10.3109/ 13813450312331342265

120. McManaman J, Palmer C, Wright R, Neville M. Functional regulation of xanthineoxidoreductase expression and localization in the mouse mammary gland: evidence ofa role in lipid secretion. J. Physiol. 2002;545:567–579. doi: 10.1113/jphysiol.2002.027185
121. Vorbach C, Scriven A, Capecchi MR. The housekeeping gene xanthine oxidoreductaseis necessary for milk fat droplet enveloping and secretion: gene sharing in the lactatingmammary gland. Genes Dev. 2002;16:3223–3235. doi: 10.1101/gad.1032702
122. Battelli MG, Polito L, Bortolotti M, Bolognesi A. Xanthine oxidoreductase-derivedreactive species: physiological and pathological effects. Oxid. Med. Cell Longev. 2016:1–8. doi: 10.1155/2016/3527579
123. Silanikove N, Shapiro F, Leitner G. Posttranslational ruling of xanthine oxidase activityin bovine milk by its substrates. Biochem. Biophys. Res. Commun. 2007;363:561–565.doi: 10.1016/j.bbrc.2007.08.188
124. Harrison R. Milk xanthine oxidase: properties and physiological roles. Int. Dairy J.2006;16:546–554. doi: 10.1016/j.idairyj.2005.08.016

CHAPTER 10

Bacterial Tick-Borne Diseases of Livestock Animals

Şükrü Kirkan, Göksel Erbaş and Uğur Parin

Department of Microbiology, Faculty of Veterinary Medicine, Adnan Menderes University, Aydin, Turkey

ABSTRACT

Bacterial tick-borne diseases (BTBDs) are very significant in practical one health medicine. In contrast to the restrictions related to diagnostic and clinical application, the control and prevention of bacterial tick-borne diseases are difficult because they require the disruption of a complicated transmission chain, involving vertebrate hosts and ticks, which interact in a constantly changing environment. Q fever, rickettsiosis, borreliosis, ehrlichiosis, anaplasmosis and tularemia are BTBDs, which are discussed in this chapter. Epidemiology, clinical symptoms, diagnosis and prevention subtopics are planning to be prepared under main topics. This chapter presents a brief background of key livestock BTBDs and ticks and reviews the general aspects of BTBDs to identify topics in knowledge and understanding of these diseases, propose areas for future research and draw attention to the need for improved tools for the diagnosis and control of BTBDs.

Keywords: tick, bacterial zoonoses, Q fever, rickettsiosis, borreliosis, ehrlichiosis, anaplasmosis, tularemia

1. INTRODUCTION

Bacterial tick-borne diseases (BTBDs) affect the productivity of livestock animals in various regions of the world, leading to a significant adverse impact on the production of resource-poor farming communities. Hence, the livestock industry has become an integral part of world economy, and the large number of dairy cattle is being imported between continents in order to meet an increasing demand of meat and dairy products, it is essential to review current status of bovine BTBDs and to identify diagnosis and prevention in the knowledge of BTBDs and their prevention. Although there has been a recent increase in the number of studies of BTBDs in various geographical regions, information on their prevalence, distribution, tick vectors and control is limited.

2. BACTERIAL TICK-BORNE DISEASES OF LIVESTOCK ANIMALS

2.1. Q fever

Q fever is a zoonosis associated with *Coxiella burnetii* that is an obligate intracellular parasite classified within the family Rickettsiaceae and which can be divided into six genomic groups based on restriction fragment length polymorphism. Unlike the other members of Rickettsiae, *C. burnetii* is quite resistant to environmental influences and is not dependent upon arthropod vectors for transmission. *C. burnetii* exhibits two antigenic phases: phase I and phase II (**Figure 1**). Phase I organisms are more infectious. The organism has worldwide distribution, although a large serological survey argues that it is not present in New Zealand [1].

C. burnetii cycles in a wide variety of wildlife species and their ectoparasites. The infection also cycles in domestic animals. Rates of infection in farm animals vary considerably between locations, between countries and with time as there appears to be cycles of infection within regions [3].

In cattle, prevalence figures range from 6 to 82% of cattle and 23 to 96% of herds seropositive depending upon location and country. Seropositivity rates in sheep and goats are similar but also vary according to year and region. There is little information on management or other factors that might influence this variation in prevalence but one study found a significantly higher prevalence in housed cattle compared to cattle at kept at pasture. The transmission of infection is spread by direct contact and inhalation. Infection of non-pregnant animals is clinically silent and is followed by latent infection until pregnancy when there is recrudescence with infection in the intestine, uterus, placenta and udder and excretion from these sites at parturition. The organism is present in high concentration in the placenta and foetal fluids, and subsequent vaginal fluids are also excreted in urine and are present in the faeces of sheep from 11 to 18 days post-partum [4, 5]. Infection can result in abortion, stillbirths or poorly viable lambs but commonly the neonates of infected, excreting, ewes are born clinically normal. Abortion usually does not occur at successive pregnancies but there can be recrudescence of infection and excretion at these pregnancies, especially the one immediately following [6].

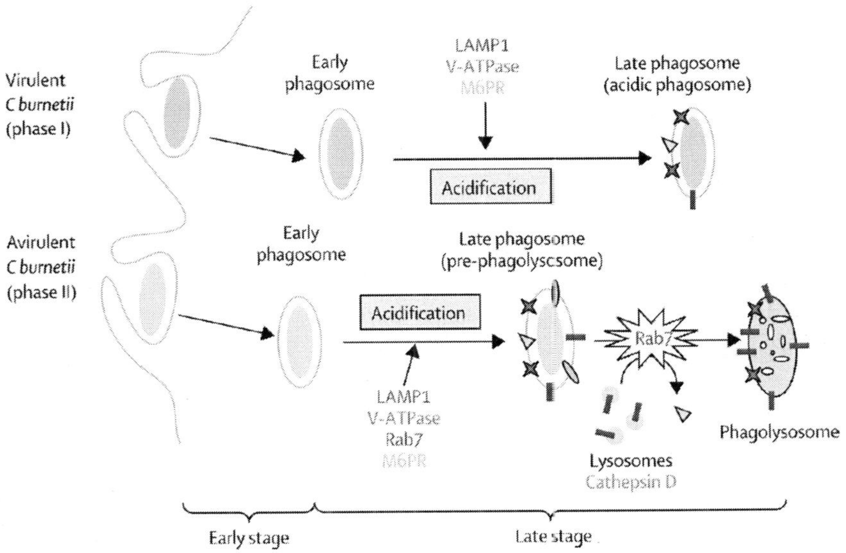

Figure 1. *Coxiella burnetii* mobilization in macrophages [2].

Goats also excrete the organism in vaginal discharges for up to 2 weeks, and it is present in goat milk for up to 52 days after kidding and also in faeces. Maximum shedding in cattle also occurs at parturition and for the following 2 weeks but cattle excretes the organism in the milk for at least several months and up to 2 years and infection is common in bulk tank milk [7–10].

There is strain variation in the organism and differences in plasmid sequence types have been correlated with differences in the type of disease occurring in humans. The organism is highly infectious, and it is estimated that the infective dose for humans approximates one organism zoonotic implications in human infection is primarily by inhalation. Sources of infection include such diverse materials such as soil, air-borne dust, wool, bedding and other materials contaminated by urine, faeces or birth products of animals. The potential for human infection from these sources is substantial; for example, ovine manure used as a garden fertilizer has been incriminated as a source. Sheep have traditionally been incriminated as the major reservoir of infection for humans, but the trend for urban populations to locate in close proximity to large dairy herds suggests that cattle could become an increasingly significant reservoir [11–13].

The organism is found in the milk of infected livestock. A significant proportion of seropositive cattle excrete the organism in milk and periods and duration of excretion are variable but may persist at least 2 years. Rates of seropositivity in humans vary markedly between surveys, but there is a higher rate of seropositivity in people (farm workers, veterinarians, livestock dealers, dairy plant and slaughter house workers, shearers, etc.) that are associated with domestic animals and their products and with farm environments [14, 15]. Several incidents of infection in humans have been linked to exposure to parturient sheep and goats [16].

Infection of ruminants can occur at any age and is usually clinically unapparent. In the experimental disease in cattle, anorexia is the only consistent clinical finding. Abortion occurs during the latter part of the lambing period in the flock and in the latter period of pregnancy in individual ewes. The dam shows no signs of impending abortion. As with sheep, infection in goats can be accompanied by abortion, but abortion in cattle is rare although it is recorded. Correlations between herd level seroprevalence and herd fertility are equivocal. There are a number of serological tests available including complement fixation, microagglutination, enzyme-linked immunosorbent assay (ELISA) and indirect immunofluorescence (IF). The IF assay is used as the sero-reference test for the serodiagnosis of Q fever. It can detect antibody to phase variants and can provide epidemiological information as phase I antibody is associated with recent and acute infections and phase II antibody with chronic infections [17].

There are seldom gross lesions in aborted foetuses, but foci of necrosis and inflammation are occasionally seen in the liver, lung and kidney microscopically. The placenta from aborting animals is usually thickened and a purulent exudates or large, red-brown foci of necrosis are typically seen in the thickened intercotyledonary areas. Microscopically, large numbers of necrotic neutrophils are usually visible on the chorionic surface and swollen trophoblasts filled with the organisms can also be found in well-preserved specimens. Examination of placental impression smears stained with Gimenez, Koster's, or other appropriate techniques provides a means of rapid diagnosis. However, care must be taken to avoid confusing Coxiella-infected trophoblasts with cells containing Chlamydophila organisms. Coxiellosis can be confirmed fluorescent antibody staining of fresh tissue or immunohistochemical staining of formalin-fixed samples. In most laboratories, culture is not attempted due to the zoonotic potential of this agent. Polymerase chain reaction (PCR) is the most accurate tool for the diagnosis of infectious abortions. In a previous study, six (4.3%) samples were detected PCR positive out of 138 samples [13]. In another research, *C. burnetii* gene was detected in 34.66% of the samples taken from 200 cattle, 200 sheep and 200 goats in the Aegean region of Turkey [19]. In a multidisciplinary research made with veterinarians, farm workers and butchers, among 92 people, 32 (34.8%) and 9 (9.8%) people were positive and equivocal by ELISA and immunoglobulin G (IgG), respectively. The ELISA positive and equivocal sera were studied further by the immunofluorescence antibody (IFA) test, and seven (7.6%) cases were confirmed with immunoglobulin M (IgM), 39 (42.4%) cases were confirmed with IgG. There was no significant difference for Coxiellosis seropositivity among the profession groups ($p > 0.05$). Only four (4.3%) cases were confirmed with PCR positive [20].

Aborting animals should be isolated for 3 weeks and aborted and placental contaminated material burnt. Ideally, manure should be composted for 6 months before application to fields. Feed areas should be increased to keep them free from contamination with faeces and urine. While Q fever has significant implications for human health, it is not significantly important enough to have generated national or regional control strategies based on control in the animal population [21–23].

Milk and milk products should be pasteurized. Veterinarians dealing with herds that provide raw milk should ensure that these herds are seronegative for *C. burnetii*. Vaccine trials with killed vaccines in animals show a good and persistent antibody response and suggest that vaccination can limit the excretion of the organism. However, there is little economic incentive for a vaccination programme involving livestock, and livestock vaccines are not available in most countries [24].

2.2. Rickettsiosis

The members of the family Rickettsiaceae have cell walls similar to those of other Gram-negative bacteria. Ultra structural studies have shown that the Anaplasmataceae family have outer membranes but lack an obvious peptidoglycan layer [25]. Organisms in the family Rickettsiaceae, referred to as rickettsiae, generally target endothelial cells. Although several new species of rickettsiae have recently been identified in domestic animals using molecular techniques, their pathogenicity is uncertain and currently the only species of veterinary importance in the family Rickettsiaceae is *Rickettsia rickettsii*, the causative agent of Rocky Mountain spotted fever. Many *Rickettsia* species including the causal agents of typhus (*R. prowazekii*), murine typhus (*R. typhi*) and scrub typhus (*R. tsutsugamushi*) are primarily human pathogens. These highly pathogenic organisms have a predilection for the endothelial cells of small blood vessels, resulting in vasculitis and thrombosis in many organs. *Rickettsia* species produce phospholipase that damages the membranes of phagosomes allowing the organisms to escape into the cytoplasm (**Figure 2**). *R. rickettsii* replicates in both the cytoplasm and the nucleus of host cells, inducing cytotoxic effects [26].

Definitive classification of the members of the Rickettsiales is based on 16S ribosomal ribonucleic acid (RNA) sequencing, lipopolysaccharide content and metabolic requirements. In diagnostic laboratories, identification of these organisms is based on the species affected, cell predilection, microscopic appearance and molecular techniques. Some members of the Rickettsiales can be cultured in embryonated eggs or tissue culture cells. These difficult procedures are usually performed only in laboratories engaged in research or vaccine production [28].

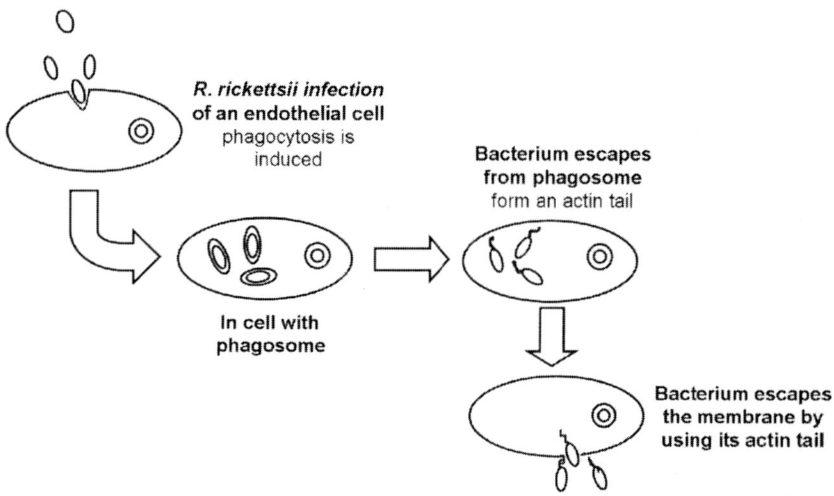

Figure 2. Infection diagram of *R. rickettsii* [27].

R. rickettsii affects mainly humans and dogs. *Rhipicephalus sanguineus* and *Amblyomma cajennense* are the main vectors in Central and South America. Ticks acquire the pathogen while feeding on infected small wild mammals [29].

An infected tick must remain attached for up to 20 hours before salivary transmission to the host occurs. The organisms, which replicate in endothelial cells of infected dogs, produce vasculitis, increased vascular permeability and haemorrhage. Rocky Mountain spotted fever should be considered in dogs with systemic diseases, which have been exposed to ticks in endemic areas. Indirect fluorescent antibody test (FAT) or ELISA demonstrating an increasing antibody titre to *R. rickettsii* is diagnostic. Antibodies are not demonstrable until at least 10 days after infection. A marked thrombocytopenia and leucopoenia may be present during the acute phase of the disease. The disease must be differentiated from acute canine monocytic ehrlichiosis. PCR detection in tick tissues has been described by a number of workers. Tetracycline therapy, which usually produces clinical improvement within 24 hours, must be continued for 2 weeks. Supportive therapy is necessary for severely debilitated dogs. Frequent removal of ticks is recommended. Because the disease is zoonotic, gloves should be worn during this procedure or a forceps should be used [30].

Ticks acquire the pathogen while feeding on infected small wild mammals. *R. rickettsii* is maintained in the tick population by transovarial and transstadial transmission and thus the tick acts as both a reservoir and a vector of the organism. An infected tick must remain attached for up to 20 hours before salivary transmission to the host occurs. The incubation period of the disease is 2–10 days and the course is usually less than 2 weeks. Clinical signs include fever, depression, conjunctivitis, retinal haemorrhages, muscle and joint pain, coughing, dyspnoea and oedema of the extremities [31].

2.3. Borreliosis

Borreliae, which are longer and wider than other spirochaetes, have a similar helical shape. In addition to a linear chromosome, which is unique among bacteria, borreliae possess linear and circular plasmids, some of which appear to be essential for growth and survival of the organism. Although these spirochaetes can cause disease in animals and humans, subclinical infections are also common. Borreliae are transmitted by arthropod vectors. Arthropod vectors are responsible for transmission of *Borrelia* species in animals. Borreliae are obligate parasites in a variety of vertebrate hosts. Although these organisms persist in the environment for short periods, they depend on vertebrate reservoir hosts and arthropod vectors for long-term survival. Associations of certain *Borrelia* species with particular arthropod vectors and reservoir hosts are important in determining the epidemiology of infections with *Borrelia* species. After entering the bloodstream of a susceptible host, borreliae multiply and are disseminated throughout the body (**Figure 3**). Organisms may be demonstrated in joints, brain, nerves, eyes and heart. Whether disease is caused by active infection or by host immune responses to the organism is unclear. Persistent infection leading to the induction of cytokines may contribute to the development of lesions [32]. There may be an association between different genotypes of *Borrelia burgdorferi* and particular clinical syndromes in humans; *B. burgdorferi* sensu stricto (s.s.) is frequently associated with arthritis, *B. garinii* with neurological disease and *B. afzelii* with skin disease [33, 34].

Chickens have been infected experimentally, and it was found that these animals quickly became immune to *B. burgdorferi* s.s. and did not show any clinical symptoms [36]. More recent studies have shown that pheasants can function as reservoir hosts of *B. garinii* and *B. valaisiana* in the United Kingdom (UK) [37], but no symptoms of disease in infected birds have been reported.

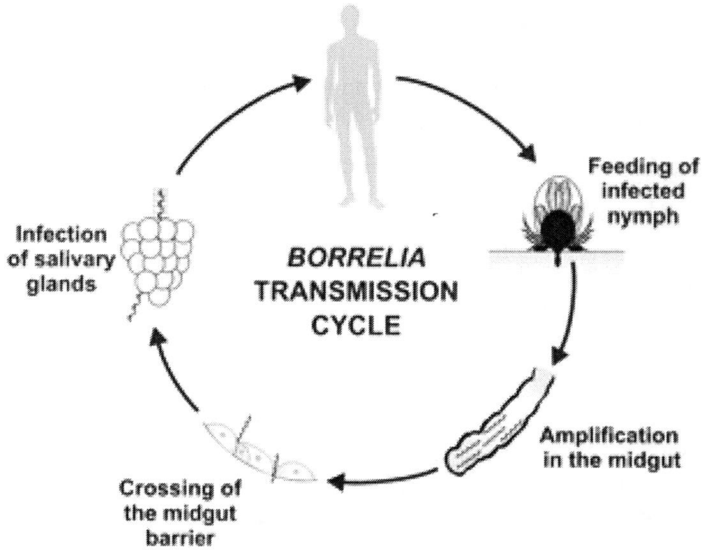

Figure 3. Life cycle of *Borrelia* spp. [35].

Most infections are subclinical. Serological surveys demonstrate that exposure is common in both animal and human populations in endemic areas. The clinical manifestations of Lyme disease are mainly related to the sites of localization of the organisms. Clinical disease is reported frequently in dogs. Symptoms include fever, lethargy, arthritis and evidence of cardiac, renal or neurological disturbance. In the United States of America (USA), arthritis is a common finding whereas neurological disturbance is the most frequent clinical feature in Europe and Japan. The clinical signs in horses are similar to those in dogs and include lameness, uveitis, nephritis, hepatitis and encephalitis. However, some authors observe that definitive evidence of clinical Lyme disease in horses is lacking [38]. Lameness in cattle and sheep associated with *B. burgdorferi* sensu lato infection has been reported.

Laboratory confirmation of Lyme disease may prove difficult because the spirochetes may be present in low numbers in specimens from clinically affected animals. In addition, the organism is fastidious in its cultural requirements. A history of exposure to tick infestation in an endemic area in association with characteristic clinical signs may suggest Lyme disease. Increasing antibody titres to *B. burgdorferi* sensu lato along with typical clinical signs are indicative of disease. Because subclinical infections are common in endemic areas, high titres alone are not confirmatory. The ELISA is extensively used for antibody detection; western immunoblotting is sometimes used for confirmation of ELISA results. It has been shown that ELISA techniques based on this antigen may be able to differentiate naturally infected and vaccinated animals [39]. Immunofluorescence assays may also be used but the results of these methods may be difficult to interpret. Culture of borreliae from clinically affected animals is confirmatory. Cultures in Barbour-Stoenner-Kelly medium should be incubated for 6 weeks under microaerophilic conditions and should be carried out in specialized laboratories. Low numbers of borreliae can be detected in samples by PCR techniques.

Acute Lyme disease responds to treatment with amoxicillin and oxytetracycline. In chronic disease, prolonged or repeated courses of treatment may be required. Acaricidal sprays, baths or dips should be used to control tick infestation. Where feasible, tick habitats such as rough brush and scrub should be cleared. Prompt removal of ticks from companion animals may prevent infection. However, because some tick species can transmit spirochetes shortly after attachment, it cannot be assumed that daily removal of ticks will prevent infection [40].

A number of vaccines, including whole cell bacterins and recombinant subunit vaccines, are commercially available for use in some countries. An outer surface protein A (OspA) recombinant vaccine stimulates the production of antibodies, which are able to kill the borreliae in the gut of the tick and thus prevent infection of the host. However, the benefit of vaccinating animals with currently available vaccines is disputed [41].

2.4. Ehrlichiosis and anaplasmosis

Ehrlichia (Cowdria) ruminantium is a Gram negative, intracellular rickettsial organism in the genus *Ehrlichia*. It occurs in colonies or morulae with a predilection for the vascular endothelium and stains blue with Giemsa stain. The organism is coccoid, 0.2–0.5 μ in diameter. It can now be cultivated *in vitro*, and it can also grow in mice. Cyclical development is believed to take place in intestinal and salivary epithelia of ticks. Although strain differences exist, all isolates possess a major antigenic protein 1 (MAP 1) that is used for diagnosis. However, the antigen cross-reacts with other *Ehrlichia* spp., including *Ehrlichia equi*, the cause of equine granulocytic ehrlichiosis. *Anaplasma* spp. is obligate intraerythrocytic parasites belonging to the order Rickettsiales and infecting ruminants. Infection occurs more sporadically in temperate climate areas. In the USA and other countries, the disease has occurred beyond the boundaries of tick-infested areas and the area distribution in Europe has been advancing northward in recent years with sporadic cases in France, Switzerland, the Netherlands, Hungary and Austria. Anaplasmosis of sheep and goats has a distribution similar to that of cattle. Disease occurs sporadically in the northern states and Canada. In Australia, infection is closely related to the distribution of *Boophilus microplus*, which is restricted to the northern areas. Differences in enzootic and epizootic areas in South America and South Africa are also largely related to tick distribution and climate [42].

Heartwater is limited in its occurrence to sub-Saharan Africa, Madagascar and three Caribbean islands of Guadeloupe, Marie Galante and Antigua. It is one of the main causes of death in imported breeds of cattle, sheep and goats in endemic areas. Heartwater has been diagnosed recently in the island of Mayotte in the Indian Ocean. Measures of disease occurrence in endemic areas, morbidity and mortality rates are low, but the percentage of sera positive titres for heartwater could be as high as 100% in adults, depending on the abundance of tick vectors [43]. Case mortality can be as high as 100% in peracute cases in sheep and goats and as low as 0–10% in cattle. The disease is less severe in indigenous breeds and related game animals reared in enzootic areas, some of which may become symptomless carriers. The N'Dama breed in West Africa is said to be well adapted to heartwater, partly because it can resist tick burdens under the traditional farming system. The method of transmission in the Caribbean, cattle egrets are suspected to spread *Amblyomma variegatum* between islands. Consequently, heartwater is considered threats to the American mainland where potential vectors are present but do not harbour the disease or where the vector may be introduced and become established. Infection in ticks is transmitted transstadially and possibly transovarially. Vertical transmission to calves in colostral milk has also been reported. Several wild ruminants can be infected and become subclinical carriers and reservoirs.

Ticks feeding on them can transmit the disease to domestic ruminants. The organism does not infect humans. Cattle are infected with *Amblyomma marginale* and *Amblyomma centrale* and sheep with *Amblyomma avis. A. marginale* will establish in sheep by experimental infection but *A. avis* will not infect cattle. A variety of species of wild ruminants in both North America and Africa can be infected and may have significance as reservoirs for *A. marginale*. In the United States, the black-tail deer in the West Coast region is believed a reservoir and a number of species of antelope play a similar role in South Africa. The prevalence of infection in cattle in endemic areas is very high with seropositivity rates exceeding 60% and often approaching 90%. Seropositivity is much lower in regions that interface between endemic and non-endemic regions. Source and methods of transmission recovery from acute infection result in persistent infection characterized by repetitive cycles of rickettsemia. Persistent carriers are the reservoir for herd infection. The level of parasitemia is often too low for detection by microscopy but can be detected by nucleic acid probe analysis. Transmission occurs biologically by ticks [44].

Heartwater is the most important rickettsial infection of ruminants in Africa and it is regarded as the most important disease of ruminants. In general, heartwater is a more serious problem where *Amblyomma habraeum* is the vector. In countries or regions where there is endemic stability, losses from heartwater are minimal until new animals are introduced. On the other hand, since most losses are in exotic animals, heartwater is a major constraint to livestock improvement in sub-Saharan Africa. Furthermore, it has the potential to spread from the Caribbean to the American mainland. Heartwater requires the vector tick to get established in any community. Therefore, there is concern about possible illegal importation of infected animals or ticks to southern United States where potential vectors exist. In ewes intra-uterine infection appears to occur with ease in experimental cases provided the ewe is exposed during the latter two-thirds of pregnancy. In sheep and goats, infection is usually subclinical but in some cases, particularly in goats, a severe anaemia may occur and a clinical picture similar to that found in cattle may be seen. Severe reactions of this type in goats are most frequent when the animals are suffering from concurrent disease. Goats may show hyper excitability and may bite at inanimate objects. The experimental disease in lambs includes fever, constipation or diarrhoea, pale, icteric conjunctivae and severe anaemia 15–20 days after inoculation. The anaemia is not completely resolved in 3–4 months. *A. avis* are usually situated at the periphery of erythrocytes but as many as 40% of infested cells may show sub-marginal protozoa [45].

The incubation period is 1–3 weeks after transmission in tick saliva. Depending on the susceptibility of individual animals and the virulence of the infecting organism, the resulting disease may be peracute, acute, subacute or mild and unapparent. Peracute cases show only high fever and death with terminal convulsions in 1–2 days. Acute cases are more common and have a course of about 6 days. A sudden febrile reaction is followed by inappetence and rapid breathing followed by the classical nervous syndrome that is characteristic of heartwater. It comprises ataxia, chewing movements. Profuse, fetid diarrhoea is frequent. Subacute cases are less severe but may terminate in death in 2 weeks or the animal may gradually recover. The mild form is often subclinical and is seen mainly in indigenous animals and wild ruminants with high natural or induced resistance. The case mortality rate in peracute cases is 100%, in acute cases 50–90% and in calves below 4 weeks of age it is 5–10%, most animals recover in mild cases [46].

Haematological changes in heartwater are not specific but there may be thrombocytopenia, neutropenia, eosinopenia and lymphocytosis. Confirmatory diagnosis is based on identifying the *Rickettsia* in capillary endothelial cells using a Giemsa stained squash preparation of brain tissue at post-mortem. The rickettsiae occur as blue to reddish-purple colonies or morulae of five to several hundred coccoid organisms (0.2–0.5 μ in diameter) in the cytoplasm of the cells. An immunohistochemical staining technique has also been described [47]. Injection of blood into sheep may also be used as a diagnostic procedure. The available serological test is an indirect fluorescent antibody test used for surveys but the close antigenic relationship with other *Ehrlichia* spp. often leads to false positives. An ELISA based on recombinant MAP l protein of *C. ruminantium* was reported to be more sensitive. In general, clinical detection of heartwater is not always easy because all serological assays so far available have poor sensitivity or specificity. Diff-Quik staining of blood smears is as accurate as Giemsa in the detection of *A. marginale* and can be completed in 15 seconds as compared to nearly an hour for Giemsa. There are no diagnostic clinical chemistry findings. A rapid card agglutination test, which tests serum or plasma for antibodies against *A. marginale*, is cheap and quick, and sufficiently accurate to be used as a herd test. Currently, in most countries, the card agglutination and complement fixation (CF) tests are routinely available. It is also an accurate test for selecting recently affected animals. A dot-ELISA with high sensitivity, specificity and predictive value is also described and could be particularly applicable to field examinations. A competitive inhibition ELISA test, with high sensitivity and specificity, has been developed that detects antibody to a major surface protein that is conserved among *Anaplasma* species; this test can be used to detect cattle persistently infected for as long as 6 years. Vaccinated animals may react to all of the serological tests for periods of over 1 year. Nucleic probe analysis can be used to detect low levels of parasitaemia. Transmission to splenectomised animals has been used to detect carriers but is expensive and is now replaced by PCR in countries where this technology is available

[48]. A polymerase chain reaction assay has therefore been suggested as the method of choice for detection of *E. ruminantium* infection [49].

Field cases of heartwater are difficult to treat successfully because available drugs are effective only in early febrile stages before neurological signs develop. In the early stages, short-acting tetracyclines at 10–20 mg/kg body weight (BW) and long-acting forms at reduced doses are effective. Sulphonamides can also be used in the early stages but are less effective. Hyperimmune serum is said to be of no curative value. Supportive therapy to reduce either the pulmonary oedema or the neurologic signs or to stabilize membranes in general is being investigated but with little success. Chemoprophylaxis involves administration of tetracyclines or subcutaneous implantation of doxycycline in susceptible animals when they are introduced into an endemic area. Results are not always predictable. Anaplasmosis treatment is with tetracyclines. Treatment of clinical disease can be with oxytetracycline, 6–10 mg/kg BW daily for 3 days, or a one dose application of long-lasting 20 mg/kg oxytetracycline intramuscularly. The convalescent period is long. Concurrent administration of estradiol cypionate (14.3 mg/kg BW intramuscularly) appears to improve the rate of recovery by promoting parasitemia during treatment. Tetracycline treatment will not eliminate infection and immunity will persist. Blood transfusions are indicated in animals with a packed cell volume (PCV) less than 15%. Rough handling must be avoided. Imidocarb (3 mg/kg BW) is also an effective treatment for clinical cases and does not interfere with the development of acquired immunity to *A. marginale*. The risk for infection in the rest of the herd should be assessed and, if necessary, temporary or prolonged protection should be provided. Protection can be provided by tetracyclines, or by vaccination [50].

Past efforts to control heartwater were based on intensive acaricide treatment in endemic areas. It involved frequent use of acaricides (plunge dipping) up to 52 times a year. This has now been shown to be environmentally unfriendly, economically unsustainable, and would invariably lead to animals that remained always susceptible. For example, it was observed in Zimbabwe that large farms applying acaricides very frequently (more than 30 times per annum) had higher morbidity and mortality than farms applying acaricides less frequently. Vaccination is based on infection and treatment regimen that was first developed more than 50 years ago. It involves an intravenous injection of virulent organisms in cryopreserved sheep blood, followed by treatment with tetracyclines at the first indication of fever. Most control programmes in enzootic areas are based on increasing the resistance of the population by immunization. In any vaccination programme, particular attention should be paid to the animals at high risk, particularly animals brought in from non-enzootic areas, those in surrounding similar areas to which infection may be spread by expansion of the vector population under the influence of suitable climatic conditions, and animals within the area are likely to be exposed to climatic or nutritional stress [51].

Vaccination may lead to some deaths, the immunity may wane in the absence of reinfection, and animals may become carriers. More recently, cattle were successfully immunized for up to 10 months with a killed vaccine from a lysate of *E. ruminantium* formulated in Freund's adjuvant. In another study, the use of inactivated vaccines from cell-cultured *E. ruminantium* combined with an adjuvant led to a reduction in mortality from heartwater in cattle, sheep and goats exposed to field challenges in Botswana, Zambia, Zimbabwe, and South Africa. Experimental studies using deoxyribonucleic acid (DNA) recombinant vaccines so far have met with only limited success. Killed *A. marginale* are usually in an adjuvant vehicle. The vaccine requires two doses, 4 weeks apart, the last dose given at least 2 weeks before the vector season. However, there is a risk for neonatal isoerythrolysis. This can be reduced by vaccinating only empty cows and avoiding unnecessary booster injections. When this vaccine is used in the face of an outbreak, tetracyclines can also be given to provide temporary protection during the period of development of immunity; tetracyclines do not interfere with the development of this immunity. Preliminary reports of the efficacy of DNA vaccines are not encouraging. A living *A. centrale* vaccine is used extensively in Australia, Africa, Israel and Latin America, but not in the USA and there is some reluctance to introduce it into areas where *A. centrale* does not already occur. A single vaccination is used in endemic areas and the immunity is reinforced by continuous challenge and considered to persist for life in tick areas. Vaccine administration is limited to the relatively resistant age group below 1 year of age, to the winter months when vectors are sufficiently rare to avoid the chance of spread to other age groups, and to circumstances where animals that react severely can be restrained and treated adequately. The method has the serious disadvantage of creating a large population of carrier animals which may subsequently spread the disease. Attenuated vaccines have been attempted by irradiation of strains and passage of the organism through sheep or deer and the use of naturally low virulence isolates [52, 53].

For tick control, flumenthrin 1% pour on at 45 days interval was found to provide effective protection of Friesian/Zebu crossbred cattle against important ticks, but it must be applied correctly at the recommended dose. Pure Zebu and N'Dama cattle would probably require less frequent applications, Flumenthrin pour-on is gradually replacing plunge dipping for the control of ticks and tick-borne diseases in general. Other than routine surveillance, there are no special biosecurity concerns with heartwater, since transmission requires the presence of the vector [54].

2.5. Tularemia

The disease causes acute septicaemia, with localization and granulomatous lesions and the organs (particularly the liver and spleen). Signs are very non-specific, as expected with bacteraemia, and include fever, anorexia, lethargy, and in some cases cough, rapid respiration or diarrhoea. Stiffness and oedema of the limbs may be seen. The incubation period of the disease is usually 2–14 days in companion animals [55].

Tularemia is a highly contagious disease occurring principally in wild animals but it may transmit to farm animals, causing septicaemia and high mortality. *Francisella tularensis* is the causative organism [56].

Tularemia is primarily restricted in its occurrence to countries in the northern hemisphere and occurs in most of them. In North America, the disease is most prevalent in farm animals in the north-western states of the USA and the adjoining areas of Canada, although in these areas it is rare and the majority of reports in livestock are historical. *F. tularensis* has a wide host range and is recorded in over 100 species of bird and wild and domestic animal. Disease is recorded among farm animals, most commonly in sheep and pigs and to a lesser extent in calves, which appear more resistant but can be infected in association with heavy tick infestation [57]. Sheep and pigs of all ages are susceptible but most losses occur in lambs, and in pigs clinical illness occurs only in piglets. There is a sharp seasonal incidence, the bulk of cases occurring during the spring months. The morbidity rate in affected flocks of sheep is usually about 20% but may be as high as 40%, and the mortality rate may reach 50%, especially in young animals. With sheep, transmission occurs chiefly by the bites of the wood tick, *Dermacentor andersoni*, and from *Haemaphysalis otophila*, the ticks becoming infected in the early part of their life cycle when they feed on rodents. In Europe *Ixodes ricinus* and *Dermacentor reticulatus* are vectors [58]. Transstadial and transovarial transmission occurs in the tick. The adult ticks infest sheep, and pastures bearing low shrubs and brush are particularly favourable to infestation. The ticks are found in greatest numbers on the sheep around the base of the ears, the top of the neck, the throat, axillae and udder. It is assumed that sheep are relatively resistant to tularemia but become clinically affected when the infection is massive and continuous. Transmission to pigs and horses is thought to occur chiefly by tick bites but mechanical transmission to laboratory animals does occur with tabanid and blackflies. Tularemia is an acute septicaemia but localization occurs, mainly in the parenchymatous organs, with the production of granulomatous lesions [59].

In the sheep, the incubation period has not been determined. A heavy tick infestation is usually evident. The onset of the disease is slow with a gradually increasing stiffness of gait, dorsiflexion of the head and a hunching of the hindquarters; affected animals lag behind the group. The pulse and respiratory rates are increased, the temperature is elevated up to 42°C (107°F), and a cough may develop. There is diarrhoea, the faeces being dark and fetid, and urination occurs frequently with the passage of small amounts of urine. Body weight is lost rapidly, and progressive weakness and recumbency develop after several days, but there is no evidence of paralysis, the animal continuing to struggle while down [60]. Death occurs usually within a few days but a fatal course may be as long as 2 weeks. Animals that recover commonly shed part or the entire fleece but are solidly immune for long periods. In pigs, the disease is latent in adult pigs but young piglets show fever up to 42°C, accompanied by depression, profuse sweating and dyspnoea. The incubation period of the disease is about 7–10 days. In horses, fever (up to 42°C) and stiffness and oedema of the limbs occur. Foals are more seriously affected and may show dyspnoea and incoordination in addition to the above signs [61]. Necropsy usually reveals ticks on the carcass. Often, reddened or necrotic areas appear in and under the skin at the site of the infected bites. Regional lymph nodes may be swollen and congested. Congestion and oedema of the lungs are common [62].

An agglutination test is available for the diagnosis of tularemia, a titre of 1:50 being regarded as a positive test in pigs. Serum from pigs affected with brucellosis does not agglutinate tularemia antigen, but serum from pigs affected with tularemia agglutinates brucellosis antigen. Cross-agglutination between *F. tularensis* and *Brucella abortus* is less common in sheep and an accurate diagnosis can be made on serological grounds because of the much greater agglutination that occurs with the homologous organism. Titres of agglutinins in affected sheep range from 1:640 to 1:5000 and may persist at levels of 1:320 for up to 7 months. A titre of 1:200 is considered as positive in sheep. In horses the titres revert to normal levels in 14–21 days. An intradermal sensitivity test using 'tularin' has been suggested as being more reliable as a diagnostic aid in pigs than the agglutination test, but is unreliable in sheep. In sheep, large numbers of ticks may be present on the hides of fresh carcasses. In animals that have been dead for some time, dark red subcutaneous areas of congestion up to 3 cm in diameter are found and may be accompanied by local swelling or necrosis of tissues [63]. These lesions mark the attachment sites of ticks. Enlargement and congestion of the lymph nodes draining the sites of heaviest tick infestation are often noted. Pulmonary oedema, congestion or consolidations are inconstant findings. In pigs, the characteristic lesions are pleuritis, pneumonia and abscessation of submaxillary and parotid lymph nodes. The organisms can be isolated from the lymph nodes and spleen, and from infected ticks. Isolation can also be effected by experimental transmission to guinea pigs. Techniques such as immunoperoxidase staining of fixed specimens and PCR of fresh tissues can circumvent the need for culture of this zoonotic agent. Samples for confirmation of diagnosis are based on

- Bacteriology: lung, lymph node, spleen (CULT—requires cysteine-enriched media, PCR).
- Histology: above tissues plus liver, fixed in formalin [64].

Treatment early in the course of infection is effective. Aminoglycosides, tetracyclines or cephalosporins all are probably beneficial initially, until results of antimicrobial susceptibility testing are available. Streptomycin, gentamicin, the tetracyclines and chloramphenicol are effective treatments in humans and companion animals. Oxytetracycline (6–10 mg/kg BW) has been highly effective in the treatment of lambs and much more effective than penicillin and streptomycin. Insecticide removal of ticks from affected animals and herdmates is important. An outbreak of tularemia in sheep can be rapidly halted by spraying or dipping with insecticide to kill the vector ticks. In areas where ticks are enzootic, sheep should be kept away from shrubby, infested pasture or sprayed regularly during the months when the tick population is greatest. An experimental live attenuated vaccine has been developed, but there is no routine vaccination of livestock [65].

3. CONCLUSION

Given that the livestock industry has become an integral part of world economy and a large number of dairy cattle are being imported between countries, in order to meet an increasing demand of meat and dairy products, it is essential to review current status of bovine BTBDs and to identify diagnosis and prevention in the knowledge of BTBDs and their control. Although there has been a recent increase in the number of studies of BTBDs in various regions and facilities, information on their prevalence, distribution, tick vectors and control is limited. This chapter provides a brief background on key bovine BTBDs and ticks and reviews the general aspects of bovine BTBDs to identify gaps in knowledge and understanding of these diseases, propose areas for future research and draw attention to the need for improved tools for the diagnosis and control of BTBDs.

REFERENCES

1. Maurin M, Raoult D. Q fever. Clin. Microbiol. Rev. 1999;12:518–553.
2. Raoult D, Marrie TJ, Mege JL. Natural history and pathophysiology of Q fever. Lancet Infect. Dis. 2005;5:219–26.

3. Mertens K, Samuel JE. Bacteriology of Coxiella. In: Raoult D, Parola P, editors. Rickettsial Diseases. New York: Informa Healthcare Inc.; 2007. pp. 257–270.
4. Marrie TJ, Raoult D. Update on Q fever, including Q fever endocarditis. Curr. Clin. Top. Infect. Dis. 2002;22:97–124.
5. Stein A, Raoult D. Pigeon pneumonia in Provence. A bird borne Q fever outbreak. Clin. Infect. Dis. 1999;29:617–620.
6. Stoker MG, Marmion BP. The spread of Q fever from animals to man. The natural history of a rickettsial disease. Bull WHO. 1955;13:781–806.
7. Marrie TJ, Langille D, Papukna V, Yates L. Truckin pneumonia-an outbreak of Q fever in a truck repair plant probably due to aerosols from clothing contaminated by contact with newborn kittens. Epidemiol. Infect. 1989;102:119–127.
8. Babudieri B. Q fever. A zoonosis. Adv. Vet. Sci. 1959;5:82–182.
9. Sanford ES, Josephson GKA, MacDonald A. Coxiella burnetii (Q fever) abortion storms in goat herds after attendance at an annual fair. Can. Vet. J. 1994;35:376–378.
10. Huebner RJ, Bell JA. Q fever studies in Southern California. Summary of current results and a discussion of possible control measures. J. Am. Med. Assoc. 1951;145:301–305.
11. Webster JP, LLoyd G, Macdonald DW. Q fever (Coxiella burnetii) reservoir in wild brown rat (Rattus norvegicus) populations in the UK. Parasitology. 1995;110:31–35.
12. Lang GH. Serosurvey of Coxiella burnetii infection in dairy goat herds in Ontario. Can. J. Vet. Res. 1988;52:37–41.
13. Marrie TJ. Epidemiology of Q fever. In: Raoult D, Parola P, editors. Rickettsial Diseases. New York: Informa Healthcare Inc.; 2007. pp. 281–289.
14. Arricau-Bouvery N, Rodolakis A. Is Q fever an emerging or reemerging zoonosis? Vet. Res. 2005;36:327–349.
15. Dupuis G, Peter O, Luthy R, Nicolet J, Peacock M, Burgdorfer W. Serological diagnosis of Q fever endocarditis. Eur. Heart J. 1986;7:1062–1066.
16. Peacock MG, Philip RN, Williams JC, Faulkner RS. Serological evaluation of Q fever in humans: enhanced phase I titers of immunoglobulins G and A are diagnostic for Q fever endocarditis. Infect. Immun. 1983;41:1089–1098.
17. Sidi-Boumedine K, Rousset E, Henning K, Ziller M, Niemczuck K, Roest HIJ, Thiéry R. Development of harmonised schemes for the monitoring and reporting of Q-fever in animals in the European Union. EFSA Scientific Report on Question No EFSA-Q-2009-00511. 2010. 48 p.
18. Kırkan Ş, Kaya O, Tekbıyık S, Parın U. Detection of Coxiella burnetii in cattle by using PCR. Turk. J. Vet. Anim. Sci. 2008;32:215–220.
19. Parın U, Kaya O. Detection of Coxiella burnetii prevalence in bovine, ovine and caprine herds. Ankara Üniv. Vet. Fak. Derg. 2015;62:177–181.
20. Eyigör M, Kırkan Ş, Gültekin B, Yaman S, Tekbıyık S, Aydın N. Detection of antibodies against Coxiella burnetii in risk groups for Q fever: a study with ELISA and IFA tests. Turk. J. Inf. 2006;20:31–36.
21. Chiu CK, Durrheim DN. A review of the efficacy of human Q fever vaccine registered in Australia. N.S.W. Public Health Bull. 2007;18:133–136.
22. Kovacova E, Kazar J. Q fever-still a query and underestimated infectious disease. Acta Virol. 2002;46:193–210.
23. Waag DM, England MJ, Tammariello RF. Comparative efficacy and immunogenicity of Q fever chloroform:methanol residue (CMR) and phase I cellular (Q-Vax) vaccines in cynomolgus monkeys challenged by aerosol. Vaccine. 2002;20:2623–2634.
24. OIE. Terrestrial Manual [Internet]. 2015. Available from: http://www.oie.int/ fileadmin/Home/eng/Health_standards/tahm/2.01.16_Q_FEVER.pdf [Accessed: 2016-07-25]
25. Cowan GO. Rickettsial infections. In: Cook GC, Zumla A, editors. Manson's Tropical Diseases. 21st ed. London Saunders: Elsevier Science, Health Sciences Division; 2003, Vol. 50, pp. 891–906.
26. Walker DH, Raoult D. Rickettsia rickettsii and other spotted fever group Rickettsiae (Rocky Mountain spotted fever and other spotted fevers). In: Mandell GL, Bennet JE, Doalin R, editors. Principles and Practice of Infectious Diseases. Philadelphia: Churchill Livingstone; 2000. pp. 2035–2042.

27. Felton B, Lacroix Z, Denisson A, Jacques J. Rocky Mountain Spotted Fever [Internet]. 2012. Available from: https://bacterialandviraldiseases.wikispaces.com [Accessed: 2016-07-03]
28. Anacker RL, McDonald GA, List RH, Mann RE. Neutralizing activity of monoclonal antibodies to heat-sensitive and heat-resistant epitopes of Rickettsia rickettsii surface proteins. Infect. Immun. 1987;55:825–827.
29. McDonald GA, Anacker RL, Mann RE Milch LJ. Protection of guinea pigs from experimental Rocky Mountain spotted fever with a cloned antigen of Rickettsia rickettsii. J. Infect. Dis. 1988;158:228–231.
30. Vishwanath S, McDonald GA, Watkins NG. A recombinant Rickettsia conorii vaccine protects guinea pigs from experimental boutonneuse fever and Rocky Mountain spotted fever. Infect. Immun. 1990;58:646–653.
31. Diaz-Montero CM, Feng HM, Crocquet-Valdes PA, Walker DH. Identification of protective components of two major outer membrane proteins of spotted fever group Rickettsiae. Am. J. Trop. Med. Hyg. 2001;65:371–378.
32. Roberts ED, Bohn RP, Lowrie RC. Pathogenesis of Lyme neuroborreliosis in the Rhesus monkey: the early disseminated and chronic phases of disease in the peripheral nervous system. J. Infect. Dis. 1998;178:722–732.
33. Van Dam AP, Kuiper H, Vos K. Different genospecies of Borrelia burgdorferi are associated with distinct clinical manifestations of Lyme borreliosis. Clin. Infect. Dis. 1993;17:708–717.
34. Gray JS. Tick-borne pathogen interactions. Infect. Dis. Rev. 1999;1:117–121.
35. Hajdušek O, ŠímaR, Ayllón N, Jalovecká M, Jan Perner J, de la Fuente J, Kopáček P. Interaction of the tick immune system with transmitted pathogens Front. Cell. Infect. Microbiol. 2013;26:1–15.
36. Piesman J, Dolan MC, Schriefer ME, Burkot TR. Ability of experimentally infected chickens to infect ticks with the Lyme disease spirochete, Borrelia burgdorferi. Am. J. Trop. Med. Hyg. 1996;54:294–298.
37. Kurtenbach K, Peacey M, Rijpkema SGT, Hoodless AN, Randolph SE Nuttall PA. Differential transmission of genospecies of Borrelia burgdorferi sensu lato by game birds and small rodents in England. App. Environ. Microbiol. 1998;64:1169–1174.
38. Butler CM, Houwers DJ, Jongejan F, van der Kolk JH. Borrelia burgdorferi infections with special reference to horses. A review. Vet. Quart. 2005;27:146–156.
39. O'Connor TP, Esty KJ, Hanscom JL, Shields P, Philipp MT. Dogs vaccinated with common Lyme disease vaccines do not respond to IR6, the conserved immunodominant region of the VlsE surface protein of Borrelia burgdorferi. Clin. Diag. Lab. Immunol. 2004;11:458–462.
40. Korenberg EI, Moskvitina GG. Interrelationships between different Borrelia genospecies and their principal vectors. J. Vector Ecol. 1996;21:178–185.
41. Littman MP, Goldstein RE, Labato MA, Lappin MR, Moore GE. ACVIM small animal consensus statement on Lyme disease in dogs: diagnosis, treatment, and prevention. J. Vet. Int. Med. 2006;20:422–434.
42. Anonymous [Internet]. 2013. Ehrlichiosis and Anaplasmosis: Zoonotic Species Available from: http://www.cfsph.iastate.edu/Factsheets/pdfs/ehrlichiosis.pdf [Accessed: 2016-08-12]
43. Parola P, Raoult D. Ticks and tickborne bacterial diseases in humans: an emerging infectious threat. Clin Infect Dis 2001;32:897–928. Erratum: Clin. Inf. Dis. 33:749.
44. Raoult D, Roux V. Rickettsioses as paradigms of new or emerging infectious diseases. Clin. Microbiol. Rev. 1997;10:694–719.
45. Anonymous [Internet]. 2012. Available from: http://www.bada-uk.org [Accessed: 2016-08-13]
46. Cunha B, Domachowske J, Keim SM, Abuhammour W, Bennett NJ. Ehrlichiosis [Internet]. 2012. Available from: http://emedicine.medscape.com/infectious_diseases
47. Dumler JS, Barbet AF, Bekker CP, Dasch GA, Palmer GH, Ray SC, Rikihisa Y, Rurangirwa FR. Reorganization of genera in the Families Rickettsiaceae and Anaplasmataceae in the order Rickettsiales: unification of some species of Ehrlichia with Anaplasma, Cowdria

with Ehrlichia, and Ehrlichia with Neorickettsia, descriptions of five new species combinations and designation of Ehrlichia equi and "HGE agent" as subjective synonyms of Ehrlichia phagocytophila. Int. J. Syst. Evol. Microbiol. 2001;51:2145–2165.

48. OIE, Terrestrial Manual. [Internet]. 2015. Available from: http://www.oie.int/fileadmin/Home/eng/Health_standards/tahm/2.04.01_BOVINE_ANAPLASMOSIS.pdf
49. Alberti A, Zobba R, Chessa B, Addis MF, Sparagano O, Pinna Parpaglia ML, Cubeddu T, Pintori G, Pittau M. Equine and canine Anaplasma phagocytophilum strains isolated on the island of Sardinia are phylogenetically related to pathogenic strains from the United States. Appl. Environ. Microbiol. 2005;71:6418–6422.
50. Uilenberg G. Note sur les babésioses et l'anaplasmose des bovins à Madagascar. IV. Note additionelle sur la transmission. Rev. Élev. Méd. Vét. Pays Trop. 1970;23:309–312.
51. Connell M, Hall WTK. Transmission of Anaplasma marginale by the cattle tick Boophilus microplus. Aust. Vet. J. 1972;48:477.
52. Leatch G. Preliminary studies on the transmission of Anaplasma marginale by Boophilus microplus. Aust. Vet. J. 1973;49:16–19.
53. Anonymous, Tick-borne Livestock Diseases and Their Vectors. [Internet]. 2016. Available from: http://www.fao.org/docrep/004/x6538e/X6538E02.htm [Accessed: 2016-06-23]
54. MERCK, Anaplasmosis. [Internet]. 2016. Available from: http://www.merckvetmanual.com/mvm/circulatory_system/blood_parasites/anaplasmosis.Html
55. Mchardy N. Immunization against anaplasmosis: a review. Prev. Vet. Med. 1984;2:135–146.
56. Kocan KM, DE LA Fuente J, Guglielmone AA, Meléndez RD. Antigens and alternatives for control of Anaplasma marginale infection in cattle. Clin. Microbiol. Rev. 2003;16:698–712.
57. Ellis J, Oyston PCF, Green M, Titball RW. Tularemia, Clin. Microbiol. Rev. 2002:15;631–646.
58. OIE, Terrestrial Manual. [Internet]. 2008. Available from: http://web.oie.int/fr/ normes/mmanual/2008/pdf/2.01.18_TULAREMIA.pdf
59. MERCK, Overview of Tularemia. [Internet]. 2016. Available from: http://www.merckvetmanual.com/mvm/generalized_conditions/tularemia/overview_of_tularemia.html
60. Bell JF. Tularaemia. In: Steel JH, editor. CRC Handbook Series in Zoonoses. Section A: Bacterial, Rickettsial, and Mycotic Diseases. 2nd ed. Florida: CRC Press; 1980. pp. 161–193.
61. Morner T, Addison E. Tularemia. In: Williams ES, Barker IK, editors. Infectious Diseases of Wild Mammals. 3rd ed. Iowa: Iowa State University Press; 2001. pp. 303–313.
62. Smith BP. Large Animal Internal Medicine. 4th ed. Missouri: Elsevier Health Sciences; 2008. 1184 p. AVMA, Tularemia. [Internet]. 2016. Available from: https://www.avma.org/News/Journals/Collections/Documents/javma_222_6_725.pdf
63. Anonymous. Tularemia, Kosovo. Wkly. Epidemiol. Rec. 2000;75:133–134.
64. Gilligan PH. Pseudomonas and Burkholderia. In: Murray PR, Baron EJ, Pfaller MA, Tenover FC, Yolken RH, editors. 6th ed. Manual of Clinical Microbiology. Washington DC: American Society for Microbiology Press; 1995. pp. 509–519.
65. Bell JF. Tularaemia. In: Steel JH, editor. CRC Handbook Series in Zoonoses. Section A: Bacterial, Rickettsial, and Mycotic Diseases. 2nd ed. Florida: CRC Press; 1980. pp. 161–193.
66. Morner T, Addison E. Tularemia. In: Williams ES, Barker IK, editors. Infectious Diseases of Wild Mammals. 3rd ed. Iowa: Iowa State University Press; 2001. pp. 303–313.
67. Smith BP. Large Animal Internal Medicine. 4th ed. Missouri: Elsevier Health Sciences; 2008. 1184 p.
68. AVMA, Tularemia. [Internet]. 2016. Available from: https://www.avma.org/News/Journals/Collections/Documents/javma_222_6_725.pdf Anonymous. Tularemia, Kosovo. Wkly. Epidemiol. Rec. 2000;75:133–134.
69. Gilligan PH. Pseudomonas and Burkholderia. In: Murray PR, Baron EJ, Pfaller MA, Tenover FC, Yolken RH, editors. 6th ed. Manual of Clinical Microbiology. Washington DC: American Society for Microbiology Press; 1995. pp. 509–519.

CHAPTER 11

Dual-Use Bioenergy-Livestock Feed Potential of Giant Miscanthus, Giant Reed, and Miscane

David M. Burner[1†], Amanda J. Ashworth[2*], Daniel H Pote[1], Jim R. Kiniry[3], David P. Belesky[4], James H. Houx III[5], Paul Carver[6], Felix B. Fritschi[5]

[1]*US Department of Agriculture, Agricultural Research Service, Dale Bumpers Small Farms Research Center, Booneville, AR, USA*

[2]*US Department of Agriculture, Agricultural Research Service, Poultry Production and Products Safety Research, Fayetteville, AR, USA*

[3]*Grassland Soil and Water Research Laboratory, USDA ARS, Temple, TX, USA*

[4]*West Virginia University, Plant & Soil Sciences, Morgantown, WV, USA*

[5]*Division of Plant Sciences, University of Missouri, Columbia, MO, USA*

[6]*New Energy Farms, Leamington, Canada*

ABSTRACT

High yielding perennial grasses could integrate bioenergy-livestock operations, thereby, offsetting diversions of cropland to lignocellulosic crops, but research is needed to determine chemical composition and digestibility of leaf and stem fractions that might affect downstream reside uses. The objective of this study was to compare feedstock quality of leaf and stem tissues of dedicated bioenergy feedstocks: giant miscanthus (Miscanthus × giganteus), giant reed (Arundo donax), and miscane (Saccharum hybrid × Miscanthus spp.) when grown with or without supplemental irrigation on an upland site. Three species were space-planted on a silt loam soil in March 2007 and harvested prior to the first freeze in plant-cane, first ratoon, and second-ratoon crops for three years. Giant miscanthus leaf tissue had greatest acid detergent lignin and cellulose, and lowest concentrations of nitrogen (N) and total nonstructural carbohydrates (TNC) in ratoon crops. Giant reed leaf tissue had greatest concentrations of *in vitro* digestible dry matter (IVDMD), TNC, and N ($P \leq 0.05$). Conversely, miscane stem tissue had greatest concentrations of IVDMD, TNC, hemicellulose, and low dry matter and combustible energy ($P \leq 0.05$). Results suggest all species' residue has positive feedstock attributes for thermochemical bioenergy conversion, and albeit giant miscanthus has very little potential value as fodder. Miscane stem and giant reed leaf tissue have potential value as livestock feed, although giant reed is not currently recommended for planting. Further research is needed on dietary composition, acceptability, voluntary intake, and live weight gain before any of these species are recommended as livestock feed sources.

Keywords: Arundo donax, Invasive Species, Miscanthus × giganteus, Agricultural Residue, Nutritive Value, Saccharum spp. Hybrid, Tissue Components, Thermochemical Conversion, Feedstock Traits

1. INTRODUCTION

Compared with their first-generation biofuel counterparts, second-generation biomass crops require fewer inputs, produce greater energy on a landmass basis, reduce greenhouse gas emissions, and do not directly compete for arable land for food production [1] [2]. As much as 21.3 million ha of existing agricultural land in the US has been projected for conversion to perennial grass feedstocks [3], therefore, second-generation feedstock crops may indirectly compete for hay and grazing landmass. Therefore, in order to transform the forage-livestock industry towards second-generation biofuels without affecting grassland resources for animal operations, identifying dual-use perennial forage feedstocks is key.

Small commercial farms, those with gross cash farm income from $10,000 to $249,999, account for 36% of all farms and 22% of total agricultural production in the US, but they are challenged to remain profitable [4]. A new market for grassland agriculture is emerging because the Energy Independence and Security Act will require 36 billion gallons of renewable fuels by 2022 [5]. To meet this goal, an estimated 1 billion $Mg \cdot yr^{-1}$ dry lignocellulosic biomass would be converted to renewable, liquid biofuels [6].

There are approximately 746,000 small farms in the US [7], and many of those could benefit from an integrated, dual-use, bioenergy-livestock production system. Considering, bioenergy-livestock feeds could produce sustainable yields of combustible energy (high thermal output) or be fed to livestock [2] [6] [8]. Producers that adopt this multifunctional practice could make fairly short-term management decisions to shift between commodities (bioenergy or livestock feed) based on markets, thereby minimizing risk and maximizing production efficiency [9]. For switchgrass (Panicum virgatum L.) and other perennial forage species, usage is primarily driven by cattle (Bos L. spp.) weight gains or hay sales, but bioenergy could provide a secondary market option [10]. It is less clear; however, if perennial grasses traditionally considered dedicated bioenergy feedstocks offer a secondary market as livestock feed.

Commercially, sugarcane (Saccharum L. spp.) leaves are mechanically separated from stems during harvest, a convenient technology, whereby leaf tissue could be diverted to bioenergy or livestock feed uses [11] as has been done with maize (Zea mays L.)-stover [12] [13]. The fibrous, lignocellulosic components of plant cell walls (cellulose and hemicellulose) can produce substantial amounts of 5 and 6 carbon sugars [6] [14]. For instance, [15] found that whole-plant (leaf and stem) conversion of giant miscanthus (Miscanthus × giganteus J.M. Greef & Deuter ex Hodkinson & Renvoize), giant reed (Arundo donax L.), and sugarcane yields ranges from 0.29 to 0.33 g glucose g^{-1} dry mass, or is about 0.1 g bioethanol g^{-1} dry mass.

Plant tissue can differ in various feedstock quality attributes including concentrations of primary metabolites, fiber, gross energy density, and digestibility; which can impact their downstream value and use. Some types of sugarcane, i.e. energy cane, differs from commercial sugarcane by having high stalk fiber concentrations [16] and some [Saccharum sp. × Miscanthus sp. (miscane)] reportedly have lower leaf ash and N concentrations than switchgrass and giant miscanthus [17], which are desirable feedstock attributes. Further, harvest timing influences tissue composition, as delayed harvests allow leaf nutrients to translocate to stems or rhizomes, or are sloughed off as leaf litter [18].

Limited information is available on the potential of giant miscanthus and giant reed to serve as livestock feed. Giant reed is invasive in many riparian areas [19] [20] [21], but could serve as an animal fodder or bioenergy resource during its eradication [22]. Fresh, whole-chopped sugarcane is similar to other roughage sources, like cottonseed (Gossypium hirsutum)-hulls, when fed to cattle [11]. Sugarcane has adequate in vitro dry matter digestibility (IVDMD) for livestock [23], and IVDMD of miscane can exceed that of giant miscanthus and switchgrass in Florida [17]. Similarly, adequate IVDMD concentration has been reported for giant reed leaves in Georgia

[24], while that of giant miscanthus and giant reed stems are quite low, even for C4 grasses [24] [25] [26].

Given the history of cultivation and use of perennial grasses [2], an existing knowledge base on cultural practices, availability of harvesting technology, and familiarity with livestock production, small farmers might transition to a dual- use system if given a bioenergy market option. Thus, research is needed on chemical composition of leaf and stem tissue of perennial grasses that might be used in dual-use, bioenergy-livestock feed practices, as concentrations of primary plant metabolites in leaf and stem tissues could affect their ultimate utilization. The objective of this study was therefore to compare feedstock quality of leaf and stem tissues of dedicated bioenergy feedstocks (giant miscanthus, giant reed, and miscane) when grown with or without supplemental irrigation on an upland site.

2. MATERIALS AND METHODS

2.1. Experimental Description and Management

Giant miscanthus, miscane, and giant reed were grown on an upland Leadvale silt loam soil (fine-silty, siliceous, thermic Typic Fragiudult) with a fragipan at 0.4 to 0.6 m depth near Booneville, Arkansas (35.08°N, 93.98°W). In the fall of 2006, giant miscanthus rhizomes (proprietary clone Q4264, Biomass Industrial Crops, Ltd., United Kingdom), axillary internode buds of miscane (sugarcane clone US84-1028; F1 hybrid of Saccharum hybrid × Miscanthus spp. bred and selected at US Department of Agriculture, Agricultural Research Service, Canal Point, Florida), and giant reed (originally obtained from a riparian area along the Little River near Temple, Texas) were transplanted in the greenhouse. Further information on the selected varieties can be found in [27]. On 29 March 2007, greenhouse-grown clones of the three species were planted as split plots in four replications under either rainfed or irrigated whole plots. Split-plots consisted of 5 clones of each species arranged in a single row and spaced 1 m apart. Rows of split-plots were spaced 2.5 m apart and were randomly arranged within main plots. Plants of the respective species were planted at the ends of each split-plot and two rows of sugarcane were planted between main plots.

Weed control was conducted annually in May with applications of atrazine [2-chloro-4-(ethylamino)-6-(isopropylamino)-s-triazine] at 1.12 kg ai ha^{-1}, and was supplemented by manual weeding as needed. During spring green-up, all plots were fertilized with 80 kg N ha^{-1} in the form of ammonium nitrate (NH_4NO_3) and 60 kg P ha^{-1} (P_2O_5), and 100 kg N ha^{-1} (NH_4NO_3) was applied in both 2008 and 2009. In irrigated whole plots, precipitation was supplemented during the growing season by additions of 719, 580, and 690 mm irrigations during 2007, 2008, and 2009, respectively. To provide irrigation, soaker hoses were installed for each irrigated row and water was applied three times per week except when a rainfall event exceeding 12 mm occurred.

2.2. Biomass Sampling and Tissue Analyses

A representative 2-stalk (giant reed and miscane) or 4-stalk (giant miscanthus) sample was taken 0.15 m above soil surface on plant-cane (establishment year) (PC, 29 October 2007), first-ratoon [or subsequent shoots sprouting from plant base (FR, 27 October 2008)], and second-ratoon (SR, 19 November 2009) crops prior to the first frost. Samples were separated (split-split plot) into leaf lamina (leaf) and stem plus leaf sheath tissue components (stem), and dried at 60°C. Plant residue tissue samples were ground to 1 mm particle size on a Wiley mill (Arthur Thomas Co., Philadelphia, Pennsylvania) and stored at −20°C until laboratory analysis. Tissue sample N analyses were conducted by combustion (Vario Macro CN, Elementar Americas, Inc., Mt. Laurel, New Jersey). The concentration of IVDMD was determined using the procedure of [28] modified for the Ankom Daisy II fiber analyzer #F200 (Ankom Technology, Macedon, New York). Source of fluid for IVDMD analysis was a ruminally-cannulated steer (Bos taurus L.) adapted for 10 d to a basal diet of cracked maize and common bermudagrass (Cynodon dactylon L.) hay (15:85 ratio of grain:hay, w:w), and allowed ad libitum access to fresh water.

Cellulose, hemicellulose, and acid detergent lignin (ADL) concentrations were determined by mass loss during sequential treatment of samples in acid detergent solution (72% v:v H_2SO_4), and combustion at 550°C following methods described by Ankom Technology for the Filter Bag Method. Specifically, bags were agitated every 30 mins by manually pushing a 2 L beaker up and down. After 3 hrs, samples were removed and rinsed with water to remove acid. Rinses were repeated until pH was neutral, then samples were rinsed with acetone for 2 - 3 mins and then air dried. Thereafter, drying was completed in a batch oven at 105°C for 2 hrs, then placed in a desiccator until cooled to ambient temperature and re-weighed. Acid detergent lignin is acid insoluble residue minus ash [28].

Cellulose was calculated as [1] and hemicellulose was calculated as [2].

$$[\text{Acid detergent fiber (ADF)} - \text{ADL}] = \text{cellulose.} \quad (1)$$

$$[\text{Neutral detergent fiber (NDF)} - \text{ADF}] = \text{hemicellulose.} \quad (2)$$

For analysis of total nonstructural carbohydrates (TNC), samples were further ground to 0.5 mm particle size on an Udy cyclone sample mill (Seedburo Equip- ment Co., Chicago, Illinois). Total nonstructural carbohydrate concentrations were determined using an automated hydrolysis method [29]. Dry samples were combusted at 550°C to measure mineral ash residue.

Gross calorific value (combustible energy, $MJ \cdot kg^{-1}$) was measured using a bomb calorimeter (IKA C2000, IKA Works; Wilmington, North Carolina). The procedure used 10 ignitions of C_6H_5COOH standard according to ASTM International [30]. Briefly, a 0.50-g sample was pelletized, weighed to 5 decimal places, and combusted in purified O_2 atmosphere. To account for HNO_3 formed during combustion, the inside of the vessel and the crucible and components were rinsed into a beaker with H_2O purified by a Labconco Water Pro PS (Labconco Corp, Kansas City, Missouri) 4-stage purification system (18.1 $M\Omega \cdot cm^{-1}$ resistivity). Two drops of methyl red titration indicator were added to the beaker and the solution was titrated with 0.035 M Na_2CO_3 solution.

2.3. Statistical Analyses

Analysis of variance tests of feedstock quality characteristics were performed using a mixed linear model procedure, Proc Mixed in SAS v.9.2 [31] [32], with irrigation treatment (whole-plot), species (split-plot), tissue component (split- split plot), year, and their interactions as fixed effects, and replication as a random effect. Species within replication and year was the repeated measure with a first-order autoregressive covariance structure and restricted maximum likelihood estimation method [31] [32]. When main effect differences were found, pair-wise post hoc comparisons were performed using Least Square Means using the Tukey Honest Significant Difference test (HSD) at a Type I error rate of 5%.

3. RESULTS AND DISCUSSION

Overall results revealed significant effects for year, plant residue component (leaf and stem), and species for all traits examined (Table 1). Irrigation main effects impacted ($P < 0.05$) TNC, ADL, cellulose, and energy density but not IVDMD, hemicellulose, ash, and N. Because a large number of two and three-way interaction effects were significant, results herein are focused on significant three-way interactions involving species as follows: year × tissue × species for IVDMD, ADL, and N; and tissue × species × irrigation for TNC, cellulose, and gross energy density.

Table 1. Probability values from analysis of variance of in vitro dry matter digestibility (IVDMD), total nonstructural carbohydrates (TNC), acid detergent lignin (ADL), hemicellulose (Hemi), cellulose (Cell), ash, nitrogen (N), and energy in response to year (Y), residue tissue component (T), species (S), and irrigation (I) fixed effects.

Effect	IVDMD	TNC	ADL	Hemi	Cell	Ash	N	Energy
Year (Y)	<0.001	<0.001	<0.001	0.050	<0.001	0.001	<0.001	0.012
Tissue (T)	<0.001	<0.001	<0.001	<0.001	<0.001	<0.001	<0.001	<0.001
Y × T	<0.001	<0.001	<0.001	0.025	<0.001	0.039	<0.001	<0.001
Species (S)	<0.001	<0.001	<0.001	<0.001	<0.001	0.001	<0.001	<0.001
Y × S	<0.001	<0.001	0.101	0.776	<0.001	0.838	<0.001	0.604
T × S	<0.001	<0.001	<0.001	<0.001	<0.001	0.002	<0.001	<0.001
Y × T × S	0.002	<0.001	<0.001	0.835	<0.001	0.810	<0.001	<0.001
Irrigation (I)	0.109	0.026	0.021	0.220	<0.001	0.205	0.071	0.024
Y × I	0.068	0.061	0.440	0.020	<0.001	0.427	0.475	0.304
T × I	<0.001	0.056	0.362	0.047	0.834	0.258	0.100	0.056
Y × T × I	<0.001	0.017	0.004	0.049	<0.001	0.050	0.053	<0.001
S × I	0.532	0.966	0.646	0.726	0.691	0.818	0.042	<0.001
Y × S × I	0.774	0.215	0.645	0.973	0.254	0.798	0.553	0.512
T × S × I	0.087	0.005	0.092	0.974	0.047	0.304	0.791	<0.001
Y × T × S × I	0.527	0.042	0.941	0.928	0.292	0.704	0.309	0.033

Since these three-way interactions were not significant for hemicellulose and ash, tissue × species interactions are highlighted herein for these two traits.

3.1. Digestibility

Across the three years, IVDMD for the three species ranged from 330 to 672 $g \cdot kg^{-1}$ for leaf tissue and 193 to 547 $g \cdot kg^{-1}$ for stem residue (**Table 2**). In general, average leaf IVDMD (539 $g \cdot kg^{-1}$) was greater than stem IVDMD concentrations (386 $g \cdot kg^{-1}$). Specifically, giant reed had greater leaf IVDMD (>642 $g \cdot kg^{-1}$) than giant miscanthus and miscane, and miscane had greatest stem IVDMD concentrations (>428 $g \cdot kg^{-1}$) among all species (**Table 2**). Giant reed leaf and stem IVDMD concentrations were stable across years, whereas giant miscanthus stem IVDMD concentration decreased with every season. Giant miscanthus also had lower ratoon leaf (<495 g·kg^{-1}) and ratoon stem IVDMD concentrations (<303 g·kg^{-1}) than other species.

Table 2. Concentrations of in vitro dry matter digestibility (IVDMD), acid detergent lignin (ADL), and nitrogen (N) as affected by the crop × tissue × species interaction.

Tissue	Species	Crop	IVDMD	ADL	N
				(g·kg^{-1})	
Leaf	Giant miscanthus	PC[a]	478.8 c[b]	91.5 c	17.5 c
	Giant reed	PC	672.2 a	107.9 b	27.6 a
	Miscane	PC	503.1 c	67.8 e	15.1 de
	Giant miscanthus	FR	494.7 c	111.5 b	11.3 f
	Giant reed	FR	667.4 a	110.9 b	22.8 b
	Miscane	FR	587.0 b	70.9 de	13.6 e
	Giant miscanthus	SR	329.9 d	125.1 a	4.4 g
	Giant reed	SR	642.4 a	108.1 b	16.1 cd
	Miscane	SR	469.7 c	80.2 d	11.1 f
Stem	Giant miscanthus	PC	396.4 bc	98.7 a	6.0 b
	Giant reed	PC	361.0 cd	72.2 cd	8.6 a
	Miscane	PC	547.4 a	56.9 e	7.2 b
	Giant miscanthus	FR	303.2 e	86.1 abc	4.0 c
	Giant reed	FR	358.8 cd	77.7 bcd	7.1 b
	Miscane	FR	545.2 a	63.4 de	6.1 b
	Giant miscanthus	SR	192.5 f	90.7 ab	1.5 d
	Giant reed	SR	330.3 de	79.6 bcd	4.5 c
	Miscane	SR	427.6 b	69.0 de	4.5 c

[a]PC = plant-cane; FR = first-ratoon; SR = second-ratoon; [b]Variable means within an irrigation treatment

In vitro dry matter digestibility is a function of soluble cellular components including TNC and crude protein (CP) as well as variable amounts of cell wall constituents (CP, cellulose, and hemicellulose) that are readily degraded by rumen microorganisms [33]. Thus, high stem IVDMD of miscane may have been due to greater soluble sugars and a less lignified parenchyma than stems of other species. Greater concentrations of stem

IVDMD in miscane than giant miscanthus and switchgrass is consistent with previous reports [17]. While primarily a livestock feed attribute, IVDMD concentration has been indirectly linked to methane production [34] and positively associated with bioethanol production [24]. Nonetheless, despite its high stem IVDMD, miscane productivity is low at this latitude because of poor ratooning [27].

Giant miscanthus leaf IVDMD concentration was comparable to that of single-cut per year, whole-plant switch grass in West Virginia [35]. Giant reed leaf and stem IVDMD compared favorably to the 541 $g \cdot kg^{-1}$ and <290 $g \cdot kg^{-1}$, respectively, reported for giant reed by [24]. Similarly, miscane leaf (520 $g \cdot kg^{-1}$) and stem (507 $g \cdot kg^{-1}$) IVDMD observed in this study are consistent with an absence of tissue differences reported by [34] in Florida. However, the observed IVDMD of giant miscanthus leaves and stems, and giant reed stems, would be among the lowest in a wide range of reported values (280 to 740 $g \cdot kg^{-1}$) for C4 grasses [25] [26].

Like many traditional livestock feeds, green-harvested tissues could be fed fresh-chopped or ensiled [11] [36]. Steer weight gain was adequate when fresh- chopped sugarcane was fed at 30% - 40% of the total diet, but performance decreased when the dietary proportion was >60% [36]. This suggests that sugarcane is similar to other roughage sources (e.g. cottonseed hulls) when fed in steer fattening diets, but it should not be a major component of the diet. Ensiling chopped sugarcane decreases total digestible nutrient concentration, due to conversion of sugar to alcohols, therefore, silage is better for maintenance rather than for increasing stocker cattle weights [11]. Incubation of sugarcane bagasse with white-rot fungi (i.e. Lentinula edodes, Ceriporiopsis subvermispora) that selectively degrade lignin may substantially increase IVDMD [37], although the effect of such biological degradation on livestock intake was not assessed and the practicality of such technology to improve livestock feed quality is unknown.

3.2. Acid Detergent Lignin

Giant miscanthus and giant reed had greater leaf ADL concentrations (>91.5 $g \cdot kg^{-1}$) than miscane (<80.2 $g \cdot kg^{-1}$) and any other feedstock source in this study (Table 2). Similarly, giant miscanthus stem ADL (>86.1 $g \cdot kg^{-1}$) was consistently greater than that of miscane (<69.0 $g \cdot kg^{-1}$). In contrast, giant reed stem ADL concentration was only greater ($P < 0.05$) than that of miscane in PC but not in ratoon crops. Averaged across treatments, leaf tissue (97.3 $g \cdot kg^{-1}$) had significantly greater ADL than stem tissue (77.0 $g \cdot kg^{-1}$). Leaf and stem ADL did not change ($P \geq 0.05$) over the course of three seasons, except for giant miscanthus leaves, which became significantly more lignified. Concentrations of ADL were within the range

(20 to 115 g·kg^{-1}) previously reported for various C4 grasses [26]. Reference [10] reported giant reed leaves (<41 g·kg^{-1}) had less ADL than stems (90 g·kg^{-1}), whereas ADL observed in this study averaged 109 g·kg^{-1} for leaves and 76.5 g·kg^{-1} for stems across all seasons.

Lignin is the major non-polysaccharide component of plant cell walls, reduces digestibility of polysaccharides through cross-linkages, and is virtually indigestible by rumen microorganisms [18] [33]. Lignin is high in combustible energy, similar to that of cellulose [38], and the values we found should not cause excessive tar production during gasification [39].

3.3. Nitrogen Concentration

Giant reed had greater ($P < 0.05$) leaf N concentrations (16.1 to 27.6 g·kg^{-1}) than the other two species (4.4 to 17.5 g·kg^{-1}) (Table 2). Second ratoon giant miscanthus had the lowest leaf N (4.4 g·kg^{-1}) and stem N (1.5 g·kg^{-1}) concentrations among species and crops. Leaf and stem N concentrations were lower in ratoon than PC for giant miscanthus and giant reed, but not miscane. Second ratoon leaf and stem N concentrations were also lower than PC for each species, suggesting a decrease in N uptake in ratoon crops and greater N-use efficiency with crop age. This was particularly evident for giant miscanthus considering its relatively high yields [27]. Decreases in N concentration in giant miscanthus from PC to ratoon crops were also previously observed [40].

Across diverse species, bioenergy feedstocks can be highly variable in N concentrations, ranging from 0.5 [Sequoia sempervirens (Lamb. ex D. Don) Endl. wood] to 30.4 g·kg^{-1} (Triticum aestivum L. dust) [41]. Tissue-N is usually considered an anti-quality attribute in combustible feedstocks, as concentrations of 10 g·kg^{-1} or greater can necessitate NOx removal at thermochemical conversion facilities [41]. On that basis alone, leaf residue would be less preferred as a combustible feedstock than stem tissue, except for SR giant miscanthus leaves. In practice, however, while coal contains more N (159 g·kg^{-1}) than switchgrass (8.7 g·kg^{-1}), blending coal with a second-generation feedstock is an approach to reduce total NOx emissions from utility boilers compared to coal-firing alone [8]. A similar approach might be taken with high-N bioenergy feedstocks for combustion.

Conversely, plant proteins are a beneficial attribute in livestock feed, and are often expressed as CP [42]. Data presented herein are consistent with previous findings [34] in that leaves of mature plants have substantially greater CP concentration than stems. Except for SR giant miscanthus (28 g·kg^{-1} CP), leaf CP concentrations (calculated based on N) ranged from 69 to 172 g·kg^{-1}, and were generally adequate for maintaining beef cattle weights of 450 to 635 kg mature weight [43]. Conversely, stem tissue CP levels were low across species and inadequate (<54 g·kg^{-1}) for cattle maintenance [43].

3.4. Total Nonstructural Carbohydrates

A tissue × species × irrigation interaction was observed for TNC ($P < 0.05$; Table 1). Within an irrigation treatment, giant reed had greater leaf TNC concentrations (I+: 98.0 $g{\cdot}kg^{-1}$; I−: 112.8 $g{\cdot}kg^{-1}$) than the other two species (<73.2 $g{\cdot}kg^{-1}$) ($P < 0.05$; Table 3). All three species differed significantly in stem TNC [sugarcane (>247.2 $g{\cdot}kg^{-1}$) > giant reed (>170.9 $g{\cdot}kg^{-1}$) > giant miscanthus (<80.6 $g{\cdot}kg^{-1}$)]. For all species, stem residues had greater ($P < 0.05$) TNC concentrations than leaf tissues. Composed primarily of sugars and starches, TNC are important energy reserves for plants and grazing animals [33], and it was not surprising that miscane had greatest levels of stem TNC among species [17]. While miscane leaf TNC concentrations were slightly lower than those reported (79 to 119 $g{\cdot}kg^{-1}$) for sugarcane in Florida [34], giant reed leaf TNC was within that range. Despite its comparatively high TNC concentrations, mature giant reed stems would likely have low acceptability and voluntary intake by livestock due to their fibrous and woody texture.

Table 3. Total nonstructural carbohydrate (TNC), cellulose, and energy concentrations of irrigated (I+) and non-irrigated (I−) giant miscanthus, giant reed, and miscane leaf and stem tissues as affected by the tissue component by species by irrigation interaction.

Irrigation	Species	Tissue	TNC	Cellulose	Energy
			($g{\cdot}kg^{-1}$)		($MJ{\cdot}kg^{-1}$)
I+	Giant miscanthus	Leaf	41.7 e[a]	431.3 a	16.8 d
	Giant reed		98.0 c	389.3 b	17.1 c
	Miscane		73.2 d	347.1 c	17. 0 c
	Giant miscanthus	Stem	70.1 d	305.6 d	17.9 a
	Giant reed		170.9 b	255.2 e	17.9 a
	Miscane		247.2 a	296.9 d	17.4 b
I−	Giant miscanthus	Leaf	46.5 e	417.1 a	17.1 bc
	Giant reed		112.8 c	381.5 b	16.7 d
	Miscane		64.9 de	329.5 c	17.0 cd
	Giant miscanthus	Stem	80.6 d	290.3 d	17.8 a
	Giant reed		176.2 b	236.7 e	17.7 a
	Miscane		269.5 a	293.4 d	17.3 b

[a]Variable means within an irrigation treatment followed by a common letter do not differ significantly at $\alpha = 0.05$ using Tukey's HSD.

3.5. Fiber Constituents

A significant tissue × species × irrigation interaction was also observed for cellulose concentrations (**Table 1**). However, within residue tissue type, species ranked similarly for cellulose regardless of irrigation treatment, with cellulose concentrations being greater in leaf tissue than stems for all species (**Table 3**). Leaf cellulose concentrations were greatest in giant miscanthus, intermediate in giant reed, and lowest in miscane (**Table 3**). In contrast, stem cellulose concentrations did not differ ($P < 0.05$) between giant miscanthus and miscane, albeit both of which had greater cellulose concentrations than stem residue of giant reed. Cellulose concentrations of both leaf and stem tissue were consistently greater in irrigated than rainfed treatments, but differences were only significant for miscane leaf tissue (347.1 vs 329.5 $g \cdot kg^{-1}$) and giant reed stem cellulose (255.2 vs 236.7 $g \cdot kg^{-1}$).

Structural carbohydrates (cellulose and hemicellulose) are largely composite mixtures of 5 and 6 carbon sugars, which comprise the backbone of cell walls and provide the primary source of energy in ruminant diets [11]. Neither tissue × species × irrigation nor year × tissue × species interactions were significant for hemicellulose, although a strong tissue × species interaction was found (Table 1). The interaction was due to greater hemicellulose concentrations in leaves but lower concentrations in stems of giant miscanthus than miscane (Table 4). Hemicellulose concentrations were not different between leaves and stems of giant reed. Similarly, previous research [10] indicated that giant reed leaf residue (<318 $g \cdot kg^{-1}$) has a greater mean hemicellulose concentration than stem tissue (<260 $g \cdot kg^{-1}$). A similar trend, i.e. greater cellulose concentrations in leaves than stems, was also previously reported for sugarcane [44], but not for giant reed [24].

3.6. Ash Concentration

For all three species, leaf tissue ash concentrations were consistently greater (>92.2 $g \cdot kg-1$) than stem ash concentration (<30.6 $g \cdot kg^{-1}$; Table 4). Giant reed had greater leaf ash (107.8 $g \cdot kg^{-1}$) than giant miscanthus (95.3 $g \cdot kg^{-1}$) and miscane (92.2 $g \cdot kg^{-1}$), but species did not differ in stem ash concentration ($P < 0.05$). Of 62 potential bioenergy feedstocks evaluated by [41], ash concentrations ranged from 1.8 $g \cdot kg^{-1}$ in redwood [Sequoia sempervirens (Lamb. ex D. Don.) Endl.] mill waste to 240 g kg^{-1} for weathered rice (Oryza sativa L.) straw. Those of giant reed reportedly range from 48 to 74 $g \cdot kg^{-1}$ [45]. The greater ash concentrations found in leaves compared to stems of giant miscanthus is consistent with findings reported by [46]. By comparison, coal has 103 $g \cdot kg^{-1}$ ash [47]. Low ash concentrations are a preferred attribute of bioenergy crops used for combustion due to inorganic elements causing slagging, corrosion, and fouling in power plants [48]. In a solid fuel combustor, internal ash depositions can occur at a faster rate for bioenergy crops than coal, as ash derived from bioenergy feedstocks may be more dense, less porous, and harder to remove than that of coal [49].

Table 4. Hemicellulose and ash concentrations of giant miscanthus, giant reed, and miscane leaf and stem tissues as affected by the species × tissue interaction.

Tissue	Species	Hemicellulose	Ash
		$(g \cdot kg^{-1})$	
Leaf	Giant miscanthus	246.6 bc[a]	95.3 b
	Giant reed	214.0 cd	107.8 a
	Miscane	211.1 d	92.2 b
Stem	Giant miscanthus	279.7 ab	26.9 c
	Giant reed	233.7 cd	28.3 c
	Miscane	300.9 a	30.6 c

[a]Variable means followed by a common letter do not differ significantly at $\alpha = 0.05$ using Tukey's HSD.

Based on ash concentrations, stem tissue would be preferred to leaf tissue for combustion purposes. Little is known regarding the importance of ash concentration in livestock feed, and availability of ash and inorganic minerals to ruminants fed forage diets [50]. Plant silicates constitute 70% of switchgrass ash [51], and silicates have been implicated in reducing tissue digestibility [26]. Therefore, even within a given feedstock, it is possible to have dual-use systems both for fodder and bioenergy (e.g. leaf tissue for animal feed and stem tissue for thermochemical conversion) and within bioenergy systems (e.g. leaf tissue converted biologically and stem tissue utilized in a thermochemical platform).

Although giant miscanthus and giant reed for bioenergy are often harvested after post-freeze senescence and natural dry-down in the field, we intentionally conducted pre-freeze harvests to examine the feed/fuel dual-use potential [52]. Green harvests probably benefited livestock feed value with respect to soluble secondary cell wall constituents, such as IVDMD, N, and TNC concentrations, compared to senesced tissues [25]. However, green harvests over time may reduce stand longevity as nutrients and carbohydrates are translocated to storage tissues during maturation for the following year's regeneration [53].

3.7. Combustible Energy Density

A significant tissue × species × irrigation interaction was observed for combustible energy (**Table 1**). Specifically, energy concentration in leaves (but not stems) of the three species was affected by irrigation (**Table 3**). Giant miscanthus leaves from rainfed plots had greater energy concentrations than those from irrigated treatments, however, the opposite was true for giant reed leaf tissue; whereas, energy concentration of miscane leaves did not differ between irrigation treatments ($P < 0.05$). For all three species, energy concentrations were greater in stem than leaf residues. Stem t energy concentrations were lower in miscane than for the other two species, which did not different from each other.

In general, combustible energy of bioenergy grasses range from 14.8 to 19.3 $MJ{\cdot}kg^{-1}$ [17] [46] [47]. Seemingly small differences in energy concentration can have a multiplying impact on energy yield per unit land area. In Europe, giant miscanthus had high yields of combustible energy with low N inputs [52]. Management practice (fertilization, plant density, or date of dormant-season harvest) also had no effect on gross energy density ($\mu = 17$ $MJ{\cdot}kg^{-1}$) of giant reed [54]. Given their high biomass yields, giant miscanthus and giant reed could serve as important sources of combustible energy [27].

4. CONCLUSION

Leaves and stems of bioenergy grasses may be an underappreciated crop residue and their use could reduce the competition of cropland for bioenergy production. Invasive giant reed is commonly found in relatively remote riparian areas, and is not a recommended species for agricultural use. However, giant reed leaf tissue had relatively high concentrations of IVDMD, TNC, and N, and stems had high TNC, suggesting potential value as a livestock feed. Giant miscanthus leaf tissue had high concentrations of ADL, cellulose, and low concentrations of N and TNC in ratoon crops, indicating greater optimal traits for combustible energy than livestock feed. Miscane stem tissue had high concentrations of IVDMD, TNC, and hemicellulose, and low combustible energy concentrations, signifying contrasting desirable attributes for forage and bioenergy feedstocks. Irrigation interacted with residue tissue (stems and leaves) per species to affect TNC, cellulose, and gross energy density, meaning intra-annual rainfall fluctuations may impact dual-use potential. Therefore, further research is needed temporally and spatially on dietary composition, acceptability, voluntary intake, and live weight gain before any of these feedstocks could be recommended for livestock feeding.

ACKNOWLEDGEMENTS

The authors gratefully acknowledge J. Brent Woolley (USDA-ARS) and Karen R. Chapman (USDA-ARS) for their technical assistance in this research. Use of the proprietary clone of giant miscanthus was made possible through a non-funded cooperative agreement (No. 58-6227-0M-F155) between Biomass Industrial Crops, Ltd. (Taunton, Somerset, UK) and USDA ARS. Peter Y.P. Tai (USDA ARS, Canal Point, Florida, deceased) provided the miscane (sugarcane clone of US84-1028). Principle funding for this research was provided by USDA ARS, Dale Bumpers Small Farms Research Center, Booneville, Arkansas through Project 6227-21310-009-00D.Mention of trade names or commercial products in this article is solely for the purpose of providing specific information and does not imply recommendation or endorsement by the USDA.

REFERENCES

1. Ashworth, A.J., Taylor, A.M., Reed, D., Tyler, D.D., Allen, F.L. and Keyser, P.D. (2015) Life Cycle Assessment of Regional Switchgrass Feedstock Production Compared to Nitrogen Input Scenarios and Legume-Intercropping Systems. Journal of Cleaner Production, 87, 227-234. https://doi.org/10.1016/j.jclepro.2014.10.002
2. Sanderson, M.A. and Adler, P.R. (2008) Perennial Forages as Second Generation Bioenergy Crops. International Journal of Molecular Sciences, .9, 768-788.
3. https://doi.org/10.3390/ijms9050768
4. McLaughlin S.B., De la Torre Ugarte, D.G., Garten, Jr., C.T., Lynd, L.R., Sanderson, M.A., Tolbert, V.R. and Wolf, D.D. (2002) High-Value Renewable Energy from Prairie Grasses. Environmental Science & Technology, 36, 2122-2129.
5. https://doi.org/10.1021/es010963d
6. Hoppe, R.A., MacDonald, J.M. and Korb. P. (2010) Small Farms in the United States: Persistence under Pressure. Economic Information Bulletin No. (EIB-63), 39 p.
7. US Department of Energy (US DOE) (2007) Energy Independence and Security Act of 2007. Renewable Fuel Standard. http://www.epa.gov/otaq/fuels/ renewablefuels/ index.htm
8. US Department of Energy (US DOE) (2006) Breaking the Biological Barriers to Cellulosic Ethanol: A Joint Research Agenda. In: Houghton, J., Weatherwax, S. and Ferrell, J., Eds., A Research Roadmap Resulting from the Biomass to Biofuels Workshop, Rockville, 7-9 December 2005. http://genomicscience.energy.gov/biofuels/b2bworkshop.shtml
9. National Agricultural Statistics Service (NASS) (2012) Farms, Land in Farms, and Livestock Operations, 2011 Summary. Agricultural Statistics Board, US Department of Agriculture, Washington, DC. https://www.agcensus.usda.gov/Publications/ 2012/Full_Report/Volume_1,_Chapter_1_US/usv1.pdf

10. Sami, M., Annamalai, K. and Wooldridge, M. (2001) Co-Firing of Coal and Biomass Fuel Blends. Progress in Energy and Combustion Science, 27, 171-214. https://doi.org/10.1016/S0360-1285(00)00020-4
11. Boody, G., Vondracek, B., Andow, D.A., Krinke, M., Westra, J., Zimmerman, J. and Welle, P. (2005) Multifunctional Agriculture in the United States. Bioscience, 55, 27-38. https://doi.org/10.1641/0006-3568(2005)055[0027:MAITUS]2.0.CO;2
12. Anderson, W., Casler, M. and Baldwin, B. (2008) Improvement of Perennial Forage Species as Feedstock for Bioenergy. In: Vermerris, W., Ed., Genetic Improvement of Bioenergy Crops, 347-376. https://doi.org/10.1007/978-0-387-70805-8_12
13. Pate, F.M., Alvarez, J.J., Phillips, D. and Eiland, B.R. (1984) Sugarcane as a Cattle Feed: Production and Utilization. Institute of Food and Agriculture Sciences, University of Florida Extension, Gainesville, BUL844.
14. http://ufdc.ufl.edu/UF00027209/00001
15. Gould, K. (2007) Corn Stover Harvesting. Cattle Call, 12, 3-4. http://beef.msu.edu/LinkClick.aspx?fileticket=q0jUXM2MIqM%3d&tabid=537
16. Graham, R.L., Nelson, R., Sheehan, J., Perlack, R.D. and Wright, L.L. (2007) Current and Potential U.S. Corn Stover Supplies. Agronomy Journal, 99, 1-11. https://doi.org/10.2134/agronj2005.0222
17. Sivakumar, G., Vail, D.R., Xu, J., Burner, D.M., Lay, Jr. J.O., Ge, X. and Weathers P.J. (2010) Bioethanol and Biodiesel: Alternative Liquid Fuels for Future Generations. Engineering in Life Sciences, 10, 8-18. https://doi.org/10.1002/elsc.200900061
18. Ge, X., Burner, D.M., Xu, J., Phillips, G.C. and Sivakumar, G. (2011) Bioethanol Production from Dedicated Energy Crops and Residues in Arkansas, USA. Biotechnology Journal, 6, 66-73. https://doi.org/10.1002/biot.201000240
19. Richard, E.J., Tew, T., Cobill, R. and Hale, A. (2008) Sugar/Energy Canes as Feed Stocks for the Biofuels Industry. In: Zalesny, R.S.J., Mitchell, R. and Richardson, J. Eds., Biofuels, Bioenergy, and Bioproducts from Sustainable Agricultural and Forest Crops: Proceedings of the Short Rotation Crops International Conference, GTR-NRS-P-31, USDA Forest Service, Northern Research Station, Delaware County, 76 p.
20. Burner, D.M., Tew, T.L., Harvey, J.J. and Belesky, D.P. (2009) Dry Matter Partitioning and Quality of Miscanthus, Panicum, and Saccharum Genotypes in Arkansas, USA. Biomass and Bioenergy, 33, 610-619. https://doi.org/10.1016/j.biombioe.2008.10.002
21. Karp, A. and Shield, I. (2008) Bioenergy from Plants and the Sustainable Yield Challenge. New Phytologist, 179, 15-32. https://doi.org/10.1111/j.1469-8137.2008.02432.x
22. Bell, G. (1997) Ecology and Management of Arundo donax, and Approaches to Riparian Habitat Restoration in Southern California. In: Brock, J.H., Wade, M., Pysek, P. and Green, D., Eds., Plant Invasions: Studies from North America and Europe, Backhuys Publishers, Leiden, 103-113
23. Moore, G.W., Watts, D.A. and Goolsby, J.A. (2010) Ecophysiological Responses of Giant Reed (Arundo donax) to Herbivory. Invasive Plant Science and Management, 3, 521-529. https://doi.org/10.1614/IPSM-D-10-00007.1
24. Spencer, D.F., Liow, P.-S., Chan, W.K., Ksander, G.G. and Getsinger, K.D. (2006) Estimating Arundo donax Shoot Biomass. Aquatic Botany, 84, 272-276. https://doi.org/10.1016/j.aquabot.2005.11.004
25. Young, S.L., Gopalakrishnan, G. and Keshwani, D.R. (2011) Invasive Plant Species as Potential Bioenergy Producers and Carbon Contributors. Journal of Soil and Water Conservation, 66, 45A-50A. https://doi.org/10.2489/jswc.66.2.45A
26. Mislevy, P., Martin, F.G., Adjei, M.B. and Miller, J.D. (1995) Agronomic Characteristics of US 72-1153 Energycane for Biomass. Biomass and Bioenergy, 9, 449-457. https://doi.org/10.1016/0961-9534(95)00050-X
27. Anderson, W.F., Dien, B.S. Brandon, S.K. and Peterson, J.D. (2008) Assessment of Bermudagrass and Bunch Grasses as Feedstock for Conversion to Ethanol. Applied Biochemistry and Biotechnology, 145, 13-21. https://doi.org/10.1007/s12010-007-8041-y
28. Coleman, S.W., Moore, J.E. and Wilson, J.R. (2004) Quality and Utilization. In: Moser, L.E., Burson, B.L. and Sollenberger, L.E., Eds., Warm-Season (C4) Grasses, American Society of

Agronomy Crop Science Society of America Soil Science Society of America, Madison, 267-308.

29. Moore, J.E. and Mott, G.O. (1973) Structural Inhibitors of Quality in Tropical Grasses. In: Matches, A.G., Ed., Anti-Quality Components of Forages, CSSA Special Publication No. 4, Crop Science Society of America, Madison, 53-98.
30. Burner, D.M., Hale, A.L., Carver, P., Pote, D.H. and Fritschi, F.B. (2015) Biomass Yield Comparisons of Giant Miscanthus, Giant Reed, and Sugarcane Grown under Irrigated and Rainfed Conditions. Industrial Crops and Products, 76, 1025-1032. https://doi.org/10.1016/j.indcrop.2015.07.071
31. Van Soest, P.J. (1994) Nutritional Ecology of the Ruminant. 2nd Edition, Cornell University Press, Ithaca.
32. Smith, D. (1981) Removing and Analyzing Total Nonstructural Carbohydrates from Plant Tissue. Bull. 2107, Wisconsin Agricultural Experiment Station, Madison.
33. ASTM International (2010) Standard Test Method for Analysis of Coal and Coke. In: ASTM Volume 5.06: Gaseous Fuels; Coal and Coke, ASTM International Standard D5865-10a, ASTM International, West Conshohocken.
34. Littell, R.C., Milliken, G.A., Stroup, W.W. and Wolfinger, R.D. (1996) SAS System for Mixed Models. AS Institute Inc., Cary.
35. SAS Institute (2009) SAS/STAT User's Guide. Release 9.2. Windows Version 5.1.2600. SAS Institute, Cary.
36. Moore, K.J. and Hatfield, R.D. (1994) Carbohydrates and Forage Quality. In: Fahey, G.C.J., Moser, L.E., Martens, D.R. and Collins, M., Eds., Forage Quality, Evaluation, and Utilization, American Society of Agronomy Crop Science Society of America Soil Science Society of America, Madison, 229-280.
37. Mislevy, P., Adjei, M.B., Martin, F.G. and Miller, J.D. (1993) Response of US 72-1153 Energycane to Harvest Management. Proceedings—Soil & Crop Science Society of Florida, 52, 27-32.
38. Balasko, J.A., Burner, D.M. and Thayne, W.V. (1984) Yield and Quality of Switchgrass Grown without Soil Amendments. Agronomy Journal, 76, 204-208. https://doi.org/10.2134/agronj1984.00021962007600020009x
39. Velik, M., Baumung, R. and Knaus, W.F. (2008) Maize Silage as an Energy Supplement in Organic Dairy Cow Rations. Renewable Agriculture and Food Systems, 23, 155-160. https://doi.org/10.1017/S1742170507002104
40. Okano, K., Iida, Y., Samsuri, M., Prasetya, B., Usagawa T. and Watanabe T. (2006) Comparison of in Vitro Digestibility and Chemical Composition among Sugarcane Bagasses Treated by Four White-Rot Fungi. Animal Science Journal, 77, 308-313. https://doi.org/10.1111/j.1740-0929.2006.00353.x
41. Gani, A. and Naruse, I. (2007) Effect of Cellulose and Lignin Content on Pyrolysis and Combustion Characteristics for Several Types of Biomass. Renewable Energy, 32, 649-661. https://doi.org/10.1016/j.renene.2006.02.017
42. Panopoulos, K.D., Fryda, L.E. and Kakaras. E. (2007) Atmospheric Fluidized Bed Gasification of Promising Biomass Fuels in Southern European Regions. Thermal Science, 11, 5-15. https://doi.org/10.2298/TSCI0701005P
43. Clifton-Brown, J.C. and Lewandowski, I. (2002) Screening Miscanthus Genotypes in Field Trials to Optimise Biomass Yield and Quality in Southern Germany. European Journal of Agronomy, 16, 97-110. https://doi.org/10.1016/S1161-0301(01)00120-4
44. Ebeling, J.M. and Jenkins, B.M. (1985) Physical and Chemical Properties of Biomass Fuels. Transactions of the ASAE, 28, 898-902. https://doi.org/10.13031/2013.32359
45. Nelson, C.J. and Moser, L.E. (1994) Plant Factors Affecting Forage Quality. In: Fahey, G.C.J., Moser, L.E., Martens, D.R. and Collins, M., Eds., Forage Quality, Evaluation, and Utilization, American Society of Agronomy Crop Science Society of America Soil Science Society of America, Madison, 115-154.
46. National Research Council (NRC) (1996) Nutrient Requirements for Beef Cattle. 7th Edition, The National Academies Press, Washington DC.
47. Kevelenge, J.E.E., Said, A.N. and Kiflewahid, B. (1983) The Nutritional Value of Four Arable Farm By-Products Commonly Fed to Dairy Cattle by Small-Scale Farmers in Kenya. I.

Organic Structural Components and in Vitro Digestibility. Tropical Animal Production, 8, 162-170.

48. Lewandowski, I. and Kicherer, A. (1997) Combustion Quality of Biomass: Practical Relevance and Experiments to Modify the Biomass Quality of Miscanthus x giganteus. European Journal of Agronomy, 6, 163-177. https://doi.org/10.1016/s1161-0301(96)02044-8
49. Lewandowski, I., Scurlock, J.M.O., Lindvall, E. and Chistou, M. (2003) The Development and Current Status of Perennial Rhizomatous Grasses as Energy Crops in the US and Europe. Biomass and Bioenergy, 25, 335-361. https://doi.org/10.1016/S0961-9534(03)00030-8
50. Aerts, D.J., Bryden, K.M., Hoerning, J.M. and Ragland, K.W. (1997) Co-Firing Switchgrass in a 50 MW Pulverized Coal Boiler. Proceedings of 59th Annual American Power Conference, Vol. 59, Chicago, 1-3 April 1997, 1180-1185.
51. Monti, A., Di Virgilio, N. and Venturi. G. (2008) Mineral Composition and Ash Content of Six Major Energy Crops. Biomass and Bioenergy, 32, 216-223. https://doi.org/10.1016/j.biombioe.2007.09.012
52. Baxter, L.L. (1993) Ash Deposition during Biomass and Coal Combustion: A Mechanistic Approach. Biomass and Bioenergy, 4, 85-102. https://doi.org/10.1016/0961-9534(93)90031-X
53. Spears, J.W. (1994) Minerals in Forages. In: Fahey, G.C.J., Moser, L.E., Martens, D.R. and Collins, M., Eds., Forage Quality, Evaluation, and Utilization, American Society of Agronomy Crop Science Society of America Soil Science Society of America, Madison, 281-317.
54. Klass, D.L. (1998) Biomass for Renewable Energy, Fuels, and Chemicals. Academic Press, San Diego, 1-651. https://doi.org/10.1016/B978-012410950-6/50003-9
55. Heaton, E., Voigt, T. and Long, S.P. (2004) A Quantitative Review Comparing the Yields of Two Candidate C4 Perennial Biomass Crops in Relation to Nitrogen, Temperature and Water. Biomass and Bioenergy, 27, 21-30. https://doi.org/10.1016/ j.biombioe. 2003.10.005
56. Wilson, D.M., Heaton, E.A., Liebman, M. and Moore, K.J. (2013) Intraseasonal Changes in Switchgrass Nitrogen Distribution Compared with Corn. Agronomy Journal, 105, 285-294. https://doi.org/10.2134/agronj2012.0233
57. Angelini, L.G., Ceccarini, L. and Bonari, E. (2005) Biomass Yield and Energy Balance of Giant Reed (Arundo donax L.) Cropped in Central Italy as Related to Different Management Practices. European Journal of Agronomy, 22, 375-389. https://doi.org/10.1016/j.eja.2004.05.004

CHAPTER 12

The effect of feeding high fat diet to beef cattle on manure composition and gaseous emission from a feedlot pen surface

Dhan Prasad Gautam[1], Shafiqur Rahman[1]
Md Saidul Borhan[1] and Chanda Engel[2]

1 Department of Agricultural and Biosystems Engineering,
North Dakota State University, Fargo, ND 58102, USA

ABSTRACT

Background

Dietary manipulation is a common practice to mitigate gaseous emission from livestock production facilities, and the variation of fat level in the diet has shown great influence on ruminal volatile fatty acids (VFA) and enteric methane generation. The changes in dietary fat levels influence rumen chemistry that could modify manure nutrient composition along with odor and gaseous emissions from manure management facilities.

Methods

A field experiment was carried out on beef cattle feedlots to investigate the effect of four levels of dietary fat concentrations (3 to 5.5 %) on the manure composition and gaseous emissions (methane-CH_4, nitrous oxide-N_2O, carbon dioxide-CO_2 and hydrogen sulfide-H_2S) from the feedlot pen surface. The experiment was carried out over a 5-month period from June to October during North Dakota's summer-fall climatic condition. Air and manure sampling was conducted five times at a 20–30 day intervals.

Results

Overall, this research indicated that fat levels in diet have no or little effect on the nutrient composition of manure and gaseous emission from the pens with cattle fed with different diet. Though significant variation of gaseous emission and manure composition were observed between different sampling periods, no effect of high fat diet was observed on manure composition and gaseous emission.

Conclusions

It can be concluded that addition of fat to animal diet may not have any impact on gaseous emission and manure compositions.

Keywords: Manure composition, Gaseous emission, Diet, Volatile, fatty acids, Concentration

1. BACKGROUND

The United States of America is one of the largest producers of livestock and number one producer of beef cattle in the world [1]. According to the USDA, as of July 2015, there are 98.4 million beef cattle in the United states [2] and approximately 1.5 billion kg of manure (according to ASABE Standard D384.2, manure production from a beef cattle is 20–34 kg of manure per day) is generated daily only from beef cattle. Livestock manure is a nutrient source for crops. At the same time, it is also a major source of pollutant gases (ammonia-NH_3, hydrogen sulfide-H_2S, etc.), greenhouse gases (GHGs), volatile organic compounds (VOCs), odor, and particulate material (PM). Emission of pollutant gases and GHGs are becoming an important issue for human and animal health, and environment [3, 4]. In a livestock production systems, the rate and amount of gaseous emissions depend on animal species, diet composition, manure management, weather, types of housing system, and topographic features [5].

In a confined livestock operation, the emission of pollutant gases can impact workers' health, livestock welfare and productivity. The exposure of pollutant gas like H_2S can cause dizziness, headache, respiratory problem, bronchitis, pulmonary paralysis, and unconsciousness and higher concentration (>100 ppm) can have lethal outcomes [6, 7, 8]. Similarly, the concentration more than 25 ppm of NH_3 can cause respiratory irritation, chemical burns to the respiratory track, skin and eyes, severe cough, and chronic lung diseases [9]. Besides the impacts on human and animal health; those pollutant gases have an impact on environment. For example, NH_3 can contribute to nutrient build up and eutrophication of surface water, acidification, and the promotion of bacterial growth that leads to weathering and corrosive damage of buildings [10, 11, 12]. Livestock production systems generate GHGs and are likely to contribute to the global warming [13, 14].

The GHGs have the potential to absorb and emit infrared radiation that increases the earth's temperature and cause global warming [15]. The principal GHGs are water vapor, ozone (O_3), carbon dioxide (CO_2), methane (CH_4) nitrous oxide (N_2O), chlorofluorocarbon, per-fluorocarbon and sulfur hexafluoride; however CH_4, N_2O and CO_2 are the major GHGs emitted from livestock production systems [13, 14]. It is estimated that 3.4 % of the total GHGs emissions in the USA is emitted from livestock [16]. In general, methane is emitted mostly from cattle production systems due to enteric fermentation in rumen and decomposition of manure in the manure treatment and management facilities. Similarly, N_2O is produced during alternate aerobic and anaerobic decomposition of livestock manure [1]. Though the reported contribution of CH_4 and N_2O are only around 9.5 and 5.3 %, respectively, to the total GHG emissions [17]; the global warming potential of these gases are 25 and 298 times of CO_2, respectively [17]. On the other hand, CH_4 and N_2O emission from livestock manure management has increased by 68 and 25 %, respectively, since 1990 [18]. Researchers around the world are seeking technologies and management practices to mitigate emission of these gases from livestock production facilities [19, 20, 21]. Among treatment options, diet manipulation is one of the prominent options for minimizing the total gaseous emission (enteric and from manure management) [22, 23, 24].

Manure management is one of the major sources of CH_4 and N_2O emission; however, a larger portion of CH_4 (25.9 % of total CH_4 emission) is also emitted during enteric fermentation in rumens [17]. Basically, the enteric CH_4 production in rumen is affected by cattle feeding practices and feed diet composition [25]. Specifically, the diet composition can affect rumen pH, carbon nitrogen ratio, nutrient composition of manure, odor, and gaseous emissions from the manure system [6, 26]. In ruminal animal diets, carbohydrate and amount of intake influence the production of individual volatile fatty acids (VFAs), which is directly related with CH_4 production. Diet with higher sugar and starch components favor propionic acid production resulting in less CH_4 production [27]. Carbohydrate has the greatest impact on pH, microbial population, and VFA concentration which influences CH_4 production. Similarly, an increase of fat levels in cattle diets increases the energy density of the diet (8.8 kcal g^{-1}) [28], and also help to decrease enteric CH_4 production [29].

The addition of supplemental fat in the cattle diets is one of the management practices adopted by farmers [30]. The fat content of commercial beef cattle feed is typically 2–5 % [31]. If the fat content in feed exceeds 6 %, it can cause digestive disturbance, diarrhea, and reduce feed intake [32]. Many researchers have conducted experiments using fat and oil in beef cattle diets and observed its impact on body performance, weight gain, cold tolerance, and gaseous emission from body and manure. Engstrom et al. [33] conducted a feeding trial on feedlot performance and carcass quality with beef cattle in Canada using 0, 2 and 4 % fat from canola oil in diets. They found an increase of 9.8 % in daily weight gain with the addition of 4 % fat in diet during the first 56 days.

The increase of fat level in the diet may affect metabolic changes in the ruminant. It may favor the production of propionic acid, which can reduce CH_4 generation. In addition, supplementary fat can also lower the digestibility of fermentable substrates in the rumen, bio-hydrogenate unsaturated fat, and decrease methanogens population in rumen; ultimately reducing CH_4 emission [27]. Mathison [34] reported 33 % reduction in enteric CH_4 production is achievable by adding 4 % canola oil in a steer diet containing 85 % concentrate. Beauchemin and McGinn [35] carried out an experiment using fumeric acid, essential oil, and canola oil in beef cattle diets to observe their effect on enteric CH_4 emission. Their results showed a reduction on CH_4 emission using canola oil; though essential oil and fumeric acid did not influence ruminal fermentation or CH_4 emission. Similarly, Beauchemin et al. [36] used the fat sources from different oil seeds like sunflower, canola and flaxseed to feed the cattle, and observed significant CH_4 reduction in all cases.

Corn based distiller's dried grain with solubles (DDGS) is a by-product from the ethanol industries and widely used in livestock diets. Usually, DDGS contains 12 to 15 % oil on a dry basis; however, partial removal of corn oil is common in the ethanol industry. Typically, 3 to 9 % corn oil levels are found in the commercially available DDGS feedstuffs [37]. In beef cattle diets, DDGS is a major ingredient comprising up to 42 % of the total diets [38]. Besides DDGS; corn grain, corn silage, hay, sunflower meal, and concentrated separator by-product (CSB) are some other common ingredients added to beef cattle diets. The desired fat levels in the diets can be achieved by adjusting the inclusion level of DDGS in the diets. However, to the best of our knowledge no studies have been reported on the effect of various fat levels from DDGS on gaseous emission and manure composition from the feedlot pen surfaces. Therefore, the objective of this study was to investigate the effect of different fat levels in beef cattle diets on manure nutrient composition and GHG emission from feedlot pen surfaces.

2. METHODS

The experimental design and procedures of this study were reviewed and approved by the North Dakota State University Institutional Animal Care and Use Committee (protocol number A13068).

2.1 Feedlot description and experimental design

The research was carried out in a research feedlot at the North Dakota State University Carrington Research Extension Center (CREC). The feedlot had 16 pens and each pen with an area of 433 m^2 (≈19 m × 23 m). The overall slope of the feedlot was around 3 %. A total of 182 fall-born ($n = 92$) and spring-born ($n = 90$) Angus-steer calves in a randomized block design. Steers were blocked by weight (four groups: light, medium light, medium heavy and heavy). After blocking, the steers were allocated to one of 16 pens (11 to 12 steers per pen) and pen was allocated to 1 of 4 dietary fat levels treatment diets (high fat; medium fat, low fat, and control). Initially, the finishing ration was provided to the heavy and medium heavy animals while the growing ration was provided to light and medium light animals. However, after June the same ration (finishing) was provided to all. This study was conducted from June to October of 2013. The information about animal number, blocking groups, feeding strategies, treatment category and weight of animals on each pen on different time period has been provided in Table 1.

2.2 Weather condition

During each sampling, the pen surface temperatures were measured using an infrared thermometer (MiniTemp-MT6 Instrument, Carlsbad, CA). Ambient temperature, wind speed, solar radiation, and rainfall were collected from the North Dakota Agricultural Weather Network - NDAWN site, NDSU Carrington Research and Extension center, located 2 km from the study site.

Table 1 Summary of animal weight, feeding stage, treatments base on fat levels in diet, and animal weight at different period

Pen#	Animal weight	Feeding stage	Treatments	# of Animals	Weight of animals (kg) (Average ± Standard Deviation)				
					7 June	17 July	14 August	11 Sept.	2 Oct.
Pen 1	Heavy	Finish	Medium fat	11	448 ± 16	528 ± 23	588 ± 37	642 ± 43	683 ± 48
Pen 2	Heavy	Finish	Low fat	12	451 ± 19	533 ± 33	595 ± 39	654 ± 48	697 ± 40
Pen 3	Heavy	Finish	High fat	11	448 ± 17	535 ± 21	608 ± 21	671 ± 26	715 ± 31
Pen 4	Heavy	Finish	Control	11	446 ± 13	529 ± 18	586 ± 19	651 ± 20	695 ± 26
Pen 5	Medium- heavy	Finish	Medium fat	11	411 ± 12	504 ± 23	569 ± 28	632 ± 33	677 ± 40
Pen 6	Medium- heavy	Finish	Control	11	413 ± 12	494 ± 24	559 ± 34	619 ± 35	661 ± 38
Pen7	Medium- heavy	Finish	High fat	11	412 ± 13	490 ± 14	556 ± 20	622 ± 27	661 ± 25
Pen 8	Medium- heavy	Finish	Low fat	12	413 ± 13	489 ± 23	548 ± 27	613 ± 33	655 ± 39
Pen 9	Medium-light	Growing/Finish	Medium fat	11	358 ± 20	426 ± 23	491 ± 21	552 ± 29	628 ± 33
Pem10	Medium-light	Growing/Finish	Low fat	11	358 ± 18	426 ± 26	487 ± 29	545 ± 31	629 ± 38
Pen 11	Medium-light	Growing/Finish	High fat	12	360 ± 21	433 ± 25	508 ± 31	572 ± 31	649 ± 34
Pen 12	Medium-light	Growing/Finish	Control	12	360 ± 21	429 ± 22	500 ± 34	554 ± 33	638 ± 39
Pen 13	Light	Growing/Finish	Medium fat	11	307 ± 18	380 ± 26	441 ± 34	505 ± 43	585 ± 47
Pen 14	Light	Growing/Finish	High fat	12	307 ± 19	384 ± 17	455 ± 22	516 ± 22	596 ± 34
Pen 15	Light	Growing/Finish	Control	11	306 ± 20	383 ± 24	448 ± 23	505 ± 28	594 ± 29
Pen 16	Light	Growing/Finish	Low fat	12	309 ± 18	386 ± 23	443 ± 29	506 ± 32	588 ± 36

2.3 Dietary composition

In this study, the effects of four different dietary fat levels (high, medium, low and control) on beef cattle performance, manure composition and gaseous emissions from feedlot pen surfaces were studied. Three different DDGS products sourced from different ethanol plants, were used to obtain different oil levels. High fat treatment group used DDGS purchased from High-water Ethanol, Lamberton, MN; and had 12.96 % corn oil (no corn removal).

Table 2 Diet ingredient and nutrient composition of growing and finishing ration

Diet ingredients	Growing Rations				Finishing Rations			
	Control	High Fat	Med. Fat	Low Fat	Control	High Fat	Med. Fat	Low Fat
Corn (%)	48.76	43.25	42.76	43.09	66.68	61.02	60.95	60.96
DDGS (%)	--	18.89	18.76	18.7	--	19.4	19.34	19.4
Sunflower meal (%)	13.24	--	--	--	13.3	--	--	--
Hay (%)	16.96	16.95	17.06	16.96	11.13	11.27	11.29	11.29
Corn silage (%)	12.85	12.9	13.18	13.04	--	--	--	--
CSB (%)	6.32	6.3	6.27	6.29	6.76	6.73	6.73	6.73
Supplement (%)	1.87	1.71	1.96	1.92	1.56	1.58	1.69	1.62
Nutrient Composition								
CP (%)	12.09	11.85	12.39	12.53	12.42	12.12	12.7	12.88
NEm (Mcal kg^{-1})	0.37	0.37	0.37	0.37	0.37	0.41	0.41	0.41
NEg (Mcal kg^{-1})	0.23	0.23	0.23	0.23	0.24	0.27	0.27	0.27
Fat (%)	3.19	5.07	4.12	3.64	3.58	5.48	4.52	4.02

Note: DDGS = Distiller's dried grains with solubles; CSB = Concentrated separator by-product; CP = Crude protein; NEm = Net energy of maintenance; NEg = net energy of gain; All feed samples were analyzed by a commercial laboratory; net energy prediction calculations were done as per NRC [58, 59, 60]

Medium fat treatment group consisted of DDGS purchased from Blue Flient Ethanol, Washburn, ND; which used 8.05 % corn oil (partial removal). Similarly, low fat treatment group consisted of DDGS purchased from POET, Groton, SD; used 5.47 % corn oil (higher removal). The control diet included sunflower meal used 2.44 % oil. Besides DDGS, other ration ingredients were chopped grass hay, dry-rolled corn grain, corn silage, condensed separator by product and a vitamins and minerals supplements. The diets were formulated to meet the nutrient requirement recommended by NRC [39]. Overall, the fat content of high, medium, low, and control diet (composite diet) were 5.07, 4.12, 3.6, and 3.19 %, respectively in the growing ration and they were 5.48, 4.52, 4.02, and 3.58 %, respectively in the finishing ration. The diet ingredients and the nutrient composition of composite diet is listed in Table 2, and the nutrient composition of each ingredient is listed in Table 3.

Table 3 Nutrient composition in each diet ingredient

Ingredient	DM %	CP %	ADF %	TDN %	NEm Mcal kg^{-1}	NEg Mcal kg^{-1}	Fat %
Corn	87.66	8.33	3.58	87	0.45	0.31	4.37
Corn silage	32.90	7.53	28.69	68	0.33	0.20	2.59
Mixed hay	85.74	7.42	44.70	52	0.22	0.10	1.88
Sunflower meal	90.63	39.44	22.77	70	0.34	0.21	2.44
DDGS medium fat	89.26	31.90	16.48	87	0.44	0.29	8.05
DDGS high fat	88.74	28.76	15.74	88	0.44	0.29	12.96
DDGS low Fat	88.83	32.69	11.93	92	0.43	0.29	5.47
CSB	71.56	10.07	0.19	86	0.42	0.28	1.28

Note: *DDGS* Distiller's dried grains with solubles, *CSB* Concentrated separator by-product, *DM* dry matter, *CP* Crude protein, *ADF* Acid detergent fiber, *TDN* Total digestible nutrients, *NEm* Net energy for maintenance, *NEg* Net energy for gain. All feed samples were analyzed by a commercial laboratory; net energy prediction calculations are from [58, 59, 60]

2.4 Gaseous sampling and analysis

Air samples from the pen surface were collected for five times during June to October 2013 with a sampling interval of 30 ± 10 days. Air samples were collected using a custom built portable wind tunnel with a foot print area of 0.32 m^2 (0.8 m × 0.4 m), Tedlar bag, and Vac-U-Chamber (SKC Inc., Eighty Four, PA) (Fig. 1). Measured air velocity over the foot-print area of tunnel immediate over the manure surface was maintained 0.35 m s^{-1} that prompted an air flow through the tunnel of 2.75 L s^{-1} (0.00275 m^3 s^{-1}). In each sampling location, a 5 L Tedlar bag was placed inside a vacuum chamber and a uniform air flow rate (2.75 L s^{-1}) was maintained inside the tunnel throughout the sampling period using a DC motor. Additional sampling protocol can be found at Rahman et al. [40]. In each pen, two samples were collected; one from the front end of the pen next to feeding area, and another one from the backside of pen. So, a total of 160 air samples (16 pens × 2 samples/pen × 5 times) were collected and they were brought back to the laboratory for H_2S, CH_4, CO_2, and N_2O analysis.

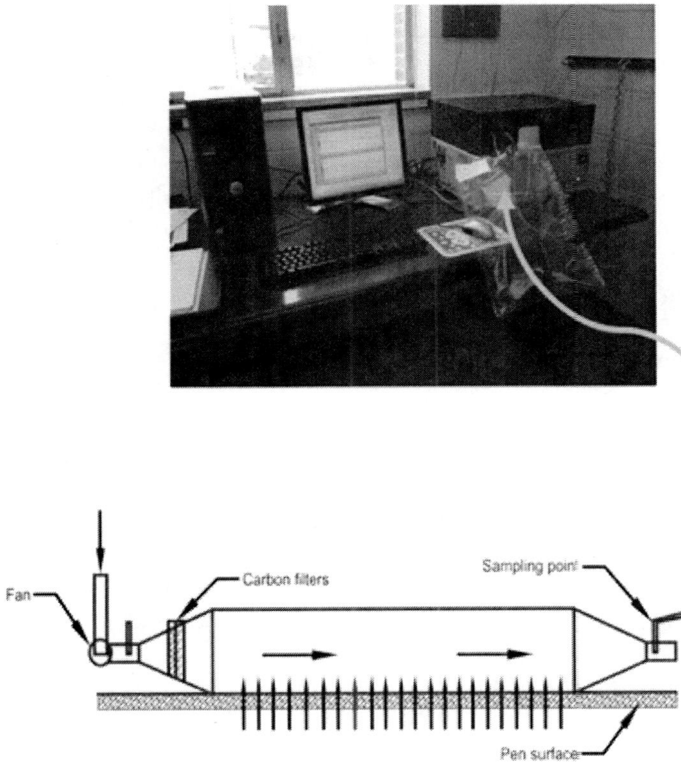

Figure. 1 Schematic diagram of gas collection using a wind tunnel and GHG analysis using a gas chromatograph (drawing not to scale)

Within 24 h of sampling, they were analyzed for GHGs concentration using a greenhouse gas GC (Model No. 8610C, SRI Instruments, and 20720 Earl St., Torrance, CA 90502), and H_2S concentration using a Jerome meter (Jerome® 631-X, Arizona instrument, Arizona, USA). The GC was equipped with a flame ionization detector (FID) to measure CO_2 and CH_4 and an electron captured detector (ECD) to analyze N_2O. GHG was analyzed following the procedure described in Rahman et al. [40].

2.5 Manure sampling and analysis

Manures on the pen surface were allowed to accumulate until animals were sold out. During a sampling event, the manure sampling was paired with air sampling and relatively fresh manures (freshly excreted or few minutes old) were sampled. The composite manure samples collected from five to seven spots in a pen, bagged and mix in a zip-locked bag, kept in ice cooled cooler in the field and during transportation; and finally stored in a refrigerator at around 4 °C until analysis. Thus, in each sampling day, a total of 16 composite manure (each approximately ~ 800 g) samples were collected. Before analysis, samples were mixed thoroughly again, divided into two sub-samples (sub-sample 1: pH, total nitrogen (TN), potassium (K), total phosphorus (TP) and total carbon (TC); sub-sample 2: volatile fatty acids (VFAs), crude protein (CP) and ammonical nitrogen), and were sent to two analytical laboratory for analysis following the standard laboratory methods (Table 4).

2.6 Emission calculation

In order to estimate the emission rate; the volumetric gas concentration was standardized at standard pressure and temperature (1 atm and 25 °C), and mass concentration of the compound was calculated from calculated volumetric concentration (Equation 1). Flux rates (g m^{-2} d^{-1}) was calculated using the average airflow through the wind tunnel, mass concentration of the target gas and the surface area covered by the wind tunnel as shown in Equation 2. Finally, emission rate was estimated using the surface area of pen, flux rate, and animal unit (AU) in the pen (Equation 3).

$$C_{mass} = \frac{C_{ppm} \times MW}{24.24} \tag{1}$$

where, C_{ppm} = Volumetric concentration of the target gas (ppm)

C_{mass} = Mass concentration of the target gas (mg m^{-3})

MW = Molecular weight of the target gas (g mol^{-1})

24.25 = Volume per mole of an ideal gas at standard temperature and pressure (L mol^{-1})

$$FR = \frac{C_{mass} \times V_{wt} \times 3600 \times 24}{A_{wt} \times 1000} \tag{2}$$

where, FR = GHG emission flux rate from pen surface (g m^{-2} d^{-1})

V_{wt} = Airflow rate through wind tunnel (m^3 s^{-1})

A_{wt} = Surface area covered by the wind tunnel (0.4 × 0.8 m^2)

$$ER = \frac{FR \times A_{Sc}}{AU} \tag{3}$$

where, ER = GHG emission rate from pen surface (g hd^{-1} d^{-1})

A_{sc} = Surface area of the source (m^2)

AU = Animal unit (total weight of animals in pen divided by 500 kg live weight)

Table 4 Method/protocol used to analyze manure samples

Parameters	Methods/protocol used
TN	Recommended methods of manure analysis, A3769 Macro-Kjeldahl method (adapted from Kane, 1998)
K	Recommended method of manure analysis, A3769
TP	Recommended method of manure analysis, A3769
TC	U.S. EPA method 415.1: Catalytic combustion and non-dispersive infrared detection (NDIR) method
CP	Official Method 2001.11, AOAC International (2005) 18th ED., AOAC[1] International Gaithersburg, MD, USA
NH_3-N	Sigma Technical Bulletin #640. Sigma Diagnostics, St. Louis, MO 63178
VFA	Method of Goetsch and Galyean, 1983. Agilent 6890 N Gas Chromatograph with a FID (flame ionization detector) and the 7683 Series auto injector and auto sampler. Column used was the Supelco brand, NUKOL Fused Silica Column, 15 m × 0.53 mm × 0.5 um

[1] *AOAC* Association of Official Agricultural Chemists

2.7 Ambient weather and feedlot pen surface temperature measurement

The daily mean air temperature, wind speed, solar irradiation, and rainfall at the sampling locations during each sampling period are listed in Table 5. The August sampling time had the highest ambient temperature, while October had the lowest ambient temperature. Likewise, the highest pen surface temperature was noted in August, which equates to the ambient temperature (Fig. 2). Similarly, the lowest pen surface temperature was observed in September. Overall, average pen surface temperatures were very consistent among pens at each sampling time. Besides temperature, solar radiation was also the highest in August, and the lowest in September. During the sampling time, no noticeable rainfall was observed, which might have some effects on gaseous emission from the manure pen surface.

Table 5 Ambient weather condition at the study site

Sampling date	Air temperature (°C)			Average wind speed ($m\ s^{-1}$)	Solar radiation ($MJ\ m^{-2}$)	Rainfall (mm)
	Average	Minimum	Maximum			
20-Jun-13	20.56	17.78	23.33	5.09	7.57	0.00
30-July-13	16.67	10.56	22.78	2.14	16.99	0.80
20-Aug-13	26.11	16.11	36.11	2.77	23.01	0.00
18-Sep-13	18.33	13.89	22.78	2.55	5.19	0.00
9-Oct-13	11.11	2.78	19.44	1.52	10.08	0.00

2.8 Statistical analysis

It is known that temperature variation on the pen surface effects gaseous emissions. Usually, higher temperatures enhance CH_4 production [41]. The temperature range of 25–30 °C is considered optimum for CH_4 production [42]. Surface temperature also influence on N_2O emission. Luo et al. [43] reported the highest N_2O emissions with moist and warm soil, and the soil exposed to freezing and thawing condition. Lang et al. [44] also observed the higher soil temperature promoted greater nitrification and N_2O emissions.

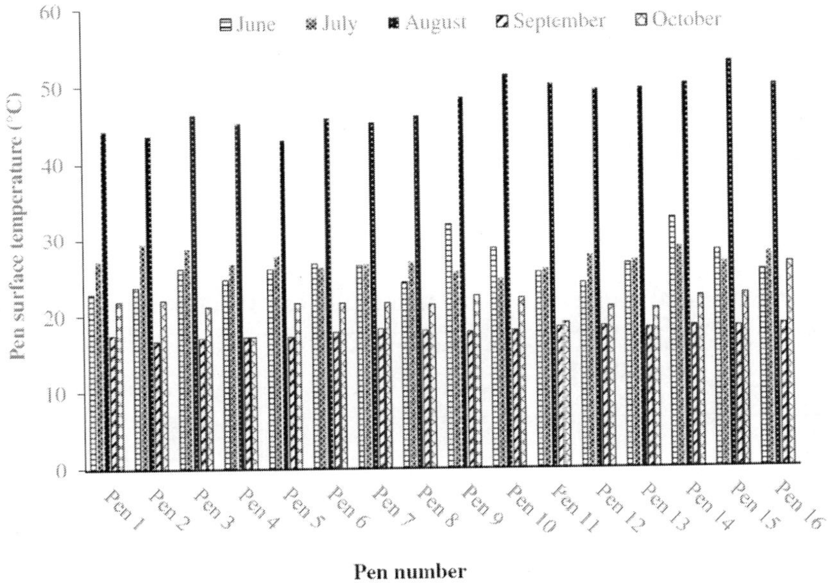

Figure. 2 Variation in feedlot pen surface temperature during the experimental period

The effect of fat levels in the diet on GHG emission and manure composition were compared using the Generalized Liner Model (GLM) procedure in SAS software (SAS 9.3, 2002-2010). Randomized complete block design was chosen for each sampling event (months) with animal weight as a block (light and heavy) for four treatments (control, low, medium and high). However, during analysis no significant difference of treatments were observed separating the animals on weight basis. Therefore, a comparative study of different treatments were carried considering the animal types as a single block. All significance tests were evaluated at $P = 0.05$. The null hypothesis was that the means value of GHGs concentrations, emission flux, emission rates, manure nutrient and VFAs concentrations were equal within and among treatments and sampling time.

3. RESULTS AND DISCUSSION

3.1 Manure composition

Overall, no significant differences in manure composition were observed among treatments in most of the sampled months, but significant differences on some manure composition were observed over the sampling period (Table 6). Likewise in August, the moisture content, crude protein (CP), and TN were also significantly lower in manure from pens with cattle fed high fat diets compared to other treatment groups.

Table 6 Average manure composition for each month based on treatment

Sampling date	Manure composition	Treatments			
		control	Low fat	Medium fat	High fat
20-Jun-13	pH	5.4 a* ± .0.2	5.6 a ± 0.4	5.5 a ± 0.4	5.4 a ± 0.2
	Moisture % wb	76.7 a ±1.9	76.9 a ± 1.0	74.2 a ± 5.5	77.4 a ± 1.6
	Ash %	11.1 a ± 1.4	13.9 a ± 4.2	10.0 a ± 0.8	10.4 a ± 1.6
	CP %	14.9 a ± 0.8	14.9 a ± 1.4	13.7 a ± 2.8	15.1 a ± 0.6
	TN %	2.4 a ± 0.1	2.4 a ± 0.2	2.2 a ± 0.5	2.4 a ± 0.1
	NH_3 (mM)	5.6 a ± 1.2	4.9 a ± 1.7	5.5 a ± 1.6	4.7 a ± 1.4
	TC (%)	43.7 a ± 1.0	43.2 a ± 1.5	44.8 a ± 0.5	43.7 a ± 1.0
	TP %	0.1 a ±0.1	0.2 a ± 0.1	0.2 a ± 0.1	0.1 a ± 0.0
	K %	0.1 a ± 0.0	0.1 a ± 0.0	0.1 a ± 0.0	0.1 a ± 0.0
30-Jul-13	pH	5.5 a ± 0.1	5.4 a ± .2	5.5 a ± .4	5.5 a ± 0.1
	Moisture % wb	77.8 a ± 0.6	77.2 a ± 1.5	76.8 a ± 1.5	75.5 a ± 2.6
	Ash %	9.1 a ± 0.8	8.2 a ± 0.3	10.2 a ± 1.3	8.5 a ± 1.6
	CP %	15.6 a ± 1.5	15.4 a ± 1.1	16.3 a ± 0.8	14.9 a ± 1.0
	TN %	2.5 a ± 0.3	2.5 a ± 0.2	2.6 a ± 0.1	2.4 a ± 0.1
	NH_3 (mM)	5.2 a ± 1.7	5.5 a ± 1.5	9.4 a ± 5.7	8.9 a ± 2.1
	TC (%)	44.4 a ± 0.4	43.6 a ± 1.8	44.3 a ± 0.4	45.4 a ± 0.8

	TP %	0.1 a ± 0.0	0.1 a ± 0.0	0.1 a ± 0.0	0.1 a ± 0.0
	K %	0.1 a ± 0.0	0.1 a ± 0.0	0.1 a ± 0.0	0.1 a ± 0.0
20-Aug-13	pH	5.2 a ± 0.1	5.1 a ± 0.3	5.1 a ± 0.1	5.0 a ± 0.2
	Moisture % wb	73.7 ab ± 1.2	73.3 ab ± 2.2	75.6 a ± 0.7	72.2 b ± 1.9
	Ash %	8.1 a ± 0.5	8.9 a ± 1.1	7.7 a ± 0.6	7.4 a ± 1.0
	CP %	15.2 ab ± 1.1	17.6 a ± 1.4	17.4 ab ± 1.8	14.9 b ± 0.6
	TN %	2.4 ab ± 0.2	2.8 a ± 0.2	2.8 ab ± 0.3	2.4 b ± 0.1
	NH_3 (mM)	12.8 a ± 1.2	17.1 a ± 9.0	16.8 a ± 6.8	11.3 a ± 3.7
	TC (%)	44.2 a ± 1.3	45.0 a ± 2.2	44.3 a ± 0.5	42.9 a ± 1.8
	TP %	0.1 a ± 0.0	0.1 a ± 0.1	0.1 a ± 0.0	0.2 a ± 0.0
	K %	0.1 a ± 0.0	0.1 a ± 0.0	0.1 a ± 0.0	0.1 a ± 0.0
18-Sep-13	pH	5.6 a ± 0.2	5.4 a ± 0.2	5.4 a ± 0.3	5.5 a ± 0.1
	Moisture % wb	75.4 a ± 1.4	74.9 a ± 2.0	75.1 a ± 2.8	75.7 a ± 1.3
	Ash %	9.5 a ± 2.1	8.0 a ± 1.3	8.7 a ± 1.0	7.9 a ± 1.0
	CP %	15.0 a ± 1.5	15.0 a ± 2.6	15.6 a ± 0.8	15.0 a ± 1.1
	TNm %	2.4 a ± 0.2	2.4 a ± 0.4	2.5 a ± 0.1	2.4 a ± 0.2
	NH_3 (mM)	10.0 ab ± 2.9	7.6 b ± 3.3	8.0 ab ± 2.1	11.5 a ± 2.3
	TC (%)	44.4 a ± 0.8	44.1 a ± 0.7	43.5 a ± 1.7	44.9 a ± 0.4
	TP %	0.2 a ±0.1	0.1 b ± 0.0	0.1 b ± 0.0	0.1 b ± 0.0
	K %	0.1 a ± 0.0	0.1 a ± 0.0	0.1 a ± 0.0	0.1 a ± 0.1
9-Oct-13	pH	5.5 a ± .01	5.6 a ± 0.1	5.9 a ± .02	5.3 a ± 0.0
	Moisture % wb	72.6 a ± 0.5	72.9 a ± 0.0	72.4 a ± 3.6	72.4 a ± 3.0
	DM %	95.6 a ± 0.2	96.7 a ± 0.0	96.6 a ± 0.5	96.1 a ± 0.3
	Ash %	8.7 a ± 1.3	8.5 a ± 1.2	19.9 a ± 10.9	6.9 a ± 0.9
	CP %	16.4 a ± 1.2	14.9 a ± 1.2	13.9 a ± 1.1	15.0 a ± 0.6
	TN %	2.6 a ± 0.2	2.4 a ± 0.2	2.2 a ± 0.2	2.4 a ± 0.1

	NH_3 (mM)	6.9 a ± 0.9	14.9 a ± 2.7	19.0 a ± 5.7	8.0 a ± 1.1
	TC (%)	44.0 a ± 0.9	39.9 a ± 5.0	37.2 a ± 7.2	44.8 a ± 0.2
	TP %	0.1 a ± 0.0	0.2 a ± 0.1	0.1 a ± 0.0	0.1 a ± 0.0
	K %	0.1 a ± 0.0	0.2 a ± 0.0	0.2 a ± 0.0	0.1 a ± 0.0

*Values followed by the same letter in row are not significantly different at $P \leq 0.05$; wb = wet basis

However, when the analysis was simply carried out on time basis (comparison among months), significant difference on most of the parameters of manure composition were observed (Table 7). Manure pH was significantly lower in August as compared to other months. Similarly, moisture content of manure was significantly lower in October as compared to June, July and September as shown in Table 7. Ash content of manure was the highest in June and the lowest in August. Crude protein, TP, and ammonical nitrogen NH_3-N content in manure were the lowest in June and the highest in August. Total carbon (TC) in manure was significantly lower in August as compared to other months. In 2012 summer, Borhan et al. [45] had also measured the nutrient composition of the manure in the same feed lot under similar condition and the values of nutrient parameter were almost comparable with this study.

Table 7 Average manure composition on monthly basis

Parameters	June	July	August	September	October
pH	5.5 a* ± 0.1	5.5 a ± 0.0	5.1 b ± 0.1	5.5 a ± 0.1	5.6 a ± 0.2
Moisture %	76.32 a ± 1.2	76.8 a ± 0.8	73.7 bc ± 1.2	75.3 ab ± 0.3	72.6 c ± 0.2
Ash %	11.3 a ± 1.5	9.0 abc ± 0.8	8.0 c ± 0.6	8.5 bc ± 0.5	11.0 ab ± 5.2
CP %	14.6 a ± 0.6	15.5 ab ± 0.5	16.2 b ± 1.2	15.1 ab ± 0.3	15.0 ab ± 0.9
TN %	2.3 a ± 0.1	2.5 ab ± 0.1	2.6 b ± 0.2	2.4 ab ± 0.0	2.4 ab ± 0.1
NH_3-N (mM)	5.2 d ± 0.4	7.3 cd ± 1.9	14.5 a ± 2.5	9.3 bc ± 1.6	12.2 ab ± 5.0
TC (%)	43.8 a ± 0.6	44.4 a ± 0.6	44.1 a ± 0.8	44.2 a ± 0.5	41.5 b ± 3.1
TP %	0.1 a ± 0.0	0.1 a ± 0.0	0.1 a ± 0.0	0.2 a ± 0.0	0.1 a ± 0.0
K %	0.1 a ± 0.0	0.1 a ± 0.0	0.1 a ± 0.0	0.1 a ± 0.0	0.1 a ± 0.0

* Values followed by the same letter in row are not significantly different at $P \leq 0.05$; *CP* Crude protein, *TN* Total nitrogen, *NH_3-N* ammonical nitrogen, *TP* Total phosphorus, *K* Potassium, *TVFA* Total volatile fatty acids, *DM* Dry matter

3.2 Effect of dietary fat level on volatile fatty acid (VFAs) composition of manure

No significant differences in any of VFAs concentration were observed among treatments during the study period except for two months. In July, isovaleric acid was significantly higher in manure from pens with cattle fed the low fat diets than the control. Likewise, in September butyric acid was significantly higher in the manure from pens with cattle fed the medium fat diets compared to the control (Table 8). Similarly in August and September, the total volatile fatty acid (TVFA) content were significantly lower in the high fat group than the others (Table 8), which may contribute to lower CH_4 emission. However, when the analysis carried out on monthly basis, the lowest acetic acid concentration and the highest propionic acid concentrations were observed in August (Table 9). Likewise, the TVFA content of manure was significantly higher in July and August compared to other months (Table 9), which is likely due to temperature effect on VFA production. Due to higher TVFA, comparatively higher CH_4 emission can be expected during July and August.

Table 8 Volatile fatty acids content in manure measured based on the treatment

Sampling date	VFAs (mM)	Treatments			
		control	Low fat	Medium fat	High fat
20-Jun-13	Acetic	49.5 a* ± 3.2	51.3 a ± 3.5	46.0 a ± 1.5	46.5 a ± 4.8
	Propionic	17.0 a ±2.1	17.3 a ± 2.5	20.2 a ± 4.1	19.4 a ± 4.0
	Isobutyric	2.1 a ± 1.3	2.0 a ± 0.6	2.0 a ± 1.4	2.1 a ± 0.6
	Butyric	24.0 a ± 2.6	21.7 a ± 3.0	25.7 a ± 3.7	23.0 a ± 1.5
	Isovaleric	4.6 a ± 1.4	3.2 a ± 1.9	4.6 a ± 1.7	3.9 a ± 1.3
	Valeric	2.8 a ± 3.0	4.5 a ± 1.3	1.5 a ± 1.5	5.1 a ± 1.8
	TVFA (mM)	120.7 a ± 27.6	101.7 a ± 37.4	105.1 a ± 8.6	127.1 a ± 24.2
30-Jul-13	Acetic	51.4 a ± 2.2	49.5 a ± 2.4	51.1 a ± 2.9	50.9 a ± 2.8
	Propionic	23.5 a ± 1.4	19.8 a ± 3.4	21.3 a ± 3.3	19.5 a ±2.3
	Isobutyric	1.4 a ± 0.3	1.5 a ± 0.4	1.7 a ± 0.6	1.7 a ± 0.2
	Butyric	21.0 a ± 1.2	24.4 a ± 2.3	22.1 a ± 1.8	22.8 a ± 1.7
	Isovaleric	1.3 b ± 0.3	3.5 a ± 0.5	2.2 ab ± 0.8	3.4 ab ± 0.3
	Valaric	1.4 a ± 1.3	1.4 a ± 1.0	1.6 a ± 1.7	1.6 a ± 0.9

	TVFA (mM)	150.2 a ± 18.9	149.0 a ± 32.7	176.4 a ± 42.2	148.7 a ±10.9
20-Aug-13	Acetic	46.0 a ± 3.7	44.7 a ± 2.8	45.6 a ± 2.4	49.0 a ± 5.6
	Propionic	25.4 a ± 1.3	25.6 a ± 3.9	24.7 a ± 2.1	22.0 a ± 3.5
	Isobutyric	1.5 a ± 0.2	1.4 a ± 0.3	1.2 ab ± 0.4	0.7 b ± 3.6
	Butyric	22.4 a ± 2.4	22.4 a ± 2.4	23.4 a ± 1.4	24.6 a ± 3.6
	Isovaleric	1.3 a ± 0.1	1.5 a ± 0.6	1.3 a ± 0.4	0.9 a ± 0.3
	Valeric	3.5 a ± 0.5	4.4 a ± 1.7	1.3 a ± 1.2	2.9 a ± 1.1
	TVFA (mM)	147.6 ab ± 34.8	178.2 a ± 19.9	142.0 ab ± 42.5	128.8 b ± 47.5
18-Sep-13	Acetic	48.9 b ± 0.6	50.8 ab ± 0.6	53.5 a ± 2.1	51.2 ab ± 2.5
	Propionic	22.7 a ± 1.7	22.4 a ± 1.1	20.9 a ± 2.3	21.9 a ± 1.0
	Isobutyric	1.1 a ± 0.3	0.8 a ± 0.1	1.0 a ± 0.5	1.2 a ± 0.3
	Butyric	23.7 a ± 1.0	22.4 ab ± 1.8	20.6 b ± 1.6	21.3 ab ± 1.0
	Isovaleric	1.0 a ± 0.3	0.8 a ± 0.2	1.0 a ± 0.5	1.2 a ± 0.2
	Valeric	2.5 a ± 0.6	2.8 a ± 1.0	3.0 a ± 1.3	3.6 a ± 1.2
	TVFA (mM)	129.2 a ± 15.0	109.0 ab ± 14.4	114.5 ab ± 17.9	105.3 b ± 11.1
9-Oct-13	Acetic	53.5 a ± 0.9	49.8 b ± 0.0	51.9 ab ± 0.3	50.2 b ±0.8
	Propionic	20.9 a ±0.1	20.6 a ± 0.6	20.2 a ± 1.7	22.8 a ± 0.7
	Isobutyric	0.9 a ± 0.0	1.2 a ± 0.0	1.4 a ± 0.4	0.4 a ± 0.5
	Butyric	21.8 a ± 0.2	23.7 a ± 0.9	21.8 a ± 0.5	24.4 a ± 1.3
	Isovaleric	0.9 a ± 0.3	1.2 a ± 0.1	1.4 a ± 0.3	0.6 a ± 0.2
	Valeric	1.9 a ± 0.2	3.5 a ± 1.5	3.3 a ± 1.5	1.5 a ± 1.5
	TVFA (mM)	144.5 a ± 18.4	108.8 a ± 0.5	113.5 a ± 16.2	123.1 a ±8.0

*Values followed by the same letter in row are not significantly different at $P \leq 0.05$

During anaerobic decomposition of manure; acetic, propionic, butyric and valeric acids are the common VFAs produced by micro-organisms. Acetic acid is the major VFA responsible for CH_4 production from anaerobic biomass which accounts more than two third of CH_4 production [46]. Propionic and butyric acids are considered as the inhibitory agents in anaerobic process [47]. Higher concentration of propionic usually inhibits the CH_4 production in the case of anaerobic digester [48]; however, some researchers have mentioned that it's the effect rather than cause for the inhibition of CH_4 production [49, 50]. The ratio of acetic acid and propionic acid is another important factor for determining the CH_4 production rate. Higher acetic acid (>800 mg L^{-1}) as well as propionic acid and acetic acid ratios greater than 1:4 is taken as the indicator for failure of anaerobic processes [51]. However in this study, the ratio of propionic acid to TVFA was <1:4 (Tables 8 and 9), which was an indicator of anaerobic process on the pen surface.

Table 9 Monthly volatile fatty acid (VFA) analysis of manure

Parameters	June	July	August	September	October
Acetic	48.3 bc* ± 2.2	50.7 ab ± 0.7	46.3 c ± 1.6	51.1 ab ± 1.6	51.4 a ± 1.5
Propionic	18.5 c ± 1.4	21.0 b ± 1.6	24.4 a ± 1.4	22.0 b ± 0.7	21.1 b ± 1.0
Isobutyric	2.0 a ± 0.1	1.6 ab ± 0.1	1.2 bc ± 0.3	1.0 c ± 0.2	1.0 c ± 0.4
Butyric	23.6 a ± 1.5	22.6 a ± 1.2	23.2 a ± 0.9	22.0 a ± 1.2	22.9 a ±1.1
Isovaleric	4.1 a ± 0.6	1.5 b ± 0.1	1.2 b ± 0.2	1.0 b ± 0.1	1.0 b ± 0.3
Valeric	3.5 a ± 1.4	2.6 a ± 0.9	3.7 a ± 0.6	2.9 a ± 0.2	2.6 a ± 0.8
TVFA (mM)	113.7 b ± 10.6	156.1 a ± 11.8	149.2 a ± 18.1	114.5 b ± 9.1	122.5 b ± 13.7

*Values followed by the same letter in row are not significantly different at $P \leq 0.05$

3.3 Effect of dietary fat level on GHG emission

Overall, no significant difference in GHGs emissions were observed from the feedlot pen surfaces with beef cattle fed four levels of fat (control, low, medium, high) in the diet (Table 10). However, some variations on GHG emission were observed when the measurement were compared between months. In July and September, the highest CO_2 efflux was observed from pen surface with cattle fed medium fat diets. The increased fat source in the diets is likely to increase dietary energy, suppress methanogens decreasing CH_4emissions (both enteric and from manure) as well as reduce nitrogen emissions from manure [27, 52]. No significant difference in the total nitrogen and ammonium nitrogen between the treatments also support less variation of N_2O emission between treatments.

Table 10 Analysis of greenhouse gas emissions based on treatment

Sampling date	Manure composition	Treatments			
		control	Low fat	Medium fat	High fat
20-Jun-13	CH_4 concentration (ppm)	2.2 ab* ± 0.0	2.2 a ± 0.2	2.3 b ± 0.2	2.1 a ± 0.0
	CO_2 concentration (ppm)	378.0 a ± 24.0	378.3 a ± 17.9	390.0 a ± 40.4	374.3 a ± 15.8
	N_2O concentration (ppm)	0.9 a ± 0.1	1.2 a ± 0.5	0.8 a ± 0.4	1.4 a ± 0.4
	CH_4 FR (g m^{-2} d^{-1})	1.1 a ± 0.0	1.1 a ± 0.1	1.1 a ± 0.1	1.0 a ± 0.0
	CO_2 FR (g m^{-2} d^{-1})	504.8 a ± 32.0	505.2 a ± 23.9	520.8 a ± 54.0	499.8 a ± 21.1
	N_2O FR (g m^{-2} d^{-1})	1.2 a ± 0.2	1.6 a ± 0.6	1.1 a ± 0.5	1.8 a ± 0.6
	CH_4 EF (g AU^{-1} d^{-1})	54.3 a ± 8.5	53.4 a ± 9.4	55.7 a ± 6.3	51.4 a ± 6.0
	CO_2 EF (kg AU^{-1} d^{-1})	25.8 a ± 2.8	24.7 a ± 2.9	27.1 a ± 2.2	25.9 a ± 1.7
	N_2O EF (g AU^{-1} d^{-1})	62.2 a ± 8 .9	78.2 a ± 26.5	54.5 a ± 20.4	93.8 a ± 35.4
30-Jul-13	CH_4 concentration (ppm)	2.8 ab ± 0.3	2.8 b ± 0.1	3.1 a ± 0.2	2.6 b ± 0.1
	CO_2 concentration (ppm)	467.9 b ± 70.3	485.4 b ± 67.0	518.0 a ± 75.0	473.5 b ± 58.8
	N_2O concentration (ppm)	1.0 a ± 0.4	0.8 a ± 0.2	1.3 a ± 0.1	1.0 a ± 0.3
	CH_4 FR (g m^{-2} d^{-1})	1.4 ab ± 0.2	1.3 b ± 0.0	1.5 a ± 0.1	1.3 b ± 0.1
	CO_2 FR (g m^{-2} d^{-1})	624.7 b ± 93.9	648.2 b ± 89.5	691.7 a ± 100.2	632.2 b ± 78.5
	N_2O FR (g m^{-2} d^{-1})	1.4 a ± 0.5	1.1 a ± 0.3	1.7 a ± 0.1	1.4 a ± 0.4
	CH_4 EF (g AU^{-1} d^{-1})	58.4 ab ± 5.6	54.8 a ± 6.3	64.3 b ± 4.2	52.8 a ± 3.6
	CO_2 EF (g AU^{-1} d^{-1})	26.2 a ± 1.6	26.1 a ± 2.9	29.6 b ± 2.2	25.9 a ± 1.7
	N_2O EF (kg AU^{-1} d^{-1})	58.6 ab ± 23.5	42.5 a ± 7.0	74.1 b ± 9.5	55.5 ab ± 13.2
20-Aug-13	CH_4 concentration (ppm)	3.2 a ±1.0	2.8 a ± 0.4	2.7 a ± 0.5	2.8 a ± 0.7
	CO_2 concentration (ppm)	431.5 a ± 48.3	471.1 a ± 93.1	487.4 a ± 131.1	447.0 a ± 64.2
	N_2O concentration (ppm)	0.9 a ± 0.4	1.4 a ± 0.6	0.9 a ± 0.4	1.2 a ± 0.3
	CH_4 FR (g m^{-2} d^{-1})	1.6 a ± 0.5	1.4 a ± 0.2	1.3 a ± 0.2	1.4 a ± 0.4

	CO_2 FR (g m^{-2} d^{-1})	576.2 a ± 64.4	629.0 a ± 124.3	650.9 a ± 175.1	596.9 a ± 85.8
	N_2O FR (g m^{-2} d^{-1})	1.3 a ± 0.6	1.9 a ± 0.8	1.2 a ± 0.4	1.7 a ± 0.8
	CH_4 EF (g AU^{-1} d^{-1})	57.9 a ± 15.3	49.0 a ± 2.6	49.7 a ± 6.0	48.2 a ± 9.0
	CO_2 EF (kg AU^{-1} d^{-1})	21.2 a ± 1.6	22.1 a ± 1.6	24.3 a ± 4.3	21.1 a ± 1.8
	N_2O EF (g AU^{-1} d^{-1})	48.5 a ± 25.6	67.3 a ± 31.6	48.5 a ± 18.7	59.1 a ± 28.6
18-Sep-13	CH_4 concentration (ppm)	3.3 a ±0.3	3.4 a ± 0.7	3.1 a ± 0.5	3.6 a ± 0.7
	CO_2 concentration (ppm)	389.3 a ± 28.1	381.1 a ± 60.6	423.3 a ± 70.3	422.2 a ± 44.7
	N_2O concentration (ppm)	0.6 ± 28.1	0.6 ± 60.6	0.6 ± 70.3	0.6 ± 44.7
	CH_4 FR (g m^{-2} d^{-1})	1.6 a ± 0.1	1.6 a ± 0.4	1.5 a ± 0.3	1.7 a ± 0.4
	CO_2 FR (g m^{-2} d^{-1})	519.9 a ± 37.5	508.9 a ± 80.9	565.2 a ± 93.9	563.8 a ± 59.7
	N_2O FR (g m^{-2} d^{-1})	0.8 a ± 0.3	0.7 a ± 0.1	0.7 a ± 0.3	0.8 a ± 0.4
	CH_4 EF (g AU^{-1} d^{-1})	54.2 a ± 6.0	51.6 a ± 5.7	50.8 a ± 4.0	54.7 a ± 7.7
	CO_2 EF (kg AU^{-1} d^{-1})	17.3 ab ± 0.6	16.1 b ± 1.1	19.1 a ± 2.1	17.9 ab ± 1.6
	N_2O EF (g AU^{-1} d^{-1})	26.9 a ± 7.8	24.3 a ± 4.5	24.4 a ± 9.2	23.9 a ± 9.4
9-Oct-13	CH_4 concentration (ppm)	4.3 a ± 1.4	3.9 a ± 1.1	2.5 a ± 0.2	3.4 a ± 0.2
	CO_2 concentration (ppm)	367.2 a ± 35.1	381.6 a ± 23.9	345.7 a ± 11.3	379.8 ± 2.9
	N_2O concentration (ppm)	0.4 a ± 0.0	0.3 a ± 0.1	0.4 a ± 0.1	0.4 a ± 0.0
	CH_4 FR (g m^{-2} d^{-1})	2.1 a ± 0.7	1.9 a ± 0.5	1.2 a ± 0.1	1.6 a ± 0.1
	CO_2 FR (g m^{-2} d^{-1})	490.3 a ± 46.9	509.5 a ± 31.9	461.6 a ± 15.0	507.2 a ± 3.9
	N_2O FR (g m^{-2} d^{-1})	0.5 a ±0.0	0.5 a ± 0.1	0.5 a ± 0.1	0.5 a ± 0.0
	CH_4 EF (g AU^{-1} d^{-1})	62.2 a ± 16.3	59.1 a ± 15.5	38.8 a ± 1.5	47.8 a ± 5.2
	CO_2 EF (kg AU^{-1} d^{-1})	14.9 a ± 0.3	15.8 a ± 1.2	15.0 a ± 0.0	14.7 a ± 0.7
	N_2O EF (g AU^{-1} d^{-1})	14.6 a ± 0.7	14.0 a ± 2.1	15.7 a ± 2.8	13.6 a ±.09

*Values followed by the same letter in row are not significantly different at $P \leq 0.05$; where, *FR* flux rate from pen surface (g m^{-2} d^{-1}), *ER* emission rate from pen surface (g hd^{-1} d^{-1})

The effect of fat on gaseous emissions depends on many factors; such as type of fat, amount of fat in feed, and environmental condition. Though the literatures [35, 36] showed that the addition of fat effects enteric CH_4 production, this study showed that different fat levels from DDGS may not greatly influence the CH_4production from the pen surface area. The emissions from the pen surface area are most likely to influence from environmental factors. The environmental conditions were almost similar in all the pen surfaces; therefore, very little variation in gaseous emissions might have been observed under the different treatments conditions. In addition, the reduction of CH_4 concentration using supplementary fat may not be applicable for corn oil in DDGS; or the application rate of corn oil used in this research may not be sufficient for significant reduction on gaseous emission from pen surface.

When the gaseous emission were compared between different months, significant differences in the gaseous parameters were observed. The CH_4 emissions were significantly higher during September and October from the pen surfaces as compared to June, July, and August. Higher emissions of CH_4 were expected due to higher temperatures in July and August [41]. Though the CH_4 concentration was observed higher in August and July compared to June; the concentration in September and October were even higher than July and August. This could be due to the accumulation of manure on the pen surface that provide anaerobic conditions for CH_4 emission. Nitrous oxide emissions were significantly lower during September and October and higher during June, July and August (Table 11). The higher temperature during June, July and August could be a reason for higher N_2O emissions [44]. Similarly, the dry and wet condition of the pen surface due to rain in summer may provide alternate aerobic and anaerobic condition on the pen surface, thus variation of N_2O emission was observed. The wet conditions of pen surfaces favors anaerobic conditions in manure, resulting in denitrification. Dry conditions favor aerobic conditions in manure resulting in nitrification. Nitrous oxide is produced during both nitrification and denitrification processes [53]. The significantly lowest nitrous oxide and carbon dioxide fluxes during October are most likely due to prevailing dry surface and ambient condition (Table 10).

In comparing the results with the previous study; in 2011, Rahman et al [40] simply measured GHG emission from the same feedlot pen surface and they found that CH_4, CO_2 and N_2O emission were 38, 26, and 17 g $hd^{-1}d^{-1}$, respectively, during the 2011 summer period. Similarly, in 2012, Borhan et al. [45] studied the effects of two dietary crude protein levels (12 and 16 %) in the GHG emission on the similar conditions. They found that CH_4, CO_2 and N_2O emission ranged from 40–61, 31–43, and 50–116 $gAU^{-1}d^{-1}$ (0.8–1.1, 593–431, and 1–1.9 g $m^{-2}d^{-1}$), respectively, during the summer months. They noticed no significant differences on gaseous emission due to different protein diet levels.

Table 11 Average manure composition on monthly basis

Parameters	June	July	August	September	October
CH_4 concentration (ppm)	2.2 d* ± 0.0	2.8 c ± 0.2	2.9 bc ± 0.2	3.3 ab ± 0.2	3.5 a ± 0.7
CO_2 concentration (ppm)	380.2 b ± 5.9	486.2 a ± 19.4	459.3 a ± 21.5	404.0 b ± 19.0	368.6 b ± 14.3
N_2O concentration (ppm)	1.1 a ± 0.2	1.0 a ± 0.2	1.1 a ± 0.2	0.6 b ± 0.0	0.4 b ± 0.0
CH_4 FR (g $m^{-2}d^{-1}$)	1.1 d ± 0.0	1.4 c ± 0.1	1.4 bc ± 0.1	1.6 ab ± 0.1	1.7 a ± 0.3
CO_2 FR (g $m^{-2}d^{-1}$)	507.6 b ± 7.9	649.2 a ± 26.0	613.3 a ± 21.5	539.4 b ± 19.0	492.1 b ± 14.3
N_2O FR (g $m^{-2}d^{-1}$)	1.4 a ± 0.3	1.4 a ± 0.2	1.5 a ± 0.3	0.8 b ± 0.0	0.5 b ± 0.0
CH_4 EF (g d^{-1} hd^{-1})	40.5 b ± 1.2	52.7 ab ± 3.8	54.2 b ± 3.8	62.4 a ± 2.3	63.7 a ± 11.6
CO_2 EF(g d^{-1} hd^{-1})	19487 b ± 624	24958 a ± 326	23584 a ± 326	20693 b ± 143	18541 b ± 431
N_2O EF (g d^{-1} hd^{-1})	55.2 a ± 11.0	53.3a ± 9.3	58.2 a ± 9.6	29.6b ± 1.4	17.7 b ± 0.9
CH_4 EF (g d^{-1} AU^{-1})	53.2 ab ± 7.9	57.6 a ± 6.7	51.2 b ± 10.2	52.8 ab ± 6.2	52.0 ab ± 14.9
CO_2 EF(kg d^{-1} AU^{-1})	25.5 a ± 2.6	26.9 a ± 2.0	22.2 b ± 2.9	17.5 c ± 1.8	15.1 c ± 0.8
N_2O EF(g d^{-1} AU^{-1})	67.0 a ± 29.0	57.7 a ± 18.5	55.8 a ± 27.7	24.9 b ± 8.1	14.5 c ± 2.0

*Values followed by the same letter in row are not significantly different at $P \leq 0.05$

Further analsysis was carried out to see the interaction of diet and time on GHG emisison. The results reveal that all CH_4, CO_2 and N_2O emisison (concentration and emission rate) varied significantly ($p < 0.05$) over the sampling period; however, diet did not have any interaction with time for the effect on GHG emissions (Table 12).

3.4 Effect of dietary fat levels on hydrogen sulfide emission

Hydrogen sulfide concentration was very low (<80 ppb) at the pen surfaces throughout the measurement period. Other researchers have also reported the concentration around 50 ppb in feedlot [54]. There was no significant difference in H_2S emission rate among dietary treatments. However, variations in H_2S emission rates were observed during different sampling periods (Fig. 2). The H_2S emission rate was fairly low (<0.18 g m^{-2} d^{-1}) in the first month since pen surfaces had a thin layer of manure on the surface. The H_2S concentration gradually increased over time and reached up to 0.7 g m^{-2} d^{-1} in August (Fig. 3). However, as the temperature started decreasing (Fig. 2), the H_2S emission rate also declined gradually (Fig. 3).

Table 12 Probability values based on the repeated measure multivariate analysis along with time and treatment interactions

Parameters		Interaction	
	Time		Diet*Time
CH_4 g m^{-2} d^{-1}	<0.01		0.68
CO_2 g m^{-2} d^{-1}	<0.01		0.43
N_2O g m^{-2} d^{-1}	<0.01		0.37
CH_4 g AU^{-1} d^{-1}	0.03		0.41
CO_2 kg AU^{-1} d^{-1}	<0.01		0.97
N_2O g AU^{-1} d^{-1}	<0.01		0.48

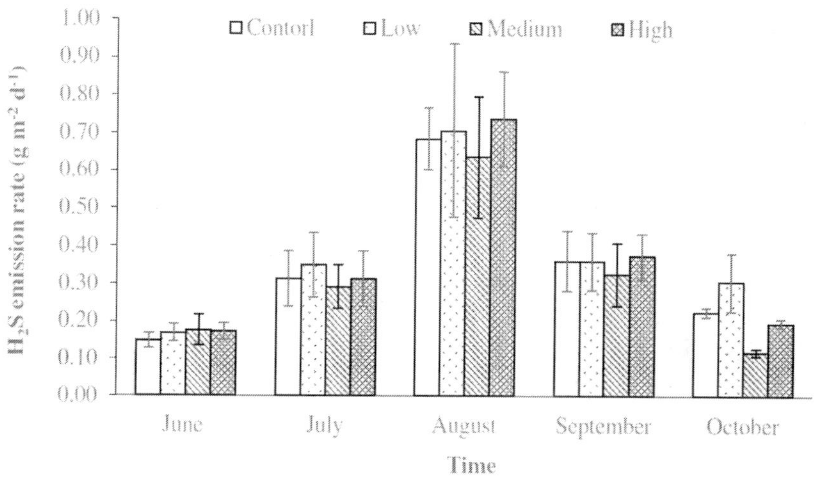

Figure. 3 Hydrogen sulfide emission from feedlot pen surfaces in different time period

This study shows that H_2S emission rate measured on the feedlot pen surfaces were correlated with temperature change ($R^2 = 0.49$), and manure accumulation (Figs. 2 and 3). Other researchers have also observed very low emission rate of H_2S from feedlot. Wood et al. [55]) reported the emission rate 103 μg m^{-2} min^{-1}. Similarly, Baek et al. [56] and Koziel et al. [57] reported an the H_2S emission rate as 1.88 μg m^{-2} min^{-1}, and 1.39 μg m^{-2} min^{-1}, respectively.

4. CONCLUSIONS

In this study the effect of four dietary fat concentrations (3 to 5.5 % in the composite sample) feed to beef cattle was evaluated in term of manure nutrient composition, VFA concentration, hydrogen sulfide and GHG (CH_4, CO_2, and N_2O) emissions. The study was conducted over a 5-month period from June to October for five ~28-day sampling periods. Overall, the fat levels in the diets showed no or little effect on the manure compositions, VFA, and H_2S and GHGs emissions. However, some variation in the above mentioned parameters was observed among different measurement periods. In this research, the variation of fat levels from 3 to 5.5 % in cattle diets did not reflect any significant difference on GHGs and H_2S emission from beef cattle feedlot pen surfaces, as well as on manure composition. It can be concluded that addition of fat to animal diet may not have any impact on gaseous emission and manure compositions.

ACKNOWLEDGMENTS

This project was funded by the North Dakota Corn Council and State Board of Agricultural Research and Education (SBARE). The authors are thankful to the Carrington Research and Extension center, NDSU; and grateful to Dr. Vern Anderson for his help during sampling.

REFERENCES

1. *Spellman FR, Whiting NE. Environmental management of concentrated animal feeding operations (CAFOs): CRC Press. 2010.*
2. *USDA. Cattle Inventory – United States: July 1. National Agricultural Statistics Service (NASS), Agricultural Statistics Board, United States Department of Agriculture (USDA). Reterieved on 9/10/2015 from* http://www.usda.gov/nass/PUBS/ TODAYRPT/ catl0715.pdf.

3. *Viguria M, Sanz-Cobena A, Lopez DM, Arriaga H, Merino P. Ammonia and greenhouse gases emission from impermeable covered storage and land application of cattle slurry to bare soil. Agr Ecosyst Environ. 2015;199:261–71.*
4. *Portejoie S, Martinez J, Landmann G. Ammonia of farm origin: impact on human and animal health and on the natural habitat. INRA Productions Animales. 2002;15(3):151–60.*
5. *Chadwick DR, Pain BF, Brookman SKE. Nitrous oxide and methane emissions following application of animal manures to grassland. J Environ Qual. 2000;29(1):277–87.*
6. *Bowman A, Mueller K, Smith M. Increased animal waste protection from concentrated animal feeding operation (CAFOs): Potential implication for public and environmental health. Omaha: Nebraska Center For Rural Health Research; 2000.*
7. *Guidotti TL. Occupational exposure to hydrogen sulfide in the sour gas industry: some unresolved issues. Int Arch Occup Environ Health. 1994;66(3):153–60.*
8. Milby TH, Baselt RC. Hydrogen sulfide poisoning: clarification of some controversial issues. Am J Ind Med. 1999;35(2):192–5.
9. Hribar, C. and Schultz M. Understanding concentrated animal feeding operations and their impact on communities. Bowling Green, OH:National Association of Local Boards of Health. Retrieved February.2010; 18: 2013.
10. Dang H, Li J, Chen R, Wang L, Guo L, Zhang Z, et al. Diversity, abundance,and spatial distribution of sediment ammonia-oxidizing betaproteobacteriain response to environmental gradients and coastal eutrophication inJiaozhou Bay, China. Appl Environ Microbiol. 2010;76(14):4691–702.
11. Menz FC, Seip HM. Acid rain in Europe and the United States: an update.Environ Sci Pol. 2004;7(4):253–65.
12. Hartung J, Phillips VR. Control of gaseous emissions from livestock buildingsand manure stores. J Agric Eng Res. 1994;57(3):173–89.
13. Owen JJ, Silver WL. Greenhouse gas emissions from dairy manure management:a review of field-based studies. Glob Chang Biol. 2015;21(2):550–65.
14. Philippe FX, Nicks B. Review on greenhouse gas emissions from pig houses:Production of carbon dioxide, methane and nitrous oxide by animals and manure. Agr Ecosyst Environ. 2015;199:10–25.
15. IPCC. Second Assessment Report, Climate Change: The Scientific BasisContribution of Working Group I to the Second Assessment Report of theIPCC. 2001.
16. USEPA. Inventory of US Greenhouse Gas Emission and Sinks: 1990-2007.Washington DC. Retrieved on 04/29/2016 from https://www3.epa.gov/climatechange/ Downloads/ ghgemissions/GHG2007entire_report-508.pdf.
17. USEPA. Inventory of U.S. greenhouse gas emissions and sinks: 1990-2013,Washington DC, USA. 2015. Retrieved on 04/29/2016 from https://www3.epa.gov/ climatechange/ Downloads/ghgemissions/US-GHG-Inventory-2015-Main-Text.pdf.
18. USEPA. US greenhouse gas inventory report: inventory of US greenhousegas emission and Sink:1990-2012. Washington DC. Retrieved on 04/29/2016 from https://www3. epa.gov/climatechange/Downloads/ghgemissions/USGHG-Inventory-2014-Main-Text.pdf.
19. Schader C, Jud K, Meier MS, Kuhn T, Oehen B, Gattinger A. Quantification of the effectiveness of greenhouse gas mitigation measures in Swiss organic milk production using a life cycle assessment approach. J Clean Prod. 2014;73:227–35.
20. Hristov AN, Ott T, Tricarico J, Rotz A, Waghorn G, Adesogan A, et al. SPECIALTOPICS-Mitigation of methane and nitrous oxide emissions from animaloperations: III. A review of animal management mitigation options. J AnimSci. 2013;91(11):5095–113.
21. Meale SJ, McAllister TA, Beauchemin KA, Harstad OM, Chaves AV. Strategiesto reduce greenhouse gases from ruminant livestock. Acta Agric Scand SectA Anim Sci. 2012;62(4):199–211.
22. Hulshof RBA, Berndt A, Gerrits WJJ, Dijkstra J, van Zijderveld SM, NewboldJR, et al. Dietary nitrate supplementation reduces methane emission in beef cattle fed sugarcane-based diets. J Anim Sci. 2012;90(7):2317–23.
23. Chuntrakort P, Otsuka M, Hayashi K, Takenaka A, Udchachon S, Sommart K.The effect of dietary coconut kernels, whole cottonseeds and sunflowerseeds on the intake,

digestibility and enteric methane emissions of Zebu beef cattle fed rice straw based diets. Livest Sci. 2014;161:80–9.

24. Osada T, Takada R, Shinzato I. Potential reduction of greenhouse gasemission from swine manure by using a low-protein diet supplemented with synthetic amino acids. Anim Feed Sci Technol. 2011;166–67:562–74.
25. Dutreuil M, Wattiaux M, Hardie CA, Cabrera VE. Feeding strategies and manure management for cost-effective mitigation of greenhouse gas emissions from dairy farms in Wisconsin. J Dairy Sci. 2014;97(9):5904–17.
26. Mirabelli MC, Wing S, Marshall SW, Wilcosky TC. Race, poverty, and potentialexposure of middle-school students to air emissions from confined swinefeeding operations. Environ Health Perspect. 2006;114:591–6.
27. Johnson KA, Johnson DE. Methane emissions from cattle. J Anim Sci. 1995;73(8):2483–92.
28. Chizzolini R, Zanardi E, Dorigoni V, Ghidini S. Calorific value and cholesterolcontent of normal and low-fat meat and meat products. Trends Food SciTechnol. 1999;10(4–5):119–28.
29. Mathison G, Okine E, McAllister T, Dong Y, Galbraith J, Dmytruk O.Reducing methane emissions from ruminant animals. J Appl Anim Res 1998;14(1):1–28.
30. Zinn R, Plascencia A. Effects of forage level on the comparative feedingvalue of supplemental fat in growing-finishing diets for feedlot cattle. J Anim Sci. 1996;74:1194–201.
31. Eastridge M. Effects of feeding fats on rumen fermentation and milkcomposition. In: Proc. 37th Annual Conf. Pacific Northwest Animal Nutrition,Vancouver, Canada. 2002.
32. Ziegler K. Nutrition and Management: Feeding Fats and Oils in Feedlot Diets.2007. Engstrom DF, Goonewardene LA, Volek R, Grimson RE, Stilborn RP, McKinnonPJ. Effects of feeding added protein and fat on feedlot performance and carcass quality in large frame steers. Can J Anim Sci. 1994;74(3):547–9.
33. Mathison G. Effect of canola oil on methane production in steers. Can JAnim Sci. 1997;77:545.
34. Beauchemin KA, McGinn SM. Methane emissions from beef cattle: Effects offumaric acid, essential oil, and canola oil. J Anim Sci. 2006;84(6):1489–96.
35. Beauchemin KA, McGinn SM, Benchaar C, Holtshausen L. Crushed sunflower, flax, or canola seeds in lactating dairy cow diets: effects on methane production, rumen fermentation, and milk production. J Dairy Sci. 2009;92(5) 2118–27.
36. Anderson V, Engel C. Effects of fat level in distillers grain on feedlot finishingperformance and carcass traits. North Dakota Beef Report - AS1736. In North Dakota Beef Report - AS1736. Fargo: North Dakota State University; 2014. p. 9.
37. Lardy G, Anderson V. Feeding Coproducts of the Ethanol Industry to BeefCattle, AS1242. NDSU extension service. Fargo, ND: North Dakota State University; 2014. p. 9.
38. NRC. Nutrient Requirements of Beef Cattle. 7th Revised Edition. Washington,D.C: National Academy of Science; 1996.
39. Rahman S, Borhan MS, Swanson K. Greenhouse gas emissions frombeef cattle pen surfaces in North Dakota. Environ Technol.2013;34(10):1239–46.
40. van Winden JF, Reichart G-J, McNamara NP, Benthien A, Damsté JSS.Temperature-induced increase in methane release from peat bogs: a mesocosm experiment. PLoS One. 2012;7(6):e39614.
41. Boeckx P, Van Cleemput O. Methane oxidation in a neutral landfill coversoil: influence of moisture content, temperature, and nitrogen-turnover. J Environ Qual. 1996;25(1):178–83.
42. Luo G, Kiese R, Wolf B, Butterbach-Bahl K. Effects of soil temperature andmoisture on methane uptake and nitrous oxide emissions across threedifferent ecosystem types. Biogeosciences. 2013;10(5):3205–19.
43. Lang M, Li P, Zhang X. Effects of land use type and incubation temperatureon soil nitrogen transformation and greenhouse gas emission. Ying YongSheng Tai Xue Bao. 2012; 23(10):2670–6.

44. Borhan MS, Gautam DP, Engel C, Anderson VL, Rahman S. Effects ofpen bedding and feeding high crude protein diets on manurecomposition and greenhouse gas emissions from a feedlot pen surface. J Air Waste Manage Assoc. 2013;63(12):1457–68.
45. Zinder SH. Conversion of acetic acid to methane by thermophiles. FEMSMicrobiol Rev. 1990;75:125–38.
46. Ward AJ, Hobbs PJ, Holliman PJ, Jones DL. Optimisation of the anaerobicdigestion of agricultural resources. Bioresour Technol. 2008;99(17):7928–40.
47. Boone DR, Xun L. Effects of pH, temperature, and nutrients on propionatede gradation by a methanogenic enrichment culture. Appl EnvironMicrobiol. 1987;53(7):1589–92.
48. Gourdon R, Vermande P. Effects of propionic acid concentration on anaerobic digestion of pig manure. Biomass. 1987;13(1):1–12.
49. Pullammanappallil PC, Chynoweth DP, Lyberatos G, Svoronos SA. Stable performance of anaerobic digestion in the presence of a high concentration of propionic acid. Bioresour Technol. 2001;78(2):165–9.
50. Marchaim U, Krause C. Propionic to acetic acid ratios in overloaded anaerobic digestion. Bioresour Technol. 1993;43(3):195–203.
51. Machmüller A, Ossowski DA, Kreuzer M. Effect of fat supplementation onnitrogen utilisation of lambs and nitrogen emission from their manure. Livest Sci. 2006;101(1–3):159–68.
52. Klemedtsson L, Svensson B, Rosswall T. Relationships between soil moisturecontent and nitrous oxide production during nitrification and denitrification.Biol Fertil Soils. 1988;6(2):106–11.
53. Sullivan J. Feedlot Air Quality Summary: Data Collection, Enforcement and Program Development: Minnesota Pollution Control Agency. 1999.
54. Wood S, Schmidt D, Janni K, Jacobson L, Clanton C, Weisberg S. Odor andair emissions from animal production systems. In: ASAE Annual InternationalMeeting Sponsored by ASAE. 2001.
55. Baek B-H, Koziel JA, Spinhirne JP, Parker DB, Cole NA. Estimation of ammoniaand hydrogen sulfide emissions from cattle feedlots in Texas. 2003.
56. Koziel JA, Parker DB, Baek B-H, Bush KJ, Rhoades M, Perschbacher-Buser Z.Ammonia and hydrogen sulfide flux from beef cattle pens: implications for air quality measurement methodologies and evaluation of emission controls. 2005.
57. NRC. National Research Council: Nutrient Requirements of Dairy Cattle. 6thRev. Ed. Washington, D.C.: National Academy of Sciences; 2001.
58. NRC. National Research Council: Nutrient Requirements of Beef Cattle. Seventh Revised Edition. Washington, D.C.: The National Academies Press; 2001.
59. Weiss W, Conrad H, Pierre NS. A theoretically-based model for predicting total digestible nutrient values of forages and concentrates. Anim Feed Sci Technol. 1992;39(1):95–110.

CHAPTER 13

Exploring Agritourism Experience and Perceptions of Pork Production

Ann M. Cummins[1], Nicole J. Olynk Widmar[1*], Candace C. Croney[2], Joan R. Fulton[1]

[1]*Department of Agricultural Economics, Purdue University, West Lafayette, IN, USA*

[2]*College of Veterinary Medicine, Purdue University, West Lafayette, IN, USA*

ABSTRACT

The interest of consumers in farming activities is evident and agricultural producers have responded to this increasing interest in a number of ways, including inviting visitors to farm operations through tourism or farm visits. Livestock industries, in particular, are interacting with the public in numerous ways, including via marketing channels for their products, interacting with community leaders and community members in regard to location siting and regulations, and inviting people onto their operations via agritourism. An improved understanding of who visits agricultural locations (particularly livestock operations) and consumers' perceptions about livestock agriculture is needed in order to begin to understand the possible relationships between on- farm experiences and consumers' perceptions. Thus, it is important to gain a deeper understanding in consumers' interests, including their levels of concern for production process attributes, such as animal welfare, relative to product attributes like price and taste. This analysis investigates the demographics of consumers who have (and have not) visited various agritourism operations, studies differences in preferences for (or perceptions of) animal welfare relative to other pork attributes between consumers who have or have not visited a livestock operation, as well as amongst consumers with varying levels of involvement in household food production. Indeed the relationships explored are between having visited a livestock operation and the perceptions. This study found that 69% of participants had visited a livestock operation (pig farm, dairy farm, and/or horse farm) at some point. The majority of respondents agreed that agriculture was an important industry, yet, those who had visited livestock operations were more likely to note concerns about the impact of livestock operations on water quality in their county.

Keywords: Agritourism, Animal Welfare, Best-Worst Scaling, Pork Production

1. INTRODUCTION

Livestock producers today are interacting with the public in many ways including through the products they sell in stores, communication and marketing of their products, interacting with their neighborhoods and communities, and inviting people onto their farms. Consumers' interest in farming activities is evident and producers have responded to this increasing interest in multiple ways, including inviting consumers to farms through tourism or farm visits. The livestock industry has displayed an increased interest in understanding consumers' perceptions and preferences for different agriculture outputs (food products, mainly) and practices used to produce those products. Thus, it is important to gain a deeper understanding of consumers' interests, including their levels of concern for production process attributes, such as animal welfare, relative to produce attributes like price and taste.

The pork industry has received attention recently regarding the welfare of the pigs raised in specific housing systems or using certain practices. Recent literature on livestock product consumption has demonstrated an increasing concern for how livestock animals are handled and raised [1] - [5]. Consumers' preferences for different production practices and consumers' willingness to pay for verified production practices have been studied at length [2] [5] - [9].

Along with an interest in understanding consumers' preferences for pork attributes, production practices or product attributes, another area of interest is in understanding relationships between preferences for production processes and other factors, such as knowledge or consumer demographics. Showing people what occurs on farms in order to better inform them about daily farming practices and activities is currently a major topic of discussion in animal agriculture. The underlying assumption is that if people can see the farm they will better understand what happens and why, resulting in reduced concern about animal welfare or the production practices used. In order to begin to understand the possible relationships between farm experiences and perceptions, better understanding of who visits agricultural locations (particularly livestock operations) and consumers' perceptions about livestock agriculture is needed. The definition of agritourism differs across much of the literature; agritourism experiences can vary with the amount of contact participants who have with nature, the authenticity of the experience, and whether or not the farm visited is a working farm [10]. As a starting point in this analysis, agritourism is investigated as a function of people's experiences with visiting a farm operation.

There has been discussion in the international literature surrounding agritourism related to tourism and rural development in recent times and especially over the past 20 years [10]. Agritourism is frequently discussed in its relation to rural development, potential impacts of diversification and stabilization of small farm income, or in regard to the analysis of key success factors for the development of an agritourism enterprise. However, little is known about the impact agritourism has on consumers' preferences for agricultural products. Consumers' preferences and perceptions in relation to their experiences with agriculture are particularly important in the discussion of production practices used in livestock production. No research that we are aware of addresses the differences in preferences for (or perceptions of) animal welfare relative to other pork attributes between consumers who have or have not visited a livestock operation.

The goal of this analysis was to first examine the differences between those individuals in the population who have and have not been to a livestock operation. Second, this analysis sought to expand understanding about how respondents who have been to a livestock operation might differ in their preferences for pork production attributes and their perspectives of livestock operation growth compared with those who have not been to a livestock operation. Overall, this study sought to link preferences for pork products with livestock farm experience. Cummins, Widmar and Croney [11] studied consumers' preferences for seven pork attributes (animal welfare, price, taste, environmental impact, pork/food safety, locally raised/farmed pigs, and locally processed pork) to gain an understanding of the relative importance of these attributes for pork. Shares of preference were estimated for each of the seven pork attributes; by design the shares must sum to 100%, thereby allowing relative ranking of each attribute against each other attribute. Cummins, Widmar, and Croney [11] found that the mean preference share for animal welfare was 16% and that animal welfare ranked the third most important (out of the seven attributes studied).

2. MATERIALS AND METHODS

2.1 Survey Instrument and Data

The data used for this analysis comes from an online survey of U.S. households which was administered July 23-August 6 of 2014. A large opt-in panel provider, Light speed GMI, was used to recruit participants who were at least 18 years of age. The survey was targeted to be representative of U.S. households in terms of age, gender, pre-tax income, and region of residency.

According to the U.S. Census Bureau the total U.S. population is 308,745,538 people (2010 Census, Revised 2014). In order to have a sufficient sample size to offer insight, the sample size needed, S, was calculated in the following way: S= X/[1+(X/P)] where P is the total size of the U.S. population and

$$X = Z \times Z \left[F \times \frac{1-F}{D \times D} \right]$$

where Z is the value associated with the confidence interval desired assuming a normal distribution. In this case the confidence interval desired is 95%, thus 1.96 is the value of Z. The value of F is 0.5 which was the frequency of the factor in the study. The variable D was defined to be the maximum difference between the sample and population means that is acceptable, D = 0.05 in this study. Thus for the U.S. population the sample size needed to offer insights into U.S. households is 385. The survey collected 1,004 responses, however, a simple validation test was used within the survey and 857 individuals (85.4%) answered this validation question correctly (or, passed the test).

According to Gao, House, and Bi [12] the use of a simple validation test is a way to improve data quality. In their study Cummins, Widmar and Croney [11] found by testing within their sample that the respondents who passed this validation test had statistically different sample mean and variance values for the size shares of preference for many attributes studied compared to those who didn't pass the validation test. Thus, only the 857 respondents who passed the validation test have been used in this analysis. Given the calculated required sample size of 385, the sample size for this analysis of 857 individuals is more than sufficient.

A recap of survey respondent demographics and level of education for the sample being analyzed is displayed in Table 1. According to the U.S. Census Bureau [13] 49% of the U.S. population is male and in this sample 50% were male. According to the U.S. Census Bureau [13] 70% of the U.S. population over the age of 18 was 25 - 64 years old; this sample had 74% of respondents indicate they were between the ages of 25 and 64 years old. The average pre-tax income in the U.S. is $73,034 [14] and this sample had $67,453 for the mean income. The four regions of residency according to the U.S. census [15] are Northeast, South, Midwest, and West with 18%, 38%, 22% and 22% of the U.S. population respectively. These numbers are very similar to the sample used in this analysis. According to the U.S. Census Bureau [16] the percentage of the population 25 years and over who have at least a high school degree is 86.9% and 30.1% have at least a bachelor's degree. This sample was slightly more educated than the U.S. population and had 99% of the population (18 years and older) with at least a high school degree and 45% with at least a bachelor's degree.

Along with basic demographic data, information was collected regarding consumers' participation in a wide variety of tourism activities, including having visited different operations, such as livestock or crop farms, and food plants or production facilities. Participants were given the list of operation types and asked when last they had visited each operation. Respondents were asked, for each operation type, to select one of the following responses: "within the last year", "1 - 10 years ago", "over 10 years ago", and "I have never visited such an operation". Individuals who selected "within the last year", "1 - 10 years ago" and "over 10 years ago" were combined and defined as the group who had ever visited the operation type. The full list of operation types and percentage of respondents who had ever visited the respective types are displayed in Figure 1. It is important to note that while participants were counted as having visited an operation type when they indicated that they had done so, the purpose of the visit or the depth of the experience was not measured. Since this analysis is focused on gaining deeper insight on the relationships between visiting agritourism operations, particularly focused on livestock, and preferences for pork products, it is of interest to study the group of individuals who have been to one of the livestock operations. Sixty-nine percent of participants indicated that they had visited a livestock operation; in other words, a total of 69% of respondents had been to a dairy farm, pig farm and/or horse farm.

In addition to questions designed to gain an understanding of demographics and tourism participation, several questions were used to identify each participant's familiarity with agriculture, their perception of agriculture and livestock production practices, and their views on livestock operation growth. Cross-tabulations were used to look at relationships between having visited a livestock operation and variables including demographics, household production involvement and views on livestock operation growth. To analyze statistical significance throu- ghout the cross-tabulations, chi-square statistics were analyzed; those presented were all significant at the 5% level. To determine statistically significant differences across categories (at the 5% level) in cross-tabulations a z-score was used.

Table 1. Sample demographics (n = 857).

Variable Description	Number of Respondents	Frequency (%)
Male	431	50
Age		
18 to 24 years	61	7
25 to 44 years	319	37
45 to 64 years	315	37
65 years and over	162	19
Household Income		
Less than $25,000	168	20
$25,000 - $34,999	100	12
$35,000 - $49,999	135	16
$50,000 - $74,999	174	20
$75,000 - $99,999	116	13
$100,000 - $149,999	119	14
$150,000 or more	45	5
Education		
Did not graduate from high School	10	1
Graduated from high School	150	17
Attended college, no degree earned	188	22
Attended college, associate or trade degree earned	124	15
Attended college, bachelor's (B.S. or B.A.) degree earned	240	28
Graduate or advanced degree (M.S., Ph.D., law school)	144	17
Other	1	0
Region of Residency		
Northeast	158	18
South	317	37
Midwest	214	25
West	168	20

Seven pork attributes (animal welfare, price, pork/food safety, taste, environmental impacts, locally farmed/ raised pigs, locally processed pork) were studied in a previous analysis by Cummins, Widmar, and Croney [11]. Allocation of total shares of preference, necessarily summing to 100% across all seven attributes, were completed; these results are referred to as preference shares for each of the attributes. The results of the mean estimated shares of preference for the seven different pork attributes are shown in Figure 2. Correlations between the calculated individual shares of preference for the seven pork attributes from Cummins, Widmar and Croney [11] and participants' responses to questions about visiting various agricultural operation types were completed using Pearson correlations and statistical significance at the 5% and 1% levels.

3. RESULTS AND DISCUSSION

3.1. Tourism Participation

Several commonly visited operation types were investigated in this analysis, including various agritourism (and food tourism) operations. It was found that people who visited any type of operation investigated were more likely to have reported visiting other location types as well. In other words, there were positive correlations amongst attraction attendance (at the 1% significance level). Being a tourist at one type of attraction was positively correlated with being a tourist at another attraction as well.

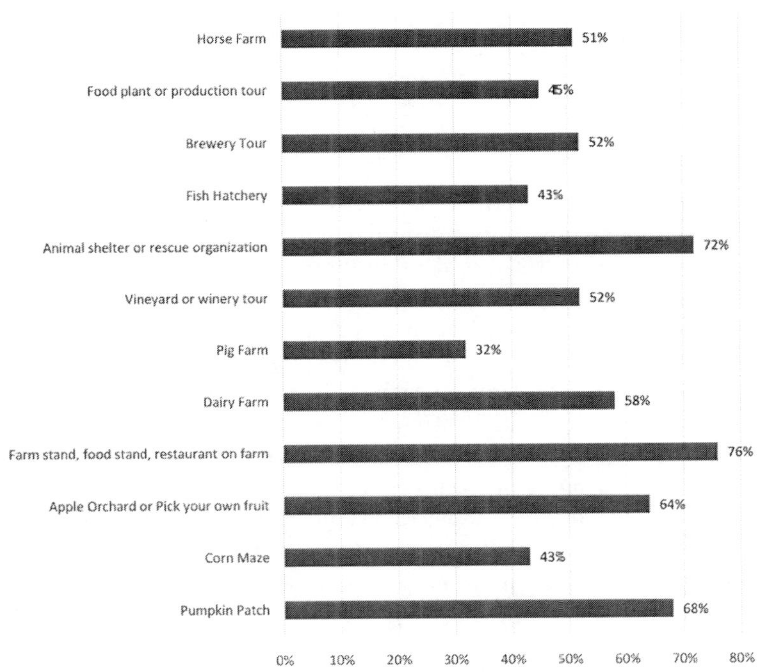

Figure 1. Percent of respondents who have ever visited the operation types investigated (n = 857).

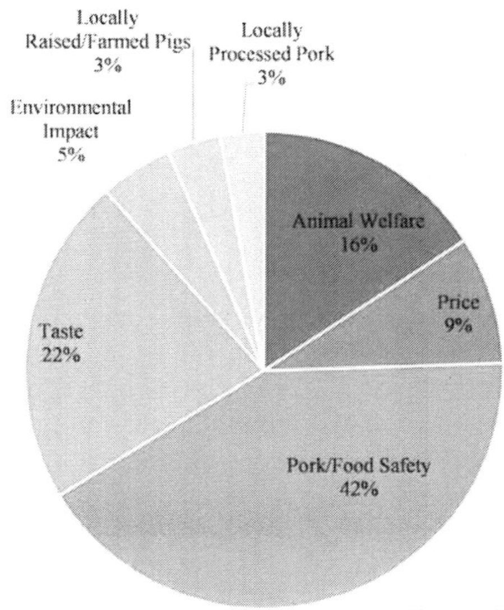

Figure 2. U.S. consumer shares of preference for pork attributes (n = 857).

Relationships between gender, age, income, and respondents who had been to each operation type were investigated using cross-tabulations (Table 2). In addition, relationships between region of residence and having visited the various operation types investigated are presented in Table 3. The values in each of the cells in the table represent the percentage of those in the corresponding demographic group that have been to the particular operation type (e.g. 70.8% of males have been to a livestock operation). Men more frequently reported having been to a dairy farm, pig farm, fish hatcher, brewery tour and food plant or production tour than did women. In general, those reporting higher income levels (up to $150,000) more frequently reported having been to each of the operations studied, with the exception of those who had attended an animal shelter or rescue organization where no statistical differences were found between income levels. Individuals living in the Midwest or West more frequently reported having been to a pumpkin patch than did those living in the South. Individuals who lived in the Midwest statistically more frequently reported having been to a pig farm than those living in the Northeast. These regional differences are not surprising in that it is reasonable to assume that those living in the vicinity of the respective operations have easier access to an operation type, and are therefore more likely to have had the opportunity to visit one.

Table 2. Cross-tabulations of basic demographics and having ever visited the operation type (n = 857).

Ever visited the following operation type	Gender[1]		Age (in years)[2]				Income[3]						
	Female (*)	Male (**)	18 - 24 (a)	25 - 44 (b)	45 - 64 (c)	65 + (d)	<$25,000 (A)	$25,000 - $34,999 (B)	$35,000 - $49,999 (C)	$50,000 - $74,999 (D)	$75,000 - $99,999 (E)	$100,000 - $149,999 (F)	$150,000 + (G)
Livestock operation	67.1	70.8	60.7	64.6	72.7	73.5	59.5 F	67.0	68.1	67.8	75	81.5A	66.7
Pumpkin patch	70.2	65.0	67.2	71.2	66.7	62.3	50.6$_{DEFG}$	63.0$_{F}$	65.2$_{F}$	71.3$_{A}$	73.3$_{A}$	82.4$_{ABC}$	80.0$_{A}$
Corn maze	43.9	42.9	67.2$_{cd}$	50.8$_{cd}$	37.8$_{ab}$	30.9$_{ab}$	25.6$_{DEFG}$	27.0$_{DEFG}$	40.0$_{F}$	47.1$_{ABF}$	51.7$_{AB}$	67.2$_{ABCD}$	57.8$_{AB}$
Apple orchard or pick your own fruit	60.8	66.4	52.5	65.5	63.8	63.6	48.2$_{DFG}$	56.0$_{F}$	63.0$_{F}$	64.9$_{AF}$	66.4	83.2$_{ABCD}$	75.6$_{A}$
Farm stand, food stand, restaurant on farm	75.4	77.0	59.0$_{cd}$	72.1	80.3$_{a}$	82.7$_{a}$	64.9$_{EF}$	68.0$_{F}$	74.8$_{F}$	78.2	83.6$_{A}$	89.9$_{ABC}$	77.8
Dairy farm	54.0$_{**}$	61.3$_{*}$	42.6$_{cd}$	52.7$_{d}$	61.3$_{a}$	66.0$_{ab}$	47.0$_{EF}$	55.0	57.0	55.7	66.4$_{A}$	70.6$_{A}$	55.6
Pig farm	24.6$_{**}$	40.1$_{*}$	29.5	35.4	30.2	32.1	26.2$_{F}$	24.0$_{F}$	30.4	32.8	37.1	48.7$_{AB}$	24.4
Horse farm	50.0	51.0	52.5	49.5	54.3	46.3	44.0$_{F}$	45.0$_{F}$	46.7$_{F}$	44.8$_{F}$	57.8	71.4$_{ABCD}$	53.3
Vineyard or winery tour	48.6	55.2	32.8$_{cd}$	49.5$_{d}$	51.4$_{ad}$	64.8$_{abc}$	26.8$_{CDEFG}$	38.0$_{EFG}$	53.3$_{AFG}$	54.0$_{AFG}$	61.2$_{AB}$	74.8$_{ABCD}$	80.0$_{ABCD}$
Animal shelter or rescue organization	74.4	69.6	67.2	71.5	73.7	71.6	66.1	69.0	71.9	75.9	72.4	79.0	66.7
Fish hatchery	36.2$_{**}$	49.7$_{*}$	26.2$_{cd}$	39.5$_{d}$	44.4$_{a}$	53.1$_{ab}$	28.6$_{EF}$	34.0$_{EF}$	40.0	44.3	55.2$_{AB}$	57.1$_{AB}$	51.1
Brewery tour	45.3$_{**}$	58.5$_{*}$	37.7$_{d}$	50.8	52.1	59.3$_{a}$	31.0$_{DEFG}$	40.0$_{FG}$	45.9$_{F}$	54.6$_{AF}$	59.5$_{AF}$	79.8$_{ABCDE}$	71.1$_{AB}$
Food plant or production tour	40.1$_{**}$	49.4$_{*}$	36.1	45.1	46.3	44.4	33.3$_{EF}$	31.0$_{EF}$	43.0$_{F}$	44.8$_{F}$	53.4$_{AB}$	63.9$_{ABCD}$	51.1

[1]Significant differences at the 5% level are marked by * and **; [2]Significant differences at the 5% level are marked by a, b, c, d'; [3]Significant differences at the 5% level are marked by A, B, C, D, E, F, G.

Table 3. Cross-tabulations between region of residency and having ever visited the operation type (n = 857).

	Region			
Ever visited one of the following tourism locations	Midwest (α)	South (β)	West (γ)	Northeast (δ)
Livestock operation	72.4	64.0	69.0	74.1
Pumpkin patch	72.9_{β}	$59.6_{\alpha\gamma}$	72.6_{β}	70.9
Corn maze	49.5	37.9	44.6	44.9
Apple orchard or pick your own fruit	70.1_{β}	$55.8_{\alpha\delta}$	60.1_{δ}	$74.1_{\beta\gamma}$
Farm stand, food stand, restaurant on farm	77.6	72.6	75.6	82.3
Dairy farm	62.6	52.1	57.7	62.0
Pig farm	39.3_{δ}	31.5	33.3	24.1_{α}
Horse farm	53.3	48.6	50.0	53.2
Vineyard or winery tour	51.9	47.0	59.5	53.8
Animal shelter or rescue organization	72.0	68.5	78.6	72.2
Fish hatchery	45.8	36.3_{γ}	54.8_{β}	39.9
Brewery tour	55.1	50.8	49.4	52.5
Food plant or production tour	45.8	40.7	50.6	45.6

Significant differences at the 5% level are marked by α, β, γ, and δ.

Six operation types were found to have differing levels of reported attendance across age groups, namely corn maze, "farm stand, food stand, restaurant on farm", dairy farm, vineyard or winery tour and brewery tour. In each of the eight, with the exception of having been to a corn maze, those in the age category 65 and older more frequently reported having been to the operation than did individuals in the age category 18 - 24. This finding is probably due in to increased experience over time. There were no statistical differences in age categories of those who had attended pumpkin patches, apple orchards or pick your own fruit locations, pig farm, horse farm, animal shelter or rescue organization or food plant/production tour.

Correlations between the size of preference shares for pork attributes and individuals who had attended each operation type were analyzed; these correlations are displayed in Table 4. Reporting having visited an animal shelter or rescue organization was positively correlated with the size of the share of preference for animal welfare when purchasing pork. This result is consistent with previous findings. The idea that individuals with connections or interactions with animals, primarily pets such as cats or dogs, tend to be more concerned about the welfare of all animals, including livestock, is not novel. Rothgerber and Mican [17] found that individuals who owned pets as children had stronger empathy for animals. McKendree, Croney and Widmar [3] postulated that human-animal interactions or relationships, particularly pet ownership, had the potential to influence people's perceptions of livestock animal welfare, and found that in U.S. households, pet owners were significantly more concerned about livestock animal welfare than those who did not own dogs and/or cat. The findings of Cummins, Widmar and Croney [11] also included that the ownership of a dog and/or cat was positively correlated with the size of the share of preference for animal welfare in the seven different pork attributes examined. Thus, finding that individuals who have visited animal shelters or rescues, which are predominantly directed towards species commonly classified as pets, are more sensitive to animal welfare concerns is supported by past studies.

Further, it is interesting to note that of all the tourism locations investigated, having visited an animal shelter/rescue organization was the only visit experience correlated with the size of the share of preference for the pork attribute of animal welfare. In particular, given the popularity of agritourism as a way to communicate with the general public by agricultural circles, it is indeed interesting to note that having visited a livestock operation was not significantly correlated with the size of the preference share for animal welfare.

Table 4. Pearson correlations between having ever visited an operation type and the size of the share of preference for pork attributes (n = 857).

	Animal welfare	Price	Pork/Food safety	Taste	Environmental impacts	Locally raised/farmed pigs	Locally processed pork
Pumpkin patch	0.004	−0.086*	0.050	0.020	0.050	0.063	0.068*
Corn maze	−0.028	−0.064	0.050	0.016	0.054	0.076*	0.095**
Apple orchard or pick your own fruit	−0.045	−0.082*	0.063	0.057	0.034	0.072*	0.098**
Farm stand, food stand, restaurant on farm	0.005	−0.079*	0.062	0.000	0.026	0.057	0.035
Dairy farm	−0.042	−0.035	0.027	0.001	0.094**	0.098**	0.112**
Pig farm	−0.011	−0.057	−0.013	0.054	0.069*	0.134**	0.173**
Horse farm	−0.029	−0.083*	0.077*	−0.002	0.042	0.114**	0.096**
Vineyard or winery tour	0.000	−0.078*	0.012	0.041	0.091**	0.112**	0.116**
Animal shelter or rescue organization	0.140**	−0.113**	−0.003	−0.028	0.060	0.062	0.046
Fish hatchery	0.000	−0.054	0.002	0.013	0.070*	0.131**	0.139**
Brewery tour	−0.021	−0.052	−0.005	0.049	0.079*	0.125**	0.130**
Food plant or production tour	−0.033	−0.055	−0.005	0.074*	0.099**	0.094**	0.121**

Note: Statistical significance (2-tailed) at the 5% and 1% level is represented by * and ** respectively.

3.2. Household Production and Involvement in Agriculture

In addition to visiting agricultural operations, people have many different ways of being exposed to agriculture and food production practices; direct ownership and/or household production of food can be among these. Participants were therefore asked, "Do you, a family member or relative own or operate a farm business (in any capacity, including a partnership or part-owner)?" Respondents selected all responses that applied from the list: "Yes, I own or operate a farm business", "Yes, I have a family member or relative who owns or operates a farm business", and "No." Eighty-eight percent of respondents did not report any familial ties (including self) to anyone who owned or operated a farm business in any capacity.

Cross-tabulations revealed that individuals who reported having been to a livestock operation more frequently also reported being an owner or operator of a farm operation and also more frequently reported having family members or relatives who owned or operated a farm business than those who had not been to a livestock operation. The cross-tabulation analysis revealed that individuals who had visited a livestock operation more frequently reported having been involved in each of the household production practices individually assessed. Individuals reporting having never been to a livestock operation more frequently reported not being involved in any of the household production practices examined in the previous three years.

To further understand households involved in home production of food products, participants were asked, "in the last three year time period, has your household been actively involved in producing food for your own family through any of the following ways?" The options included producing fruits and berries, growing produce in a garden at home or in a community garden, raising chickens for eggs or meat and raising other animals for meat or milk as well as the option of "none of the above." The results revealed that in the previous three year period, 13% of participants' households had been involved in cultivating fruit trees and/or berries, 33% of their households grew produce of some kind in a personal garden at home, 5% grew produce of some kind in a personal garden not at home, 6% raised chickens primarily for eggs, 4% raised chickens primarily for meat, and 4% raised animals (other than chickens) for meat or milk. Sixty-five percent of households reported not being involved in any household production. Participation in any type of household production was positively correlated with being involved in all other household production practices (at the 1% significance level) and negatively correlated with having self or familial ties to owning or operating a farm operation (at the 1% significance level). In other words, those individuals who produced one type of food for their household use were more likely to produce other types of food. In addition, those who owned or operated a farm were less likely to grow food for their own household consumption.

Individuals who reported having visited a livestock operation more frequently participated in household cultivation practices than those who had not visited a livestock operation. Correlations between the size of the preference shares for pork attributes and involvement in household production (in the past three years) as well as non-familial ties to farm ownership or operators is displayed in Table 5. Having indicated that the respondent and/or their relatives did not own or operate a farm business was negatively correlated (at the 1% significance level) to the size of the shares for environmental impacts, locally raised/farmed pigs, and locally processed pork. Involvement in any household production in the previous three year time period was statistically significant and negatively correlated to the size of the share of preference for price and positively correlated with the size of the shares of preference for locally raised/farmed pigs and locally processed pork. Involvement in all household production activities except for growing produce in a personal garden not at home" was positively correlated with the size of the share of preference for environmental impacts.

3.3. Perceptions of Agriculture and Growth

Participants were shown a series of 10 statements about agriculture or livestock growth and asked to respond with how much they agreed or disagreed on a Likert-scale of one through seven, where one was "very strongly disagree" and seven was "very strongly agree." The statements provided and the mean responses received are as follows: "I would oppose the building of new livestock operations in my county" (3.36), "I believe that livestock farms are environmentally harmful" (3.67), "I would oppose the growth of livestock operations in my county" (3.21), "I am concerned about the impacts of water quality from livestock operations in my county" (4.18), "I have experienced negative impacts from livestock operations located near my home or work" (2.54), "I am supportive of the growth of livestock agriculture in my county" (4.83), "I am supportive of the growth of livestock agriculture in my state but would prefer growth outside of my county/region" (3.84), "agriculture is an important industry in my state" (5.32), "odor/smell from livestock operations is a major concern for me" (3.99), and "I feel that livestock operations make good neighbors" (3.91).

Table 5. Pearson correlations between involvement in agriculture and size of the share of preference for pork attributes (n = 857).

	Animal welfare	Price	Pork/Food safety	Taste	Environmental impacts	Locally raised/farmed pigs	Locally processed pork
Respondent, a family member or relative do NOT own or operate a farm business in any capacity, including a partnership or part-owner)	−0.024	0.056	0.044	−0.029	−0.126**	−0.161**	−0.169**
Household Production (In the last three years)							
Cultivating fruit trees and/or berries	0.021	−0.080*	0.022	−0.002	0.072*	0.129**	0.135**
Growing produce of any kind in a personal garden at home.	0.011	−0.069*	0.014	0.033	0.023	0.099**	0.111**
Growing produce of any kind in a personal garden not at home (in a garden plot or community garden)	0.043	−0.097**	−0.011	−0.003	0.135**	0.220**	0.207**
Raising chickens primarily for eggs	0.031	−0.082*	−0.066	0.076*	0.158**	0.174**	0.280**
Raising chickens primarily for meat	0.038	−0.091**	−0.028	0.022	0.171**	0.182**	0.186**
Raising animals (other than chickens) for meat or milk	0.059	−0.117**	0.017	0.023	0.079*	0.104**	0.156**
None of the above household production	−0.017	0.079*	−0.009	−0.037	−0.045	−0.118**	−0.139**

Note: Statistical significance (2-tailed) at the 5% and 1% level is represented by * and ** respectively.

Cross-tabulations between participant's responses to the agriculture and livestock growth statements and whether they had been to a livestock operation were assessed and are reported for a subset of those statements (Table 6). Respondents who indicated that they had been to a livestock operation more frequently selected response "6" and less frequently selected "4" in response to the statement, "I am concerned about impacts on water quality from livestock operations in my county" than did those who had not been to a livestock operation. The most significant differences in responses between those who had and had not been to a livestock operation were responses to the statements "I am supportive of the growth of livestock agriculture in my county." In response to this statement, those who had been to a livestock operation more frequently selected response options "5", "6", and "7"and less frequently selected options "2" and "4" than those who had not been to a livestock operation. Thus, respondents who had been to a livestock operation more frequently indicated agreement that they were supportive of the growth of livestock agriculture in their county.

When asked to respond with their level of agreement or disagreement with the statement "agriculture is an important industry in my state" those who had been to a livestock operation more frequently selected options "6" and "7" (levels of agreement), and less frequently selected options "1" (very strong disagreement) or "4" (neutral). This implies that those who had been to a livestock operation felt that agriculture is an important industry in their state.

4. CONCLUSIONS

The focus of this analysis was to examine individuals who have visited livestock operations and determine how they might differ in their preferences for pork attributes, their involvement in agriculture, and their perceptions about agriculture overall (and specifically growth in the agriculture sector). The results showed that 69% of participants had visited a livestock operation (pig farm, dairy farm, and/or horse farm) at some point. Results also showed that those who had visited livestock operations did not differ in gender, age, or region of residency, but they tended to have higher incomes. Those who had been to a livestock operation were more likely to have also owned or were related to someone who owned or operated a farm business than those who had not been to a livestock operation. Similarly, those who had visited a livestock operation more frequently participated in each of the home production practices examined. This suggests that those who had been to a livestock operation had more familiarity with agricultural practices (whether for production or personal purposes).

Table 6. Cross-tabulations between perspective on agriculture and livestock growth with having visited a livestock operation (n = 857).

	I would oppose the building of new livestock operations in my county.[1]		I am concerned about impacts on water quality from livestock operations in my county.[3]		I have experienced negative impacts from livestock operations located near my home or work.[4]		I am supportive of the growth of livestock agriculture in my county.[5]		Agriculture is an important industry in my state.[6]		I feel that livestock operations make good neighbors.[7]	
Ever visited a livestock operation	No (*a*)	Yes (*b*)	No (*)	Yes (**)	No (α)	Yes (β)	No (•)	Yes (••)	No (□)	Yes (□□)	No (ϕ)	Yes (θ)
Very strongly disagree 1	20.7_b	21.8_a	10.9	10.5	45.1	41.1	4.9	2.5	$6.4_{\square\square}$	$1.7_{\square}$	12.0_θ	7.3_ϕ
2	12.4	18.1	9.0	9.8	11.7_β	22.0_α	$6.0_{\bullet\bullet}$	$2.9_{\bullet}$	3.4	2.4	13.9	10.3
3	9.0	13.7	10.2	12.5	10.5	8.6	7.9	6.8	6.4	4.4	15.8	13.9
4	33.1_b	21.5_a	29.3_{**}	19.8_{*}	19.5_β	11.0_α	$39.8_{\bullet\bullet}$	$24.2_{\bullet}$	$27.1_{\square\square}$	$15.6_{\square}$	38.7	32.3
5	12.0	10.0	21.4	19.5	7.9	6.8	$16.2_{\bullet\bullet}$	$22.3_{\bullet}$	15.4	19.8	10.5	14.9
6	6.0	9.3	10.9_{**}	16.2_{*}	3.8	6.1	$14.3_{\bullet\bullet}$	$24.7_{\bullet}$	$16.2_{\square\square}$	$23.4_{\square}$	6.8_θ	13.7_ϕ
Very Strongly Agree 7	6.8	5.6	8.3	11.7	1.5_β	4.4_α	$10.9_{\bullet\bullet}$	$16.6_{\bullet}$	$25.2_{\square\square}$	$32.8_{\square}$	2.3_θ	7.6_ϕ

[1]Significant differences at the 5% level are marked by *a* and *b*; [2]Significant differences at the 5% level are marked by A and B`; [3]Significant differences at the 5% level are marked by * and **; [4]Significant differences at the 5% level are marked by α and β; [5]Significant differences at the 5% level are marked by • and ••; [6]Significant differences at the 5% level are marked by □ and □□; [7]Significant differences at the 5% level are marked by ϕ and θ.

While the majority of participants indicated that they agreed that agriculture was an important industry in their state, those who had been to a livestock operation stated stronger levels of agreement than those who had not been to a livestock operation. Despite the belief that agriculture is important in their state, those who had been to a livestock operation more frequently agreed with the statements "I would oppose the growth of livestock operations in my county", and "I am concerned with the impacts on water quality from livestock operations in my county". While this study sought to measure levels of concern and agreement with various statements about animal agriculture, a limitation of this analysis is that there was not data collected specific to why respondents may or may not be concerned. Additional insights into the factors (beyond agritourism involvement) that may be influencing levels of agreement with beliefs or perceptions of agriculture should be explored in future studies.

ACKNOWLEDGEMENTS

Researchers at Purdue University developed the concept for this study in collaboration with Fair Oaks Farms, Belstra Milling Co., Indiana Pork and Indiana Soybean Alliance. Indiana Pork and Indiana Soybean Alliance provided the funding for the study. Researchers at Purdue University conducted the study and analysis without input, collaboration, sharing of survey design or participation in data collection by the funders, Fair Oaks Farms or Belstra Milling Co. in order to avoid biases or the perception of biases arising from working with industry groups.

REFERENCES

1. Carpio, C.E., Wohlgenant, M.K. and Boonsaeng, T. (2008) The Demand For Agritourism in the United States. Journal of Agricultural and Resource Economics, 33, 254-269. http://www.jstor.org/stable/41220626
2. McKendree, M.G.S., Croney, C.C. and Widmar, N.J.O. (2014) Effects of Demographic Factors and Information Sources on United States Consumer Perception of Animal Welfare. Journal of Animal Science, 92, 3161-3173.
3. McKendree, M.G.S., Croney, C.C. and Widmar, N.J.O. (2014) Current Factors Influencing Perceptions of Animals and Their Welfare. Journal of Animal Science, 92, 1821-1831.
4. Olynk Widmar, N.J. and Ortega, D.L. (2014) Comparing Consumer Preferences for Livestock Production Process Attributes Across Products, Species, and Modeling Methods. Journal of Agricultural and Applied Economics, 46, 375-391.
5. Olynk, N.J., Tonsor, G.T. and Wolf, C.A. (2010) Consumer Willingness to Pay for Livestock Credence Attribute Claim Verification. Journal of Agricultural and Resource Economics, 35, 261-280. http://www.jstor.org/stable/41960517
6. Lusk, J. and Briggeman, B. (2009) Food Values. American Journal of Agricultural Economics, 91, 184-196.

7. McKendree, M.G.S., Widmar, B.O., Ortega, D.L. and Foster, K.A. (2013) Consumer Preferences for Verified Pork-Rearing Practices in the Production of Ham Products. Journal of Agricultural and Resources Economics, 38, 397-417. http://ageconsearch.umn.edu/bitstream/165935/2/JARE,Dec2013,%237,McKendree,pp397G%C3%87%C3%B4417.pdf
8. Tonsor, G., Nicole J Olynk, and Christopher Wolf. (2009) Consumer Preferences for Animal Welfare Attributes: The Case of Gestation Crates. Journal of Agricultural and Applied Economics, 41, 713-730.
9. Sacket, H.M., Shupp, R. and Tonsor, G. (2013) Consumer Perceptions of Sustainable Farming Practices: A Best-Worst Scenario. Agricultural and Resource Economics Review, 42, 275-290.
10. Flanigan, S., Blackstock, K. and Hunter, C. (2014) Agritourism from the Perspective of Providers and Visitors: A Topology-Based Study. Tourism Management, 40, 394-405. http://dx.doi.org/10.1016/j.tourman.2013.07.004
11. Cummins, A.M., Widmar, N.J.O. and Croney, C.C. (2016) Understanding Consumer Pork Attribute Preferences. Theoretical Economic Letters, 6, 166-177. http://dx.doi.org/10.4236/tel.2016.62019
12. Gao, Z.F., House, L.A .and Bi, X. (2012) Finding True Consumer Attitudes: Do Validation Questions Help? Food and Resource Economics Department, University of Florida, Gainesville. http://ageconsearch.umn.edu/bitstream/123922/1/aaea-poster-05-18.pdf
13. U.S. Census Bureau (2014) DP-1 Profile of General Population and Housing Characteristics: 2010, 2010 Demographic Profile Data. http://factfinder.census.gov/faces/tableservices/jsf/pages/productview.xhtml?pid=DEC_10_DP_DPDP1&prodType=table
14. U.S. Census Bureau (2008-2012) DP01: Selected Economics: 2008-2012 American Community Survey 5-Year Estimates. http://factfinder.census.gov/faces/ tableservices/jsf/pages/productview.xhtml?pid=ACS_12_5YR_DP03&prodType=table
15. U.S. Census Bureau (2013) Annual Estimates of Housing Units for the United States, Regions, Divisions, States and Counties: April 1, 2010 to July 1, 2013: 2013 Population Estimates. http://factfinder.census.gov/faces/tableservices/jsf/pages/productview.xhtml?pid=PEP_2013_PEPANNHU&prodType=table
16. U.S. Census Bureau (2014) Educational Attainment: 2014 American Community Survey-One Year Estimates. http://factfinder.census.gov/faces/tableservices/ jsf/pages/productview.xhtml?pid=ACS_14_1YR_S1501&prodType=table
17. Rothgerber, H. and Mican, F. (2014) Childhood Pet Ownership, Attachment to Pets, and Subsequent Meat Avoidance. The Mediating Role of Empathy towards Animals. Appetite, 79, 11-17.http://dx.doi.org/10.1016/j.appet.2014.03.032

CHAPTER 14

Do Organic Livestock Farms Differ from Low-Input Conventional Ones? Insights Based on Beef Cattle in Southern Europe

Alfredo J. Escribano[1], Paula Gaspar[2], Francisco J. Mesías[3] and Miguel Escribano[4]

[1]*C/Rafael Alberti, Cáceres, Spain*

[2]*Department of Animal Production and Food Science, Faculty of Agriculture, University of Extremadura, Ctra. Cáceres s/n, Badajoz, Spain*

[3]*Department of Economy, Faculty of Agriculture, University of Extremadura, Ctra. Cáceres s/n, Badajoz, Spain*

[4]*Department of Animal Production and Food Science, Faculty of Veterinary Sciences, University of Extremadura, Avda.de la Universidad s/n. Cáceres, Spain*

ABSTRACT

The objective of this study is to determine whether there are clear differences between conventional (but low-input) and organic beef cattle farms located in the Southwest of Spain. Thirty-three conventional and 30 organic farms were compared in terms of structure, technical management, and performance. The results showed that organic farms ('All Organic') mainly focus on the production of calves at weaning age, which are fattened in conventional holdings ('Organic 1'; $n = 22$). The remaining organic farms ('Organic 2'; $n = 11$) showed to participate in almost all stages of the agri-value chain. 'Conventional' farms were mainly dedicated to producing calves at weaning age (similarly to Organic 1). Organic 1 had the smallest herd size (80.18 livestock units (LU), $p < 0.05$). Organic 2 showed greater presence of indigenous breeds (62.08%, $p < 0.05$). Conventional farms proved to bear higher feed and veterinary costs per area (161.59 and 17.87 €/ha; $p < 0.01$ and $p < 0.05$, respectively), but Organic 2 had higher feed costs per LU. Therefore, Conventional and All Organic were quite similar, and differences depended mainly on farm structure. Hence, being either conventional or organic does not seem to be a valid criterion for drawing conclusions regarding the benefits or characteristics of each system.

Keywords: semiarid, production systems, sustainability, sustainable agriculture, mediterranean, drylands

1. INTRODUCTION

Organic livestock farm numbers have sharply increased in recent years [1] as an adaptive response for increasing farm profitability (through subsidies and price premiums). However, before implementing any production systems, an analysis of the similarities and differences between both the current and the potential new farm configurations should be carried out, since it will provide a wider view of the chances of success after the change.

For this purpose, the methodological process of farm characterisation is essential as it allows in-depth understanding of the operation of livestock production systems, which is key to improving their management, economic performance and overall sustainability. Thus, Rodríguez et al. [2] stated that farm viability relies on specific management practices that are suitable for the specific socioeconomic and environmental context of the farm, and this should be based on the knowledge of the characteristics and performance of the production systems.

Subsequently, several researchers have conducted studies for characterising farms according to various parameters. Some authors have focused on livestock species reared on the farms. Thus, different authors [3–10] have studied and characterised beef cattle farms by means of descriptive and/or cluster analysis on the basis of technical, structural, economic and/or social indicator. However, to our knowledge, there are no available studies that comparatively characterise organic and pasture-based or low-input conventional beef cattle farms, contextualizing such analysis within the evolution of the production systems under study. We therefore believe that this is a particularly appropriate time to conduct the present study. This would be of even more interest if the farms studied were located in complex agro-ecosystems with great value and externalities from the socio-economic and environmental points of view. The present study was carried out with the following aims: (i) to shed light on the gap of knowledge existing due to the lack of studies that compare the characteristics of conventional and organic beef cattle farms and (ii) to find similarities and differences between organic and low-input conventional beef cattle farms. For this purpose, a characterisation (technical management, structure and economic performance) of the farms located in the '*dehesa*' was carried out.

2. MATERIAL AND METHODS

2.1. Study area

The study area was the *dehesa* located in the region of Extremadura (Southwest of Spain). From a climatic point of view, it enjoys annual average temperatures of 16–17°C, with mild winters (average temperature of 7.5°C) and hot and dry summers (the average mean temperature is greater than 26°C, exceeding 40°C in the hottest months, which correspond to a Mediterranean continental semiarid climate. Its rainfall pattern is irregular (300–800 mm). Soils are shallow, acidic and of low fertility. Due to these characteristics, the availability of grazing resources is reduced and unstable [11–13].

2.2. Sample selection

A sample of farms in the beef cattle sector located in the *dehesa* of Extremadura was selected. Due to a lack of official statistics on figures and locations of *dehesa* farms in Extremadura, the sampling was non-probabilistic by quotas. The number of farms surveyed was 63, in line with other studies analysing livestock production systems [4, 14–17]. Apart from the number of farms, various criteria (already explained by Escribano et al. [17, 18]) were used to select the farms with the aim of obtaining an indicative sample of the various beef cattle production systems located in *dehesas*. The criteria used are summarised below:

- Predominant species and productive orientation: beef cattle.
- Herd size: over 25 adult cows, in order to differentiate between small and commercial farms.

- Geographical and forest-related aspects: the study includes farms located in different areas within the *dehesa* (geographical criteria) with different woodland densities (forest criteria similar to that followed in previous studies in the area [14]). Figure 1 shows farms' spatial distribution and type of *dehesa* in which they were based on.
- Organic farms: all organic farms had already finished their conversion period to the organic system.

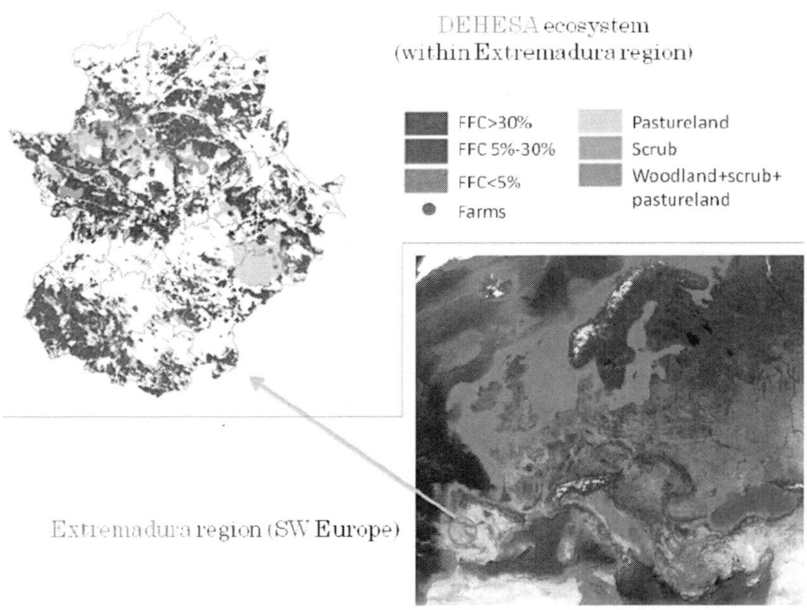

Figure 1. Dehesa location and different land cover characteristics. Forest fractional cover (FFC): Fraction of the land covered by the vertical projection of the tops of trees.

Finally, 63 farms (30 Conventional and 33 Organic farms were selected, thus achieving: a sample size similar to that of other studies characterising livestock farms [4, 5, 10, 14]; similar sample sizes of organic and conventional farms, thus allowing an adequate comparative analysis of both sectors.

2.3. Selection of parameters

In order to select the most appropriate indicators to analyse the farms under study, two main steps were followed. Firstly, the scientific literature addressing the structural and technical-economic points of view was reviewed. The selection of consistent and similar indicators allowed carrying out comparisons with studies on the topic. Moreover, economic parameters were created following the economic accounts for agriculture in the community [19] and the adaptation to dehesa livestock farms already carried out in previous studies [14, 15, 20]. As a consequence, the discussion of the results was consistent and the achievement of the aims of the present study was possible.

Finally, the selected set of indicators were confirmed to be in agreement with the recommendations of Lebacq et al. [21]: relevance, representativeness of the system, measurable, value to the end user, no ambiguity, no redundancy, and predictive.

2.4. Data collection

Data were collected from farms by means of a questionnaire in the year 2010. The questionnaire was developed according to selected indicators. These included information on structure (farms and herd characteristics: sizes, infrastructure, etc.), technical management, production results, economic data and social aspects. Subsequently, data were collected by the first author directly at the farms, followed by structured and semiclosed interviews with farm managers. Farmers' answers were the sources of information for all indicators. All these processes were carried out in accordance with the methodology used by several authors who analysed similar aspects of livestock farms [2, 6, 8, 10, 14, 15, 17, 22–27].

2.5. Analysis

The statistical analyses included descriptive statistics for the full sample of farms. Subsequently, an ANOVA test was applied to all parameters, as all of them are quantitative ones. This allowed comparing all farms following two approaches. First, conventional farms were compared to organic farms in order to compare the two production systems as a whole (Conventional vs. All Organic). Secondly, farms were compared based on three classifications that are explained in the next section: (i) Conventional farms; (ii) Organic 1 farms; (iii) Organic 2 farms. This approach offered insight into each of them, so that more valuable and precise conclusions about the organic beef cattle sector could be made. Statistical analyses were performed using SPSS v. 20.

3. RESULTS

3.1. Farm types

After collecting data and creating the database, it was noted that, based on the aspects studied, organic farms could clearly be subdivided into two production systems, so it was decided that a classification of the farms selected needed to be made, with the resulting following groups:

- Conventional; found as "Conv." in the tables (n = 30): This grouped conventional farms. With regards to the situation of the beef cattle sector in the *dehesa*, these farms were mostly focused on calf rearing (calf fattening was almost nonexistent, so these farms mainly sold their calves at weaning age (5–6 months old and 160–220 kg of live weight; see Table 1).
- Organic 1; found as "Org. 1" in the tables (n = 22): These farms were producing under the organic system, but none of them fattened their calves. On the contrary, they were marketed with almost the same characteristics (age and live weight) and price as the conventional ones (check Table 1 to observe the similarity with conventional farms).
- Organic 2; found as "Org. 2" in the tables (n = 11): Organic farms that fattened and sold their calves under the organic system.

Furthermore, organic farms were also analysed as a whole in a group called "All Organic" (n = 33).

3.2. Farm structure and management

3.2.1. Farm and herd characteristics

The average size of farms was 268.44 ha utilized agricultural area (UAA) (Table 2). Organic 2 farms were larger than 'All Organic', but the high variability within the sample did not allow identifying significant differences between this group and Organic 1. With regard to herd size, All Organic farms were also similar to Conventional farms, and an important variation within farms was identified in relation to the mean cattle herd size (140.95 livestock units (LU) in Organic 2 vs. 80.18 in Organic 1, $p < 0.05$).

In relation to the various land uses and the type of ownership, it was seen that 59% of land was in property (owned area/UAA in percentages). Fifty-two percent of UAAs had tree presence. Moreover, crop areas were almost inexistent.

Table 1. Herd and farm structure. Mean values, standard deviation and level of significance.

Parameters	Conv. (n = 30)	Org. 1 (n = 22)	Org. 2 (n = 11)	Sig. 1[4]	Sample (n = 63)	SD	All Organic (n = 33)	Sig. 2[5]
UAA[1]	275.80	223.72	337.84	0.378	268.44	223.34	261.76	0.806
Owned area/UAA	0.64	0.54	0.55	0.541	0.59	0.44	0.55	0.390
Wooded land/UAA	0.46	0.47	0.77	0.101	0.52	0.43	0.57	0.336
Crop area/UAA	0.00	0.00	0.00	0.576	0.00	0.01	0.00	0.334
Bovine LU[2]	104.92^{ab}	74.33^{a}	124.83^{b}	0.016*	97.72	52.14	91.16	0.299
Ovine LU	6.78	5.36	15.37	0.496	7.78	30.38	8.69	0.805
Swine LU	0.00	0.50	0.77	0.445	0.31	1.55	0.58	0.138
Total LU	111.70^{ab}	80.18^{a}	140.95^{b}	0.024*	105.80	63.33	100.44	0.485
Bovine LU/Total LU	0.98	0.96	0.92	0.452	0.96	0.13	0.85	0.369
Total stocking rate[3]	0.73	0.50	0.44	0.312	0.60	0.64	0.48	0.131

a, b, c Mean values with different letters in the same row are significantly different. * $p<0.05$, ** $p<0.01$, *** $p<0.001$. SD: standard deviation. [1]UAA: Utilized Agricultural Area. [2]LU: Livestock Units. 1 cow = 1 LU; 1 sheep = 0.12 LU; 1 sow = 0.37 LU. [3]Total Stocking rate = LU/ha UAA. [4]Analysis of Variance of the groups Conventional, Organic 1 and Organic 2. [5]Analysis of Variance of the groups Conventional vs. All Organic.

Table 2. Reproductive management and performance, and productive orientation. Mean values, standard deviation and level of significance.

Parameters	Conv. (n = 30)	Org. 1 (n = 22)	Org. 2 (n = 11)	Sig. 1[6]	Sample (n = 63)	SD	All Organic (n = 33)	Sig. 2[7]
Replacement rate (%)[1]	11.98	11.92	13.29	0.922	12.19	1.23	12.38	0.875
Cows/bull (N°)	31.01	30.67	28.29	0.844	30.42	1.68	29.88	0.740
Estrous synchronisation (%)[2]	6.70	0.00	0.00	0.321	3.20	–	0.00	0.132
Artificial insemination (%)[3]	6.70	4.50	0.00	0.592	4.80	–	3.00	0.658
Length of mating period (months)	10.40	10.70	10.14	0.922	10.46	0.29	10.52	0.846
Fertility rate (%)[4]	85.15	77.70	81.49	0.187	81.91	1.82	78.97	0.091
Age at first calving (month)	30.68	33.45	33.68	0.197	32.17	0.79	33.53	0.074
Calving interval (days)[5]	346.50	33500	343.64	0.165	341.98	2.74	337.88	0.117
Calves born/cow/year (N°)	0.85	0.78	0.81	0.187	0.82	0.02	0.78	0.091
Weaned calves/cow/year (N°)	0.81	0.71	0.65	0.061	0.75	0.03	0.69	0.025*
Age at weaning (months)	5.86	5.82	6.00	0.886	5.87	0.13	5.88	0.944
Live weight at calving (kg)	202.33^{a}	190.91^{b}	193.18^{ab}	0.037*	196.75	2.12	191.67	0.011*
Calves sold at weaning age/cow/year (N°)	0.63^{a}	0.66^{a}	0.27^{b}	0.000***	0.58	0.03	0.53	0.000***
Fattened calves sold/cow/year (N°)	0.07^{a}	0.00^{a}	0.45^{b}	0.000***	0.11	0.03	0.15	0.000***
Fattened calves/total calves sold	0.09^{a}	0.00^{a}	0.64^{b}	0.000***	0.15	0.30	0.21	0.119

a, b, c Mean values with different letters in the same row are significantly different. * $p<0.05$, ** $p<0.01$, *** $p<0.001$. [1]Calculated as the annual average proportion of heifers bred for reproduction/number adult. [2]Annual average proportion of cows synchronized/total adult cows in the farm. [3]Average annual proportion of inseminated cows/total adult serviced cows in the farm. [4]Annual average proportion of: live births/serviced cows. [5]Annual average number of days from calving to calving in the adult cows of the farms. [6]Analysis of Variance of the groups Conventional, Organic 1 and Organic 2. [7]Analysis of Variance of the groups Conventional vs. All Organic.

3.2.2. Reproductive management and performance, and production results

Estrous synchronisation was only carried out in the 3.20% of farms held. This practice was only observed in conventional farms, since it is not permitted in organic farming. Accordingly, only 4.80% of farms opted for artificial insemination, with all of them also carrying out natural mating, such that the use of either one or another technique was not exclusive. This scarce use of these reproductive techniques is typical in low-input beef cattle farms.

Calves weaned in All Organic had lower weights than those belonging to the conventional group, and Organic 2 farms sold less weaned calves per cow in total, thus showing a lower productivity in this regard. However, Organic 2 and All Organic sold more fattened calves per cow and also sold a higher proportion of fattened calves/total calves sold. These differences were due to the fact that the Organic 2 group was composed entirely of fattening farms, while all Organic 1 farms solely marketed calves at the age of weaning. Similarly to the Organic 1 group, 83.33% of the farms belonging to the conventional group did not carry out the fattening of any of the calves that they produced. These facts about the composition of the groups also influenced the differences between these indicators for yearlings sold per cow and calf weight at weaning.

3.2.3. Breeds

The breed distribution of organic farms is also an important issue, as autochthonous breeds are preferable for this production model, as indicated by Regulation 834/2007 [28]. Table 3 shows the composition by breed of the farms.

The percentage of purebred autochthonous cows reached 41.83%, with this percentage being higher in All Organic and Organic 2 than in the conventional group. Also, All Organic and Organic 2 showed a higher presence of these purebred cows; either autochthonous or foreign ones. The main reason for this is that Organic 2 farmers were market oriented (they had contracts with supermarkets) so that they knew that more productive breeds that allow them to produce carcass of better conformation, mainly Limousine.

3.3. Economic parameters

3.3.1. Analysis of fixed capital

This analysis allowed identification of similarities between organic and conventional systems (Table 4), with regard to infrastructure, land and animals.

Table 3. Farm breed structure. Mean percentage values, standard deviation and level of significance.

Parameters	Conv. (n = 30)	Org. 1 (n = 22)	Org. 2 (n = 11)	Sig. 1[1]	Sample (n = 63)	SD	All Organic (n = 33)	Sig. 2[2]
Purebred autochthonous cows (%)	20.11^{a}	30.76^{ab}	62.08^{b}	0.015^{*}	41.83	31.16	41.20	0.045^{*}
Purebred foreign cows (%)	8.86	11.25	6.74	0.854	22.31	9.33	9.75	0.877
Purebred cows (%)	28.97^{a}	42.01^{ab}	68.82^{b}	0.027^{*}	42.85	40.48	50.95	0.041^{*}
Purebred autochthonous bulls (%)	13.02	15.91	31.36	0.264	32.06	17.23	21.06	0.324
Purebred foreign bulls (%)	86.98	76.82	68.64	0.254	33.94	80.23	74.09	0.133

a, b Mean values with different letters in the same row are significantly different. * $p<0.05$. [1]Analysis of Variance of the groups Conventional, Organic 1 and Organic 2. [2]Analysis of Variance of the groups Conventional vs. All Organic.

Table 4. Fixed capital according to farm groups.

Parameters	Conv. (n = 30)	Org. 1 (n = 22)	Org. 2 (n = 11)	Sig. 1[5]	Sample (n = 63)	SD	All Organic (n = 33)	Sig. 2[6]
Land fixed capital[1]	5,630.07	5,194.52	5,695.62	0.788	5,489.42	310.43	5,361.56	0.669
Buildings fixed capital[2]	660.75	606.48	546.10	0.935	621.78	114.32	586.35	0.748
Machinery fixed capital[3]	215.51	138.79	107.77	0.449	169.91	35.00	128.45	0.217
Livestock fixed capital[4]	624.81	416.66	329.10	0.217	500.49	69.69	387.47	0.089
Total fixed capital	7,131.14	6,356.45	6,678.59	0.443	6,781.59	430.59	6,463.83	0.443

Note: all these indicators were measured in terms of €/ha. [1]Value of the land at market prices. This depended on the quality of the plots (grazing resources, location and tree density, among other parameters). [2]Value of infrastructure at market prices. Years of use and level of conservation/maintenance were taken into account. [3]Value of machinery (cars, trucks, etc.) at market prices. Years of use and level of conservation/maintenance were taken into account. [4]Value of all livestock present at market prices. [5]Analysis of Variance of the groups Conventional, Organic 1 and Organic 2. [6]Analysis of Variance of the groups Conventional vs. All Organic.

Table 5. Intermediate consumption and other costs.

Feed/ha UAA[1]	109.69[a]	17.55[b]	96.63[ab]	75.24	112.18	0.009**	43.91	0.019*
Feed/LU[2]	161.59[a]	38.27[b]	220.54[a]	128.82	165.70	0.003**	99.03	0.136
Seeds and fertilisers[3]	7.51	3.10	1.00	4.84	12.61	0.252	2.40	0.108
Veterinary and medicines/ha UAA[4]	17.87[a]	4.51[b]	4.84[b]	10.93	21.16	0.043*	4.62	0.012*
Veterinary and medicines /LU[5]	20.32[a]	7.45[b]	11.64[ab]	14.31	15.14	0.006**	8.84	0.002**
Maintenance of fixed capital[6]	15.74	18.95	22.60	18.06	22.68	0.681.	20.17	0.444
Energy[7]	24.24	22.44	18.27	22.57	22.84	0.765	21.05	0.584
Other expenditure[8]	24.32	20.87	21.88	22.69	33.67	0.934	21.20	0.717
Intermediate consumption[9]	199.38	87.42	165.22	154.32	177.86	0.077	113.36	0.054
Remuneration of employees[10]	60.29	42.48	61.24	54.24	100.69	0.799	48.73	0.653
Fixed capital consumption[11]	54.59	44.20	38.08	48.08	66.62	0.744	42.16	0.464
Land rented[12]	30.56	30.96	23.25	29.42	38.47	0.846	28.39	0.825

a, b Mean values with different letters in the same row are significantly different. * $p<0.05$, ** $p<0.01$. [1]Expenditures (purchases) made in external feedstuffs/hectare of UAA (€/ha). [2]Expenditures (purchases) made in external feedstuffs/LU (€/LU). [3]Expenditure in seeds and fertilisers/hectare of UAA (€/ha). [4]Expenditures made in veterinary and medicines/ha UAA (€/ha). [5]Expenditures made in veterinary and medicines /LU (€/LU). [6]Expenditures made in maintenance of fixed capital/ha of UAA (€/ha). [7]Expenditures made in energy/ha UAA (petrol and electricity) (€/ha). [8]Expenditures/ha UAA made in farmers' associations, lawyers, etc. (€/ha). [9]Sum of the following costs (€/ha of UAA: External feedstuffs + Veterinary services and medicines + Energy + Maintenance of machinery and infrastructure + Other goods and Services (lawyers, farmers' associations, etc.). [10]Expenditures made in salaries/ha of UAA (€/ha). [11]Amortization of machinery and infrastructure = Sum of (((1/20 years amortization) × Value of infrastructures) + ((1/10 years) × Value of machinery)). [12]Cost of the land rented (€/ha). [13]Analysis of Variance of the groups Conventional, Organic 1 and Organic 2. [14]Analysis of Variance of the groups Conventional vs. All Organic.

It is worth highlighting the high average value of land fixed capital 5489.42 €/ha that accounted for the 81% of total fixed capital (Table 5).

Table 6. Economic and productive performance and subsidies.

Parameters (€/ha)	Conv. (n = 30)	Org. 1 (n = 22)	Org. 2 (n = 11)	Sample (n = 63)	SD	Sig.[6]	All Organic (n = 33)	Sig.[7]
Livestock sales	291.23	151.90	215.92	229.43	223.45	0.081	173.24	0.035*
Other sales	1.49	11.34	9.75	6.37	23.05	0.276	10.81	0.110
Gross output[1]	635.24	464.09	578.52	565.57	406.81	0.328	502.24	0.197
Subsidies for livestock	158.54	153.75	185.72	161.61	118.93	0.759	164.41	0.847
Total subsidies	165.70	159.91	193.85	168.59	123.39	0.752	171.22	0.861
Total income	458.41	323.15	419.52	404.39	316.03	0.313	355.27	0.198
Total subsidies/total income (%)	0.39	0.45	0.47	0.42	0.18	0.353	0.45	0.160
Net value added[2]	388.43	338.63	383.35	370.15	249.94	0.769	353.53	0.584
Net operating surplus[3]	320.98	289.98	313.99	308.94	223.74	0.886	297.98	0.687
Net entrepreneurial income[4]	290.43	259.02	290.74	279.52	214.31	0.861	269.60	0.703
Profitability rate (%)[5]	4.39	4.18	4.35	4.31	2.69	0.961	4.24	0.819

Note: Those parameters whose unit is not showed in the table are measured per ha of UAA ((€/ha). [1]Value of all the products of agricultural activities. All agricultural output was recorded except that which was solely produced by units for their own consumption. [2]It measures the value created by all agricultural output after the consumption of fixed capital. That output is valued at basic prices and intermediate consumption is valued at purchase prices. It was calculated as follows: (Gross output – Intermediate consumption – Amortisation) + (Those subsidies not related to livestock farming). [3]It measures the yield from land, capital and unpaid labour. It is the balance of the generation of income account which indicates the distribution of income between the factors of production and the general government sector. [4]Obtained by adding the interest received and then deducting rent (i.e., farm and land rents) and interest payments, measuring compensation of unpaid labour, remuneration from land belonging to units and the yield arising from the use of capital. [5]Ratio between net surplus and average capital assets, estimated from the value of total fixed capital and the value of capital. [6]Analysis of Variance of the groups Conventional, Organic 1 and Organic 2. [7]Analysis of Variance of the groups Conventional vs. All Organic.

Table 6 shows the economic and productive performance of the farm groups, as well as aspects related to subsidies, where the Organic 1 group can be seen to have lower livestock sales per hectare of UAA and lower gross production.

Conventional farms proved to sell more calves per hectare and year, which is due to their shorter productive cycle and the low productivity of Organic 1. No differences were found for the remaining indicators, but some interesting results were found and therefore comments are necessary. Organic farms (especially Organic 2) revealed higher numerical values for other sales, which reflect a higher level of business diversification, something that is key in the farms' flexibility and adaptability to the changing market environment. Moreover, organic farms (especially Organic 2) tended to be more dependent on subsidies.

4. DISCUSSION

4.1. Structure

4.1.1. Farm and herd characteristics

All Organic farms were much smaller than the average farm size found by Perea et al. [10] in organic cattle farms located in seven regions of Spain (261.76 vs. 425 ha UAA). With regard to herd size, All Organic farms were also quite similar to Conventional farms and again smaller than the farms studied by [10], with 100.44 vs. 154 livestock units (LU).

The scarce association between land and animals continues to be an unsolved concern [6, 10]. Similarly, the integration of different livestock species is beneficial. In the farms analyzed, the proportion of cattle has been really high – 96%, in line with the findings of Perea et al. [9, 10]. This situation responds to the trend of specialisation and intensification already described [15, 17, 18], with increasing total stocking rates in beef cattle farms from 0.40 to 0.43 LU/ha ([6]—conventional farms; and [9, 10]—organic farms) to the current 0.60 LU/ha. The higher mean values observed in this study came from conventional farms (0.70). Both Organic 1 and Organic 2 farms complied with the regional organic rules [29] setting a maximum stocking rate allowed of 0.5 LU/ha.

4.1.2. Reproductive management and performance, and production results

No major differences were found between farm's groups regarding the reproductive management among groups, since most of arms followed the typical technical reproductive management in extensive ruminants production systems located in semiarid areas, where the low fertility rates compared to other breeds and systems. This is due to the fact that heats are not detected by farmers, there is no heat synchronization, and natural service is the predominant technique used for conception. Only some organic farms showed to apply artificial insemination. Average replacement rate of the sample was close to 12%, similar to that found in *dehesa* beef cattle farms, either conventional: with values ranging from 10 to 12.4% in Extremadura [30–32] or organic: 10.65% in Andalusia [9]. However, values found in the study of Milán et al. [6] were higher: 19.2%. The number of cows per bull was 30.42, lower than the 38.4 found by Milán et al. [6] and similar to the 27 found by López de Torre et al. [31] in conventional cattle farms in the *dehesas* of Extremadura. The implementation of reproductive techniques, such as artificial insemination, was even lower than that found by Milán et al. [6]: 8.5 vs. 4.80%. This divergence in results is due to the fact that they analysed farms rearing autochthonous purebred beef cattle cows. In these cases, livestock is usually registered in the Stud Book of the breeds, and the use of artificial insemination is more widespread, with the aim of rearing offspring of more appreciated genetic potential, and thus obtaining higher incomes through both selling animals as breeding animals and public subsidies.

Despite the lack of significant differences among groups, it is necessary to discuss some topics such as the reproductive calendar due to its importance in the context of uncertain availability of pastures in pasture-based systems, such as those of the Mediterranean basin. In this sense, it is recommended to avoid continuous mating and make it coincide with spring and autumn, the seasons where the availability of local feed resources allows fulfilling an important percentage of animals' nutritional needs at more affordable prices, due to a lower dependence on external feedstuff, whose prices are high and subjected to great volatility. However, also positive externalities can be found from this organization: reduced seasonality in marketing their products, thus obtaining better prices for them at certain times. Many of the farms analysed showed a distribution of mating throughout a year. Thus, the average duration of mating was 10.46 months.

Calves weaned in organic farms had lower weights than those belonging to the Conventional group. This could be due to the following aspects: Firstly, in some of the studies discussed, farms reared only local breeds, whose growing rates are lower. However, in the farms analysed in this study, many cows were either crossed or more efficient breeds, mainly the Limousine breed. Secondly, increased livestock pressure led to intensification and guidance to higher productivity which, among other adaptations, led to the inclusion of more efficient breeds. Thirdly, the rising prices of feed led to the weaning of animals at a younger age (therefore at lower weights), in order to use less feedstuff and thus reduce production costs. Finally, the next link in the food chain prefers younger animals because of their better conversion rates in feedlots. Moreover, less time grazing is usually associated with meat tenderness and lighter colour, which is in line with butchers' preferences. Thus, Organic 2 farms were those that sold more fattened calves per cow, and the age of weaning of these was lower. The latter was due to the fact that calves in Organic 2 farms were weaned before starting the fattening period, which shortened the length of the production cycle at the farm level (period between weaning and sale).

The results relating to calves weaned and sold per cow clearly show how the production of beef cattle in Southwest Europe and in semiarid areas, such as the Mediterranean basin, is mainly focused on the sale of calves at weaning. As a result, the percentage of fattened calves sold has been reduced. This is due to both the lack of infrastructure and the traditions of finishing and slaughtering animals in the Extremadura region [33]. Currently, this fact might have increased due to high feed prices and low farm profitability.

The existence of organic farms without organic products (Organic 1 in the present study) has been reported for more authors in dairy cattle [34], in a mixture of livestock and crop farms [35] and beef cattle [10, 18]. Specifically, Perea et al. [10] reported that only 40.6% of the surveyed organic beef cattle farms marketed calves as organically certified, and to the organic market. Thus, they also noticed that in different areas of Europe (from Norway to the Mediterranean area) the marketing of organic livestock is focused on the sale to conventional feedlots, and their organic stamp does not have market implications (there is a scarce market for these weaned organic animals, and they are not sold at a higher price; see [27]).

4.1.3. Breeds

The use of autochthonous breeds is a contemporary issue and usually promoted in organic farming. However, the low productivity of the rustic local cows makes it necessary to make use of other breeds that, despite not being autochthonous, are both well adapted to the local conditions and more productive. Thus, in the case of males, the racial distribution was mainly based on Limousine and Charolais breeds. This is a growing trend that responds to the need for productivity and competitiveness that requires specialisation [36]. In the *dehesas* of Extremadura, there has also been a change from Charolais towards Limousine, probably aimed at avoiding problems related to dystocia and the ability of calves to suckle, since farmers perceive that these problems are more frequent when the Charolais breed of animals are reared.

4.2. Economic parameters

4.2.1. Analysis of fixed capital

No significant differences were found between the groups of farms studied.

4.2.2. Costs, production and incomes

It is important to note that expenditure on feedstuff was lower in Organic 1 than in Conventional group when studied per hectare, while differences were found between Organic 1 and the rest of groups when these expenditures were measured per livestock unit. The expenditure on veterinary services and veterinary drugs were also lower in Organic 1 group both per area of land and per livestock unit. However, these differences only were found between Organic 1 and the Conventional group. All Organic group showed to also rely less on these external resources (feedstuff, veterinary services and drugs). However, the expenditure on feed per livestock unit was not statistically different between All Organic and the Conventional group. In general terms, these higher reliance on external resources, and in particular feed and veterinary services and drugs, is consistent with the organic production method, since the use of inputs such as feed must come from the farm itself (or the immediate surroundings), and veterinary drugs are limited to two treatments per adult cow per year, according to [28] and subsequent amendments.

When comparing Conventional and All Organic farms, one can observe very low feed costs in Organic 1 and very high feed costs in Organic 2 farms. This is due to the fact that Organic 2 farms fattened all their calves, and Organic 1, none of theirs. This increases the organic feedstuff purchased, whose price is high: around 30% above the conventional one.

The cost related to veterinary services and medicines shows that in extensive livestock systems of semiarid areas and conditions it is possible to reduce reliance on drugs with no major problems. In fact, conventional low-input farms in this area do not rely significantly on these products due to low stocking rates and dry climate. Also, as the prevalence of infectious diseases is low, it must be mentioned that the health management of organic beef cattle farms in this area is very similar to that carried out in Conventional farms, and it is not based on alternative medicine. In fact organic beef cattle farms also used some veterinary drugs as a preventive measure [17]. Organic 2 farms had higher veterinary costs than Organic 1 farms due to the fact that the transition to the fattening period usually provokes some respiratory and/or intestinal disorders.

Regarding incomes, it is necessary to increase the market orientation of Organic 1 farms, as they are not providing organic goods to the market, which influences their low economic results. Conversely, the longer productive cycle in Organic 2 farms did not allow them to clearly stand out in terms of income. Finally, the dependence on agricultural subsidies must be addressed as it is a key point for both the organic sector and the extensive beef cattle farms of Mediterranean Europe. The high dependence of this aspect makes it unstable and fragile. In the case of the organic sector, the contribution of the agro-environment subsidies makes them numerically more dependent, which is in contrast with other studies, regardless whether or not they were receiving agricultural subsidies [35, 37, 38].

The lower livestock sales per hectare of UAA and lower gross production in the Organic 1 group can be due to the fact that farms belonging to this group only sold calves at weaning age, and their prices were lower than those of fattened calves. Despite the price of organic fattened calves (marketed by the Organic farms 2 group) being greater, income from the sale of livestock per hectare of UAA was higher in Conventional farms. This was probably due to an extension of the productive cycles in Organic 2 compared to Conventional farms which, in turn, led to a reduction in the number of calves sold per cow per year. On the other hand, organic farms (especially the Organic 1 group) had higher incomes in relation to other sales (those not related to livestock). This could be a consequence of the greater degree of diversification in organic farming over conventional.

4.2.3. Other aspects worthy of discussion: workforce, agro-environmental management and marketing strategies

In addition, other aspects came up from the interviews with farmers during the farm visits which point to additional interesting aspects and open up perspectives which would be interesting to research. In this sense, Escribano et al. [17, 18] carried out a comparative sustainability assessment which showed that organic farms did not carry out so many agro-ecological practices as would be desirable to increase farm environmental protection, nutrient cycling and self-sufficiency. Moreover, these authors found that in terms of workforce both production systems are also very similar. Additionally, short marketing channels, which are commonly associated with organic production, were noted to also be very similar in various studies [17, 18, 27, 39, 40]. Profound discussion and review about these aspects can be found in other studies [40].

5. CONCLUSIONS AND FUTURE PERSPECTIVES

The present study integrated structure, technical, productive and economic parameters that allowed for a deep understanding of the organic beef cattle farms of Southern Europe, as well as their similarities and differences with conventional ones. Organic farms have proved to be very similar to Conventional farms (but pasture-based or low-input). Accordingly, the differences were based on the structure of the farms, more than the condition of their being organic.

According to the results discussed, it is worth mentioning that there was little orientation towards a different concept of farming, namely, environmental sustainability and self-sufficiency. However, the organic farm has been defined as a production system based on the principles of Health, Ecology, Fairness and Care. In this sense, consumers expect organic products to be based on these principles, and citizens support this system through taxes. All these aforementioned aspects shape the necessity to increase the implementation of sustainable agricultural practices, self-sufficiency and sales of organic products. Otherwise, the current production systems will hinder their sustainability due to high global competition, the increasing cost of agricultural inputs and reduced grazing resources in the Mediterranean area due to global warming. To do so, the education level of farmers, public support and farmer cooperation are essential. Moreover, further research is needed to study different production systems and strategies in order to improve the situation of the sector and the differential externalities of the organic livestock sector above the conventional one.

ACKNOWLEDGEMENTS

Research funded by project INIA-RTA2009-00122-C03-03 of the Spanish Ministry Economy and Competitiveness. The authors would also like to acknowledge the support provided by Junta de Extremadura and FEDER funds. Special thanks to the farmers, practitioners and experts that contributed to this research. The first author also acknowledges the pre-doctoral financial support of the Fundación Fernando Valhondo Calaff

REFERENCES

1. FiBL, IFOAM. The world of organic agriculture. Statistics & Emerging Trends 2015. 2015; http://www.organic-world.net/yearbook-2015.html. Accessed 04 August 2015.
2. Rodríguez L, Bartolomé DJ, Tabernero de Paz MJ, Posado R, García JJ, Bodas R, Garrido C, Vázquez JM, Mazariegos V, Vicente M, Olmedo S. Características técnico-económicas de las explotaciones de ovino lechero con reproducción asistida de Castilla y León: Sistemas y tipos de explotación. ITEA. 2013;109:201-214.
3. Olaizola AM, Manrique E, Maza MT. Tipos de sistemas de producción y rendimientos económicos en las explotaciones de vacuno de montaña. ITEA Producción Animal. 1995;91:47-58.

4. Serrano EP, Lavín P, Giradles FJ, Bernués A, Ruiz A. Classification variables of cattle farms in the mountains of León, Spain. Spanish Journal of Agricultural Research. 2004;2:504–511.
5. Serrano E, Lavín P, Giradles FJ, Bernués A, Ruiz A. The identification of homogeneous groups of cattle farms in the mountains of León, Spain. Spanish Journal of Agricultural Research. 2004;2:512-523.
6. Milán MJ, Arnalte E, Caja G. Economic profitability and typology of Ripollesa breed sheep farms in Spain. Small Ruminant Research. 2003;49:97-105.
7. Nieto MI, Olaizola AM, Manrique E. Los sistemas de vacuno de carne de las sierras de La Rioja: I Caracterización de las explotaciones. ITEA Producción Animal. 2005;26:183-185.
8. Ruiz R, Igarzabal A, Mandaluniz N, Amenazar M, Oregui LM. Caracterización de los principales sistemas de producción de ganado vacuno en el País Vasco. ITEA. 2005;26:195-197.
9. Perea J, García A, Acero R, Valerio D, Rodríguez V. Caracterización productiva del vacuno ecológico en Andalucía. Archivos de Zootecnia. 2007;56:517-521.
10. Perea J, Blanco–Penedo I, Barba C, Angón E, García A. Organic beef farming in Spain: Typology according to livestock management and economic variables. (Producción ecológica bovina de carne en España: tipología de acuerdo a variables técnicas y económicas). Revista Científica, FCV-LUZ. 2014;XXIV:347-354.
11. Granda M, Moreno V, Prieto M, (Eds.) Pastos naturales en la dehesa extremeña. SIA-Junta de Extremadura. 1991.
12. Hernández CG, (Ed.) La dehesa, aprovechamiento sostenible de los recursos naturales. Editorial Agrícola Española. 1998.
13. Espejo M, Espejo AM. Los sistemas tradicionales de explotación y la aplicación de innovaciones tecnológicas de la dehesa. In: Espejo M, Martín M, Matos C, Mesías FJ, (Eds.) Gestión ambiental y económica del ecosistema dehesa en la Península Ibérica. Junta de Extremadura, Consejería de Infraestructuras y D.T. 2006.
14. Gaspar P, Mesías FJ, Escribano M, Rodríguez De Ledesma A, Pulido F. Economic and management characterization of dehesa farms: Implications for their sustainability. Agroforestry Systems. 2007;71:151-162.
15. Gaspar P, Mesías FJ, Escribano M, Pulido F. Sustainability in Spanish extensive farms (Dehesas): An economic and management indicator-based evaluation. Rangeland Ecology and Management. 2009;62:153-162.
16. Franco JA, Gaspar P, Mesías FJ. Economic analysis of scenarios for the sustainability of extensive livestock farming in Spain under the CAP. Ecological Economics. 2012;74:120-129.
17. Escribano AJ, Gaspar P, Mesías FJ, Escribano M, Pulido F. Comparative sustainability assessment of extensive beef cattle farms in a high nature value agroforestry system. In: Squires VR, (Ed.) Rangeland Ecology, Management and Conservation Benefits. Nova Science Publishers. 2015.
18. Escribano AJ, Gaspar P, Mesías FJ, Pulido AF, Escribano M. A sustainability assessment of organic and conventional beef cattle farms in agroforestry systems: The case of the dehesa rangelands. ITEA Información Técnica Económica Agraria. 2014;110:343-367.
19. Regulation (EC) No 138/2004 of the European Parliament and of the Council of 5 December 2003 on the economic accounts for agriculture in the Community.
20. Gaspar P, Escribano M, Mesías FJ, Ledesma AR, Pulido F. Sheep farms in the Spanish rangelands (dehesas): Typologies according to livestock management and economic indicators. Small Ruminant Research. 2008;74:52-63.
21. Lebacq T, Baret PV, Stilmant D. Sustainability indicators for livestock farming. A review. Agronomy for Sustainable Development. 2013;33:311-327.
22. Nieto MI, Olaizola AM, Manrique E. Los sistemas de vacuno de carne de las sierras de La Rioja: I Caracterización de las explotaciones. ITEA. 2005;26:183-185.
23. García A, Perea J, Acero R, Angón E, Toro P, Rodríguez V, Gómez Castro AG. Structural characterization of extensive farms in andalusian dehesas. Archivos de Zootecnia. 2010;59:577-588.

24. Gaspar P, Escribano AJ, Mesías FJ, Escribano M, Pulido AF. Goat systems of Villuercas-Ibores area in SW Spain: Problems and perspectives of traditional farming systems. Small Ruminant Research. 2011;97:1-11.
25. Toro-Mújica P, García A, Gómez-Castro AG, Acero R, Perea J, Rodríguez-Estévez V, Aguilar C, Vera R. Organic dairy sheep farms in south-central Spain: Typologies according to livestock management and economic variables. Small Ruminant Research. 2012;100:89-95.
26. Riveiro JA, Mantecón AR, Álvarez CJ, Lavín P. A typological characterization of dairy Assaf breed sheep farms at NW of Spain based on structural factor. Agricultural Systems. 2013;120:27-37.
27. Escribano AJ. Beef cattle farms' conversion to the organic system. Recommendations for success in the face of future changes in a global context. Sustainability. 2016;8:572.
28. Regulation (EC) No 834/2007 of 28 June 2007 on organic production and labelling of organic products and repealing Regulation (EEC) No 2092/91.
29. DECRETO 9/2016, de 26 de enero, por el que se regula el régimen de ayudas agroambientales y climáticas y agricultura ecológica para prácticas agrícolas compatibles con la protección y mejora del medio ambiente en la Comunidad Autónoma de Extremadura. DOE Numero 18. Jueves, 28 de enero de 2016.
30. López de Torre G.. Metodología y esquema de selección de razas maternales de vacuno de carne: Aplicación a la raza Retinta. Especial Vacuno de Carne. Exclusivas ONE. 1986:75-93.
31. López de Torre G, García Barreto L, López Márquez JA. Encuesta sobre la explotación y mejora del ganado vacuno en la Dehesa extremeña. In: Campos P, Martín M, (Eds.) Conservación y Desarrollo de las Dehesas Portuguesas y Española. Ministerio de Agricultura, Pesca y Alimentación. 1987.
32. Martín M, Escribano M, Mesías FJ, Rodríguez de Ledesma A, Pulido F. Sistemas extensivos de producción animal. Archivos de Zootecnia. 2001;50:465-489.
33. Mateos A. 2008. Caracterización del cebo de bovino en Extremadura. Doctoral thesis. Departamento de Producción Animal, Universidad de Extremadura.
34. Flaten O, Lien G, Ebbesvik M, Koesling M, Valle PS. Do the new organic producers differ from the 'old guard'? Empirical results from Norwegian dairy farming. Renewable Agriculture & Food Systems. 2006;21:174-182.
35. Argyropoulos C, Tsiafouli MA, Sgardelis SP, Pantis JD. Organic farming without organic products. Land Use Policy. 2013;32:324-328.
36. Ligios S, Revilla R, Nardona A, Casu S. Cattle husbandry systems in Mediterranean mountains: Situation and prospects. In: Georgoudis A, Rosati A, Mosconi C, (Eds.) Animal Production and Natural Resources Utilisation in the Mediterranean Mountain Areas EAAP. 2005.
37. Hrabalová A, Zander K. Organic beef farming in the Czech Republic: Structure, development and economic performance. Agricultural Economics UZPI. 2006;52:89-100.
38. Gillespie J, Nehring R. Comparing economic performance of organic and conventional U.S. beef farms using matching examples. Australian Journal of Agricultural and Resource Economics. 2013;57:178-192.
39. Lobley M, Butler A, Winter M. Local organic food for local people? Organic marketing strategies in England and Wales. Regional Studies. 2013;47:216-228.
40. Escribano AJ. Organic livestock farming: Challenges, perspectives, and strategies to increase its contribution to the agrifood system's sustainability – A review. In: Konvalina P, (Ed.) Organic Farming – A Promising Way of Food Production, InTech. 2016.

CHAPTER 15

Molecular Characterization and Prevalence of Trypanosoma Species in Cattle from a Northern Livestock Area in Côte d'

IvoireIsidore Kpandji Kouadio[1], Didier Sokouri[1], Mathurin Koffi[2,3*], Ibrahim Konaté[2], Bernadin Ahouty[1], Alain Koffi[4], Simon Pierre N'Guetta[1]

[1]*Laboratoire de Génétique, UFR Biosciences, Université Félix Houphouet Boigny, Abidjan, Côte d'Ivoire*

[2]*Laboratoire des Interaction Hôte-Microorganisme-Environnement et Evolution, Université Jean Lorougnon Guédé, Daloa, Côte d'Ivoire*

[3]*Centre Suisse de Recherches Scientifiques en Côte d'Ivoire, Abidjan, Côte d'Ivoire*

[4]*Institut Pierre Richet, Unité de Recherche Trypanosomoses, Bouaké, Côte d'Ivoire*

ABSTRACT

Background: African animal trypanosomiasis (AAT) is caused mainly by Trypanosoma congolense, T. vivax, and T. brucei brucei and is the major constraint for livestock productivity in Sub-Saharan African countries. Information about animal trypanosomiasis status in Ivory Coast is missing, especially regarding molecular epidemiology. Therefore, this study intended to apply molecular tools to identify and characterize trypanosomes in Ivory Coast for sustainable control. Methods: 363 cattle blood samples were collected from Ferkessedougou Region in northern Ivory Coast in 2012. Buffy coat technique (BCT) and species-specific PCR assays were used to detect trypanosome species. Results: Out of 363 cattle examined with BCT, 33 were found positive with all trypanosomes species accounting for an average of 9.09% prevalence whereas polymerase chain reaction (PCR) using species-specific primers showed that 81 out of 363 cattle were infected with trypanosomes with an overall prevalence of 22.31%. Trypanosoma congolense savanah type, T. Vivax and T. brucei sl. accounted for 28.39%, 49.38% and 23.45% of the infection rate respectively. No infection with T. congo forest type was detected. T. vivax infection was the most prevalence in the area investigated compared to the two other trypanosome species. Mixed infections with different trypanosomes species were detected accounting for 7.32% of prevalence. Regarding sex- related prevalence, male cattles were slightly more infected than female but the difference was not significant. Conclusion: Our results showed that there was a high prevalence of AAT in livestock in Ferkessedougou Area. There is therefore a need to strengthen control policies and institute measures that help prevent the spread of the parasites for sustainable control of animal trypanosome in this area.

Keywords: Animal African Trypanosomiasis, Molecular Diagnosis, Species-Specific PCR, Côte d'Ivoire

1. INTRODUCTION

Animal health and livestock production in Sub-Saharan Africa suffer from high prevalence of trypanosomiasis with estimated annual losses due to direct and indirect consequences of the disease nearby billions of dollars [1] .

Regarding the burden and lost occasioned by animal African trypanosomiasis (AAT), the African Union initiated a regional control strategy, the pan Africa tsetse and trypanosomiasis eradication campaign (PATTEC) [2] . Among others, the success of this aim lies on the accurate diagnosis of the disease in each country but this is limited by use of parasitological diagnostic techniques as microscopy due to low sensitivity [3] . Fortunately polymerase chain reaction (PCR) diagnostic assays overcome the low sensitivity limitations of parasitological techniques. PCR assays are powerful marker tools, species-specific and sensitive, for molecular identification and diagnosis of trypanosome species in their mammal hosts and vectors to assess the real prevalence and correlate burden in a given country or region.

Despite campaigns towards trypanosome eradication in Sub-Saharan Africa, there is a lack of research using molecular screening of trypanosomes in Ivory Coast and the latest information about the prevalence of AAT dates about more than two decades and further drug resistance is described [4] [5] . Information cn the status of animal trypanosomiasis in Ivory Coast needs to be updated especially in the Northern part which is an important livestock area. This work was therefore designed to investigate the prevalence of trypanosomiasis in Ferkessedougou, a district in northern Ivory Coast, well known for its large population of cattle.

2. METHODS

2.1. Study Site

The study was carried out from July to August 2012 in five farming villages of the department of Ferkessedougou (9°5'N - 4°75'W) in the northern part of Ivory Coast at 650 km from Abidjan, the economical capital (Figure 1). Ferkessedougou is covered by arboreous savannah and the climate is Sudanese type (hot and dry).

2.2. Blood Sample Collection and Filed Analysis

363 cattle from both sexs were randomly selected and bled in five cattle herds. 5 ml of blood were aseptically collected from the jugular vein of each animal into an Ethylene diamine tetra acetic acid (EDTA) tube which was then labeled and placed in ice pack. In the field, trypanosomes were check based on Buffy Coat Technique (BCT) [6] : Blood samples from the EDTA tubes were transferred into capillary tubes. One end of each of the capillary tube was sealed with plasticine and spun in a microhaematocrit centrifuge at 1500 rpm for 5 minutes. The haematocrit tubes were then taken and cut at the buffy coat level to release the contents on a clean grease- free microscope glass slide to which a cover slip was placed for examination at 40× objective magnification for motile trypanosomes.

2.3. Molecular Analysis

Further to field analysis, two aliquots of each blood sample were transferred in nunc[R] cryotubes and taken to the Laboratoire des Interactions Hôte-Microorganisme-Environnement et Evolution (LIHME) at Jean Lorougnon Guédé University for molecular analysis.

Genomic DNA for the screening of trypanosomes was extracted from the blood samples using the salt out method. Briefly, 500 µl of each blood sample were washed four times. Then, 500 µl of the washing buffer, 100 µl of cell lysis buffer, 10 µl of SDS 10% and 10 µl of Proteinase K (concentration 10 mg/ml) were added to pellet and incubated at 56°C for 30 minutes. Afterwards, DNA was precipitated using ethanol 70%. Pellets were air-dried for about 4 hours, and DNA was re-suspended in 75 ml of Tris-EDTA (TE) buffer and stored at −20°C until it use for PCR.

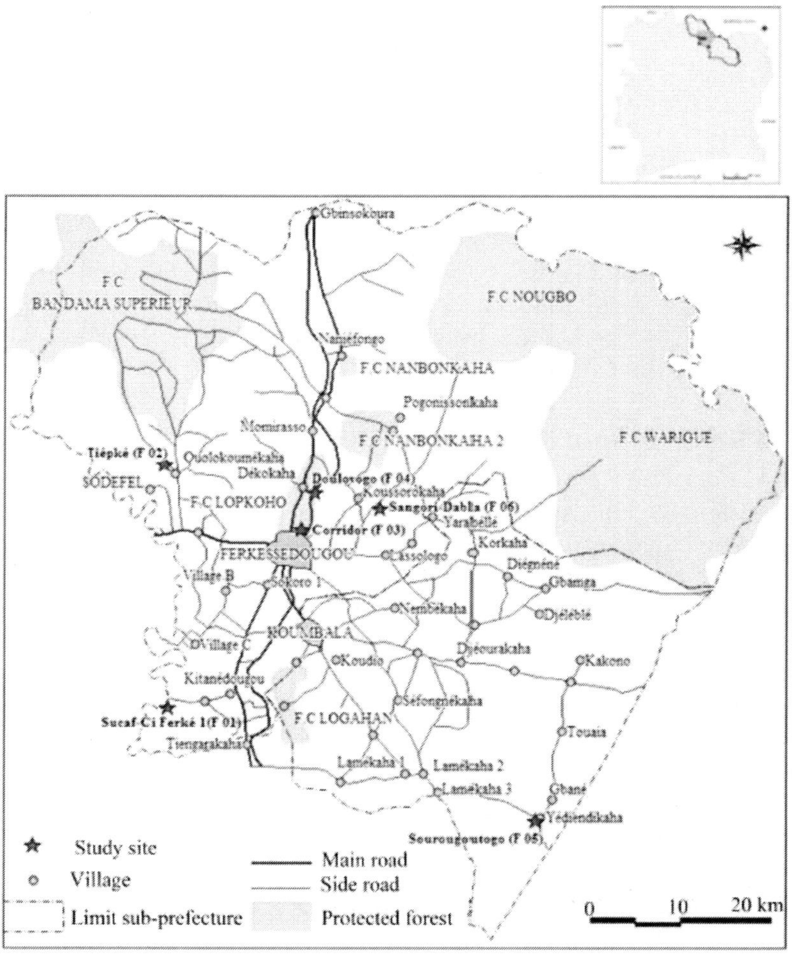

Figure 1. Study locations

Amplification of trypanosome DNA was conducted using species-specific primer pairs for Trypanosoma congolense, Trypanosoma vivax and Trypanosoma brucei sl. (Table 1).

Table 1. Species-specific primers for PCR amplification.

Trypanosome species/subgroup	Primer code and sequence 5'-3'	Product size (bp)
T. c. s.	TCS1 CGA GAA CGG GCA CTT TGC GA TCS2 GGA CAA AGA AAT CCC GCA CA	316
T. c.f.	TCF1 GGA CAC GCC AGA AGG TAC TT TCF2 GTT CTC GCA CCA AAT CCA AC	350
T. brucei	TBR1 GAA TAT TAA ACA ATG CGC AG TBR2 CCA TTT ATT AGC TTT GTT GC	177
T. vivax	TVW1 CTG AGT GCT CCA TGT CCC AC TVW2 CCA CCA GAA CAC CAA CCT GA	150

T. c. s. = *Trypanosoma congolense* savanah; *T. c. f.* = *Trypanosoma congolense* forest.

The amplifications were conducted in a total volume of 50 µl containing 5 µl of PCR buffer 10× (10 mM Tris-HCl (pH 9.0), 50 mM KCl, 3 mM $MgCl_2$), 15 picomoles of each primer, 200 µM of each of the four deoxynucleotide-triphosphate (dNTP), one unit of Taq DNA polymerase (Appligene-Oncor, USA), sterile water and 5 µl DNA extract. Amplification involved pre-denaturation at 94°C for 3 min followed by 30 cycles of denaturation at 94°C for 1 min, hybridization of primers at 60°C and elongation at 72°C for 1 minute, then final elongation at 72°C for 15 min. PCR products were separated by electrophoreses on a 2% (w/v) agarose gel for 30 min at 100 V. The gel was stained in an ethidium bromide solution for 10 min and visualized under UV light.

2.4. Statistical Analysis

The proportions of animals infected by different trypanosome species were compared between cattle herds and location using Chi-square (X^2) or Fisher exact test according to the Stata 9.2 software [7] .

3. RESULTS

3.1. Buffy Coat Technique-Based Diagnosis

Out of 363 cattle examined with BCT, comprising 73 males and 290 females (**Table 2**), 33 were found positive with trypanosomes strains (**Table 3**) accounting for an average of 9.09% prevalence.

Table 2. Number of animals sampled by site and sex.

Sampling site	Male	Female	Total
Sucaf ci ferke 1	18	42	60
Tiekpe	10	49	59
Corridor ouangolo	8	37	45
Doulovogo	8	52	60
Souroutogo	17	55	72
Sangori-Dabla	12	55	67
Total	73	290	363

Table 3. Microscopy versus PCR diagnosis.

	Number of animal tested (N = 363)			
Diagnostic technique	Positive (%)	Negative	X^2	p-value
BCT	33 (9.09)	330	21.2	<0.0001
PCR	81 (22.31)	282		

3.2. Molecular Diagnosis and Statistic Inferences

PCR using species-specific primers for Trypanosoma congolense, T. vivax and T. brucei sl. showed that 81 out of 363 cattle were infected with trypanosomes with an overall prevalence of 22.31% in the total bovine investigated. Molecular diagnosis was very significantly different to microscopic observation based on buffy coat technique ($p < 0.001$) (**Table 3**). It appears that PCR was 2.5 times more sensitive compare to BCT. Regarding sex-related prevalence, male cattle were slightly more infected than female (**Table 4**) but the different was not significant ($p > 0.05$). Out of the 81 animals tested positive, Trypanosoma congolense savanah type, T. Vivax and T. brucei sl. accounted respectively for 28.39%, 49.38% and 23.45% (**Table 5**).

Table 4. PCR-based prevalence of trypanasomiasis related to sex.

		Results of PCR			
Sex	Number of animals examined	Positive (%)	Negative (%)	X^2	p-value
Male	73	18 (26.02)	55 (73.98)		
Female	290	63 (21.72)	227 (78.28)	0.62	0.43
Total	363	81 (22.31)	282 (77.79)		

Table 5. Prevalence of trypanosomes infection related to species.

Sampling sites	Number of animals examined	Number of positive PCR (N = 81)		
		TCS	TV	TBR
Sucaf ci ferke 1	60	0	3	9
Tiekpe	59	3	6	0
Corridor ouangolo	45	12	1	7
Doulovogo	60	0	5	10
Souroutogo	72	4	2	0
Sangori-Dabla	67	4	23	0
Total	363	23	40	19
Prevalence		28.39	49.38	23.45

TCF: *Trypanosoma congolense* savannah type; TV: *Trypanosoma vivax*, TBR: *Trypanosoma brucei* sl.

The site of Corridor ouangolo was the most infected location with 37.78%, followed by Sangori-Dabla with 32.84% and less infected site was Sourougoutogo with 8.33% of prevalence (**Table 6, Figure 1**). No infection with T. congo forest type was detected. Mixed infections with different trypanosomes species were detected accounting fcr 7.32% of prevalence.

Table 6. Prevalence of trypanosomes infection related to sampling sites.

Sampling sites	Number of animals examined	Number of positive PCR (%)
Sucaf ci ferke 1	60	12 (20.00)
Tiekpe	59	9 (15.25)
Corridor ouangolo	45	17 (37.78)
Doulovogo	60	15 (25.00)
Sangori-Dabla	72	6 (8.33)
Sangori-Dabla	67	22 (32.84)
Total	363	81 (22.31)
X^2		20.37
p-value		<0.001

4. DISCUSSION

Achieving the goal of sustainable control of animal trypanosomiasis requires sensitive and specific molecular tools to conduct a rigourous epidemiological study for accuracy assessing the disease prevalence and burden. Still today, several epidemiological studies are based on microscopic observation of trypanosomes [8] -[10] . Our study just showed again that micoscopic observation underestimates the true prevalence of infection. Indeed, only about third of trypanosomes expected are detected by the BCT. Such evidence should help researchers to preferably use molecular diagnosis.

The overall molecular-based prevalence of AAT recorded in the present study, 22.31% is higher compared to what was found 3 decades before [11]. A study conducted in 2010 in Mali, a country which shares the cotton growing areas with the northern part of Ivory Coast gives 30.86% as prevalence [12] which is almost similar to the prevalence we had but in the same country another team found 15.7% as global prevalence [13] . The overall prevalences in those studies were obtained with the BCT and could be revised up if the molecular diagnosis was applied. In addition, the difference in prevalence observed in the same country shows that variations in trypanosome infection rates in different ecological zones and time are common, as previously reported [14] . Therefore, the climatic variations across different geographical regions and type of husbandry practices are factors that can influence survival of both vector and parasite. A statistically significant difference was observed ($p < 0.05$) in prevalence between the 3 trypanosomes species investigated. T. vivax was the most prevalent species contrary to studies that have reported T. congolense as the predominant species in cattle in various parts of Africa [15] . Our study is in consonant with study conducted in western Kenya [16] were T. vivax was found the most prevalent species. The high prevalence of T. vivax is consistence with other findings in domestic animals [10] [17] and may result from the level of pathogenicity of this trypanosome, which is generally low and better controlled by animals [18] . The lower prevalence of T. congolense with respect to T. vivax in animals may result from higher parasitemia in T. congolense infections, accompanied by serious anaemia, which leads to the rapid death of animals [19] [20] . Species-specific PCR did not detect T. congolense forest type in Ferkesedougou Area and this seems natural considering that it is thought to be a forest sub-groups whereas Ferkessedougou is a savanah region.

The presence of mixed infections emphasizes the need for epidemiologists to focus beyond single species to multiple infections with different parasites and different genotype within species which are a norm under field conditions [21] . Lower trypanosome prevalence found in Sourougoutogo compared to others sites could be due to recent trypanocidal treatment stated by the owners (date not shown).

Even though the prevalence of trypanosomiasis was relatively higher in male cattle than female, there was no significant difference between sex prevalence ($p > 0.05$) and this was in consonant with study conducted in 2011 in south-west Ethiopia [9] contrary to Abenga et al. [22] study were sex dimorphism in trypanosomiasis has been reported. Gender should not play a relevant role in influencing the susceptibility of the animals to infection regarding several environment interactions [10] .

This molecular epidemiology work confirmed the occurrence of mixed infections in the field, which could not have accurately been detected by the classical parasitological methods although it could influence the severity of AAT.

5. CONCLUSIONS

Animal trypanosome prevalence is not insignificant in Ferkessedougou, one of the most cattle production areas in Ivory Coast. Intervention to control the disease by the various stakeholders is therefore highly recommended.

The Pan African Tsetse and Trypanosomiasis Eradication Campaign, PATTEC set up by the African Union has been run for several years in countries that share the same borders with Ivory Coast including Mali, Burkina Faso and Ghana whereas today the Ivory Coast has not yet initiated PATTEC program. Updating data on the epidemiology of animal trypanosomiasis with molecular tools in high potential agropastoral areas for advocacy could help the Department of Animal Production first, to relieve the farmers but also in a context of international dynamic of tsetse populations, to avoid the phenomenon of re-invasion of other countries working to eliminate AAT in their countries infested.

ACKNOWLEDGEMENTS

This work was funded by the International Foundation for Science (IFS), Karlavägen 108, 5th floor, SE-115 26 Stockholm, Sweden (Fellow ship No. AB/21683R). We are very grateful to the farmers who gave their consent for this study.

REFERENCES

1. Torgerson, P.R. and Macpherson, C.N.L. (2011) The Socio-Economic Burden of Parasitic Zoonoses: Global Trends. Veterinary Parasitology, 182, 79-95. http://dx.doi.org/10.1016/j.vetpar.2011.07.017
2. Schofield, C.J. and Kabayo, J.P. (2008) Trypanosomiasis Vector Control in Africa and Latin America. Parasit Vectors, 1, 24. http://dx.doi.org/10.1186/1756-3305-1-24
3. Picozzi, K., Tilley, A., Fevre, E.M., Coleman, P.G., Magona, J.W., Odiit, M., Eisler, M. and Welburn, S. (2002) The Diagnosis of Trypanosome Infections: Applications of Novel Technology for Reducing Disease Risk. African Journal of Biotechnology, 1, 39-45.

4. Coulibaly, L., Hecker, P.A., Krebs, H.A., Rowlands, G.J., d'Ieteren, G.D.M., Leak, S.G.A., Peregrine, A.S. and Nagda, S.M. (1997) Failure of Trypanocidal Drugs to Cure Trypanosome Infections in Cattle in Boundiali District, Northern Côte d'Ivoire. In OUA/CSTR, 21, 248-249.
5. D'Ieteren, G., Coulibaly, L., Atse, P.A., Hecker, P.A., Krebs, H.A., Rowlands, G.J., Leak, S.G.A. and Nagda, S.M. (1997) Trypanocidal Drug Resistance in Four Regions of Côte d'Ivoire: Importance and Possible Impact on Sustainability of Integrated Strategies for Trypanosomiasis Control. In OUA/CSTR, 21, 233-247.
6. Murray, M., Murray, P.K. and McIntyre, W.I. (1977) An Improved Parasitological Technique for the Diagnosis of African Trypanosomiasis. Transactions of the Royal Society of Tropical Medicine and Hygiene, 71, 325-326. http://dx.doi.org/10.1016/0035-9203(77)90110-9
7. Corp, S. (2006) Stata Statistical Software: Release 9.2. Stata CORPoration, Texas.
8. Bocum, Z., Diarra, M., Maiga, H.M., Sanogo, M., Sylla, M.S.M. and Diall, O. (2012) Prevalence in the Bovine Trypanosomiasis Kadiolo Circle. Journal of Community Medicine & Health Education, 2, 9.
9. Gebreyohannes, M. and Legesse, F. (2014) Epidemiological Study of Bovine Trypanosomiasis in Woliso Woreda, Ethiopia. Journal of Animal Science Advances, 4, 833-838.
10. Fasanmi, O.G., Okoroafor, U.P., Nwufoh, O.C., Bukola-Oladele, O.M. and Ajibola, E.S. (2014) Survey for Trypanosoma Species in Cattle from Three Farms in Iddo Local Government Area, Oyo State. Sokoto Journal of Veterinary Sciences, 12, 57-61. http://dx.doi.org/10.4314/sokjvs.v12i1.9
11. Camus, E. (1981) Epidemiology and Clinical Incidence of Bovine Trypanosomiasis in the North of the Ivory Coast. Revue d'Elevage et de Médecine Vétérinaire des Pays Tropicaux, 34, 289-295.
12. Bocoum, Z., Diarra, M., Maiga, H.M., Diall, O., Guindo, M., et al. (2010) Study of the Evolution of Bovine Try- panosomosis Risk in Madina Diassa Ranch Dry Season Fresh. Review of Malian Science and Technology No. 12-CNRST 5, 119-123.
13. Mungube, E.O., Vitouley, H.S., Allegye-Cudjoe, E., Diall, O., Boucoum, Z., Diarra, B., Sanogo, Y., Randolph, T., Bauer, B., Zessin, K.H. and Clausen, P.H. (2012) Detection of Multiple Drug-Resistant Trypanosoma congolense Populations in Village Cattle of South-East Mali. Parasites & Vectors, 5, 155. http://dx.doi.org/10.1186/1756-3305-5-155
14. Ogunsanmi, A.O., Ikede, B.O. and Akpavie, S.O. (2000) Effects of Management, Season, Vegetation Zone and Breed on the Prevalence of Bovine Trypanosomiasis in South Western Nigeria. Israel Journal of Veterinary Medicine, 55, 1- 7.
15. McDermott, J., Woitag, T., Sidibe, I., Bauer, B., Diarra, B., Ouédraogo, D., et al. (2003) Field Studies of Drug- Resistant Cattle Trypanosomes in Kenedougou Province, Burkina Faso. Acta Tropica, 86, 93-103. http://dx.doi.org/10.1016/S0001-706X(03)00019-6
16. Thumbi, S.M., McOdimba, F.A., Mosi, R.O. and Jung'a, J.O. (2008) Comparative Evaluation of Three PCR Base Diagnostic Assays for the Detection of Pathogenic Trypanosomes in Cattle Blood. Parasites & Vectors, 1, 46. http://dx.doi.org/10.1186/1756-3305-1-46
17. Nimpaye, H., Njiokou, F., Njine, T., Njitchouang, G.R., Cuny, G., Herder, S., Asonganyi, T. and Simo, G. (2012) Molecular Epidemiological Studies on Animal Trypanosomiases in Ghana. Parasites & Vectors, 5, 217. http://dx.doi.org/10.1186/1756-3305-5-217
18. Authié, E., Bringaud, F., Bakalara, N., Tétaud, E. and Baltz, T. (1999) Trypanosomoses humaines et animales: Maladie du sommeil et Nagana. Annales de l'Institut Pasteur/Actualités, 10, 27-50. http://dx.doi.org/10.1016/S0924-4204(99)80021-3
19. Sidibe, I., Bengaly, Z., Boly, H., Ganaba, R., Desquesnes, M. and Sawadogo, L. (2002) Differential Pathogenicity of Trypanosoma congolense Subgroup: Implication for the Strategic Control of Trypanosomiasis. Newsletter on Integrated Control of Pathogenic Trypanosomes and Their Vectors (I.C.P.T.V), 6, 33-35.
20. Bengaly, Z., Sidibe, I., Boly, H., Sawadogo, L. and Desquesnes, M. (2002) Comparative Pathogenicity of Three Genetically Distinct Trypanosoma congolense-Types in Inbred Balb/c Mice. Veterinary Parasitology, 105, 111-118. http://dx.doi.org/10.1016/S0304-4017(01)00609-4

21. Cox, F.E. (2001) Concomitant Infections, Parasites and Immune Responses. Parasitology, 122, S23-S38. http://dx.doi.org/10.1017/S003118200001698X
22. Abenga, J.N., Enwezor, F.N.C., Lawani, F.A.G., Osue, H.O. and Ikemereh, E.C. (2004) Trypanosome Prevalence in Cattle in Lere Area of Kaduna State, North Eastern Nigeria. Journal of Veterinary Research, 57, 45-48.

CHAPTER 16

Livestock-associated risk factors for pneumonia in an area of intensive animal farming in the Netherlands

Gudrun S. Freidl[1,2*], Ineke T. Spruijt[1], Floor BorleÂe[3,4], Lidwien A. M. Smit[3], Arianne B. van Gageldonk-Lafeber[1], Dick J. J. Heederik[3], Joris Yzermans[4], Christel E. van Dijk[4], Catharina B. M. Maassen[1], Wim van der Hoek[1]

[1]*Centre for Infectious Diseases, Epidemiology and Surveillance, Centre for Infectious Disease Control,National Institute for Public Health and the Environment (RIVM), Bilthoven, the Netherlands,*

[2]*European Programme for Intervention Epidemiology Training (EPIET), European Centre for Disease Prevention and Control (ECDC), Stockholm, Sweden,*

[3]*Institute for Risk Assessment Sciences (IRAS), Division of Environmental Epidemiology, Utrecht University, Utrecht, the Netherlands,*

[4]*Netherlands Institute for Health Services Research (NIVEL), Utrecht, Netherlands*

ABSTRACT

Previous research conducted in 2009 found a significant positive association between pneumoniain humans and living close to goat and poultry farms. However, as this result mighthave been affected by a large goat-related Q fever epidemic, the aim of the current studywas to re-evaluate this association, now that the Q-fever epidemic had ended. In 2014/15,2,494 adults (aged 20±72 years) living in a livestock-dense area in the Netherlands participated in a medical examination and completed a questionnaire on respiratory health, lifestyleand other items. We retrieved additional information for 2,426/2,494 (97%) participants from electronic medical records (EMR) from general practitioners. The outcome was sel f reported, physician-diagnosed pneumonia or pneumonia recorded in the EMR in the previous three years. Livestock license data was used to determine exposure to livestock. We quantified associations between livestock exposures and pneumonia using odds ratios adjusted for participant characteristics and comorbidities (aOR). The three-year cumulative frequency of pneumonia was 186/2,426 (7.7%). Residents within 2,000m of a farm with at least 50 goats had an increased risk of pneumonia, which increased the closer they lived to the farm (2,000m a OR 1.9, 95% CI 1.4±2.6; 500m aOR 4.4, 95% CI 2.0±9.8). We found no significant associations between exposure to other farm animals and pneumonia. However, when conducting sensitivity analyses using pneumonia outcome based on EMR only, we found a weak but statistically significant association with presence of a poultry farm within 1,000m (aOR: 1.7, 95% CI 1.1±2.7). Living close to goat and poultry farms still constitute risk factors for pneumonia. Individuals with pneumonia were not more often seropositive for *Coxiella burnetii*, indicating that results are not explained by Q fever. We strongly recommend identification of pneumonia causes by the use of molecular diagnostics and investigating the role of non-infectious agents such as particulate matter or endotoxins.

1. INTRODUCTION

In the Netherlands, the number of intensive livestock farms doubled within the first decade of the 21st century [1]. Although the total number of farms has decreased over the past decades, the number of farm animals has increased [2]. This trend has raised concern about potential negative health effects on residents living close to intensive livestock farms. The debate between civil groups opposed to intensive livestock farming, the farming sector and policy makers was further fueled by the recent Q fever epidemic that occurred between 2007 and 2009 in the southern part of the Netherlands [3]. Caused by the zoonotic bacterium *Coxiella burnetii*, this epidemic resulted in more than 4000 notified human cases, mostly presenting as pneumonia [4]. Aborting dairy goats and dairy sheep were found to be the main cause of infection in humans, who were infected through inhalation of contaminated dust or aerosols distributed via the airborne route. Human cases started to decrease in 2010 in the Netherlands, coinciding with the introduction of veterinary interventions comprising of culling of pregnant goats and sheep on Q fever positive farms and vaccination of dairy goats and dairy sheep [3, 5]. Since the start of this vaccination campaign in early 2009, farms with at least 50 sheep or goats are obliged to vaccinate, which is strictly reinforced [6].

The health risks for residents living in the vicinity of livestock farms in the Netherlands were first addressed within the ªIntensive Animal Husbandry and Healthº study [7]. Within this project Smit et al. [8] studied the relationship between living in the vicinity of livestock farms and Q fever or pneumonia diagnoses in 70,142 adults. Outcomes were retrieved from electronic medical records from general practitioners (GP) located in an area with Q fever positive farms during 2009. This study found that a high number of goats within 5km of the home address quartile (Q) 4: 17,191±20,969 versus Q1: 0±2,250 goats] was indeed associated with Q fever Adjusted Odds ratio (aOR) 12.03, 95% confidence interval (CI) 8.79±16.46] and pneumonia (aOR 1.86, 95% CI 1.28±2.21). Presence of poultry farms within 1km of the home address was also identified as a risk factor for pneumonia among adults (OR 1.25, 95% CI 1.06±1.47), which was hypothesized to be potentially linked to exposure to pathogens or air pollutants. Potential exposure to other pathogens, such as influenza viruses, and higher susceptibility to community-acquired pneumonia due to exposure to high levels of fine dust and endotoxins emitted by poultry farms were considered likely explanations of this finding [8, 9]. Residual confounding by goat exposure could not be fully ruled out in explaining the high incidence of pneumonia close to poultry farms.

Therefore, the aim of the current study was to re-investigate previously found associations between pneumonia and goat/ poultry exposure (as well as other types of animals) in the same study area in a period in which the Q-fever incidence among humans has dropped to pre-epidemic levels.

2. METHODS

2.1 Study population and study design

As part of the ªLivestock Farming and Neighbouring Residents' Healthº (VGO) project, a population- based cross-sectional study was conducted to investigate the relationship between adverse health effects in humans living close to livestock farms. A detailed description of the study design can be found in a project report [10].

In short, a questionnaire study was conducted among 14,882 adults living in the east of North-Brabant and the north of Limburg, an area characterized by a high density of livestock farms [11].

In total, 9220 (62%) gave consent to be contacted for further studies, of which 7180 (72%)met the following inclusion criteria and were invited to participate: (i) not working or living on a farm, and (ii) living within a 10 km radius of one of the 12 temporary research centers. Of these, 2494

(34.7%) participated in the medical examination which was conducted between March 2014 and February 2015 and included, among others, filling in an extended questionnaire comprising questions on demographics, respiratory health and lifestyle and providing a serum sample. We also investigated a possible association between having children in the household and pneumonia, as was shown in previous research [12, 13].

Electronic medical record (EMR) data were used, if (i) GPs registered according to certain quality criteria [8, 11] and (ii) if participants granted access to their EMR. EMR data were available through the NIVEL Primary Care Database of the Netherlands Institute for Health Sciences Research. Of the 2,494 participants, we excluded 68individuals from the analysis. Of these, 66 did not provide consent to access their EMR data, whereas for two other individuals EMR data were unavailable. The final study population amounted to 2,426 individuals.

2.2 Ethical aspects

The VGO study protocol was approved by the Medical Ethical Committee of the UniversityMedical Centre Utrecht (protocol number 13/533). All 2,494 subjects signed informed consent.Patients' privacy was ensured by keeping medical information and address records separated at all times by using a Trusted Third Party.

2.3 Data availability statement

In consultation with the Medical Ethical Committee that approved the study protocol, datafrom the VGO study are not publicly available due to privacy protection of participants. Thestudy's privacy regulations stated that only researchers from NIVEL, IRAS, and RIVM (consortium partners) have access to the study database. Sharing an anonymized and de-identified dataset is not possible as it would still contain Electronical Medical Records and the personal data of participants, which could potentially lead to the identification of subjects. Researchers may reach a privacy agreement to access the data by contacting Prof. Dr. Dick Heederik (d. heederik@uu.nl) or Dr. L.A.M. Smit (l.a.smit@uu.nl).

2.4 Data collection

Questionnaire. Study participants provided information on personal characteristics and lifestyle factors through the questionnaire, including age, gender, smoking habits, and education level. Body mass index (BMI) was calculated based on weight and height measured during the medical examination. In addition, information on respiratory diseases (i.e. pneumonia,COPD and asthma) and related risk factors and determinants, e.g. growing up on a farm or inthe study area, keeping goats or poultry as a hobby, or reception of yearly influenza vaccination was collected.

Pneumonia outcome. `Having had pneumonia in the three years preceding the medical examination' (i.e. between 2012 and 2014/15) was defined as the outcome variable. We defined the outcome using two sources of information: (i) self-reported, physician-diagnosed pneumonia over the past three years as reported in the questionnaire, or (ii) having had at least one pneumonia episode recorded in the electronic medical record (EMR) during the three years preceding the medical examination (Fig 1). The benefit of combining data sources in order not to miss cases was described previously for COPD and asthma [14]. By asking for physician- diagnosed pneumonia in case of self-reporting we aimed to avoid misclassification bias. Information on pneumonia episodes was extracted from EMR using International Livestock-associated risk factors for pneumonia in the Netherlands we also conducted sensitivity analysis using `Having had a pneumonia episode registered in the EMR in the three years preceding the medical examination' as an outcome.

Risk factors extracted from electronic medical records. We identified co-morbiditiesregarded as risk factor for pneumonia from the literature. Information on comorbidities ofparticipants was similarly extracted from EMRs using ICPC codes. Individual risk factors weregrouped into the following categories: chronic lung disease, chronic cardiovascular disease,cerebrovascular disease, chronic liver disease, chronic nephropathy, malignancies, autoimmunedisease and neurological comorbidities (S2 Table).

Exposure to livestock farms. We obtained livestock data from provincial databases of mandatory environmental licenses for keeping livestock for 2012 to determine exposure to livestock [11]. Using a Geographic Information System (ArcGis), exposure to farms was quantified for each study participant individually based on geocodes of home address and farms(centroid of stable complex), as described previously [8]. The following variables were analyzed:(i) presence of livestock farms with a minimum number of animals (in increments of 500m) around the residence, (ii) distance between residence and poultry and goat farms expressed in quartiles, (iii) number of farms expressed in tertiles and (iv) number of animals expressed in tertiles (cattle, goats, horses, pigs, poultry, sheep) within 1000m of the home. As the distribution of number of goats within 1000m of their home address was highly skewed (S1 Fig), the construction of tertiles for this variable was not possible. Therefore, we decided to set the minimum number of goats to 50 in line with the threshold applied during the compulsory vaccination campaign during the

Q-fever epidemic [6]. Farms were considered `Q-fever positive' if they had experienced Q-fever-induced abortion waves or if they tested positive in bulk tank milk monitoring (data from GD Animal Health and the Food and Consumer Product Safety Authority).

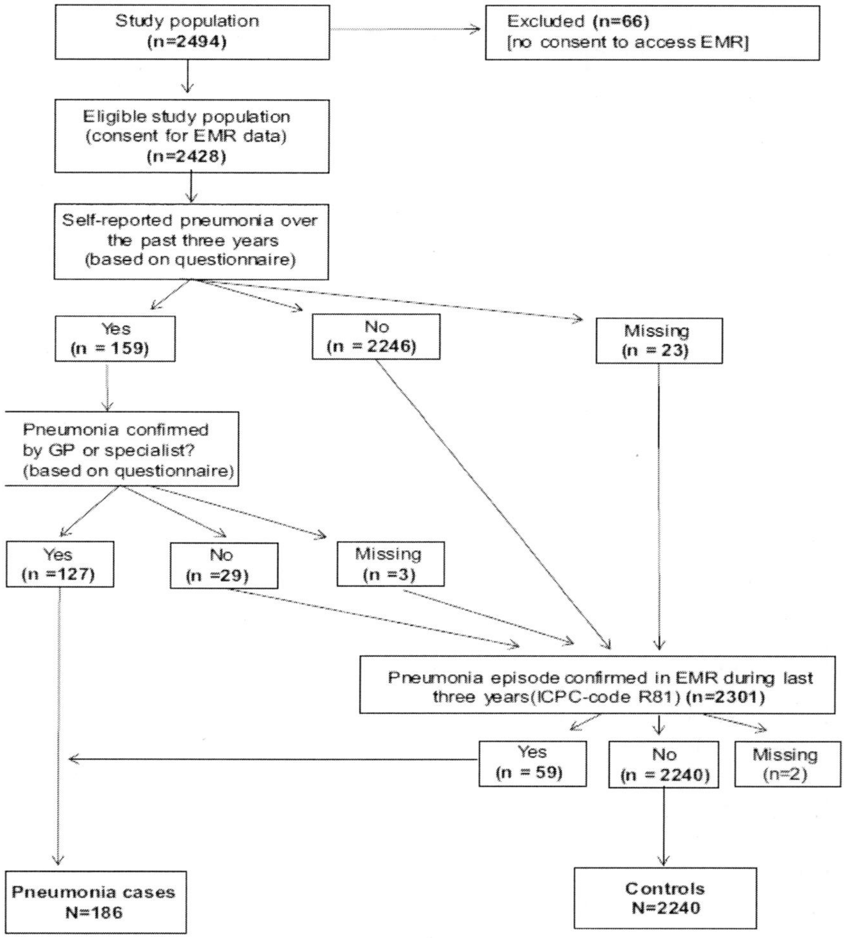

Figure 1. Construction of outcome variable `Having had pneumonia in the past three years' based on information from questionnaires and electronic medical records (EMR).

Serology for Q fever. Serological analyses were performed using a commercial enzymelinked immunosorbent assay (Serion ELISA classic, Virion/Serion, WuÈrzburg, Germany) to test for IgG to C. burnetii phase II antigen. A titer of <20 IU/ml was considered negative, between 20 to 30 IU/ml borderline, and >30 IU/ml was classified seropositive. More details on the serological findings are described elsewhere [10].

Non-response analyses. To assess whether self-selection was present in this study, we conducted non-response analyses on different subpopulations. We first compared characteristics of individuals who were invited and responded (responders) or did not respond to the short questionnaire (non-responders), as well as individuals who were invited for the medical examination and participated (responders) or did not participate (non-responders).

2.5 Statistical analysis

We used univariate logistic regression analysis (Wald Chi Square test statistic) to explore associations between risk factors related to lifestyle and livestock exposure and the occurrence of pneumonia, respectively. To adjust p-values retrieved from the univariate analyses for multiple testing, we used the Benjamini-Hochberg correction with a false discovery rate set to 10% [15].We decided *a priori* to include age and gender in the multiple logistic regression models. Additional determinants with a p-value of less than 0.15 were also included in multivariable analyses investigating associations between livestock exposure and pneumonia. To retrieve the most accurate estimates for the odds ratio, multiple logistic regression models were used with three incremental sets of risk factors included in addition to livestock exposure variables: model A (age and gender), model B (age, gender, smoking, education level and BMI) and model C(age, gender, smoking, education level, BMI, chronic lung diseases and other comorbidities).Univariate and multiple logistic regression analyses were conducted in SAS 9.4 (SAS Institute Inc., Cary, NC, USA). For the multiple logistic regression models, co-morbidities described above±with exception of chronic lung disease±were combined in one dichotomous variable expressed as `Having had at least one registered episode of any listed comorbidity in the three years preceding the medical examination'. As ORs retrieved from the three adjusted models did not substantially differ in magnitude, this article focuses on results of the most parsimonious model, model A. Results of model B and C are shown in S3 Table. To rule out that associations between goat exposure and pneumonia were not due to Q fever, we analyzed associations between seropositivity against *C. burnetii* and pneumonia by creating a dichotomous variable in which seropositive and borderline results were grouped together.

To assess how many cases would be avoided if exposures were to be removed we used the following formula to calculate the population attributable fraction (PAF):

PAF=P_E*[(adjusted OR -1)/ adjusted OR], where P_e is the proportion of cases that is exposed.

3. RESULTS

3.1 Characteristics of study population

Among the 2,426 eligible study participants, we identified 186 pneumonia cases and 2,240 individuals without a history of recent pneumonia (Fig 1). Among the twelve research centers where medical examinations were conducted, location Heeswijk-Dinther had the highest percentage of pneumonia cases (16.3%), whereas the percentage was lowest in location Horn (3.7%, Table 1).

Table 1. Three-year cumulative frequency of pneumonia cases (defined as having had at least one pneumonia episode during the three years preceding the medical examination) including 95% confidence intervals (CI), overall and per location of the research center where the medical examination took place.

Research center	Total number of participants	Number of pneumonia cases	Percentage (95% CI)
Heeswijk-Dinther	369	60	16.3 (14.8–17.8)
Heusden	72	7	9.7 (8.5–10.9)
Deurne	129	11	8.5 (7.4–9.6)
Afferden	48	4	8.3 (7.2–9.4)
Boxtel	165	11	6.7 (5.7–7.7)
Asten	289	19	6.6 (5.6–7.6)
Stramproy	227	15	6.6 (5.6–7.6)
Bakel	305	18	5.9 (5.0–6.8)
Someren	169	10	5.9 (5.0–6.8)
Budel	186	10	5.4 (4.5–6.3)
St. Anthonis	386	18	4.7 (3.9–5.5)
Horn	81	3	3.7 (3.0–4.5)
Total	2,426	186	7.7 (6.6–8.8)

Study participants were between 20 and 72 years old. Pneumonia cases were slightly older than non-cases, with a median age of 61.8 [interquartile range (IQR) 54.9–66.9] versus 58.6 years (IQR 48.7–65.4), respectively (p-value<0.001). A comparison of other characteristics between cases and non-cases is shown in Table 2. In addition to gender, age and smoking, a low educational level, being under- and overweight, having chronic lung diseases and having other comorbidities were identified as risk factors

(Table 2). Compared to non-cases (30.9%), pneumonia cases were more often vaccinated against influenza (46.2%) (Table 2). Influenza vaccination was strongly correlated with increasing age (Chi squared test, p-value <0.001) and was therefore excluded from multivariable analyses to avoid multicollinearity. Having grown up on a farm or outside the study area, or keeping goats or poultry for a hobby over the past five years did not constitute a risk factor (Table 2).

Table 2. Characteristics of the study population (n = 2,426) and risk factors for pneumonia (186 pneumonia cases and 2,240 non-cases) based on univariate logistic regression analyses.

Characteristic		Pneumonia cases n (%)	Non-cases n (%)	Crude odds ratio (95% CI)	p-value[a]
Age categories	≤49	34 (18.3)	573 (25.6)	Ref	0.004[d]
	>49 & ≤59	36 (19.4)	588 (26.3)	1.0 (0.6–1.7)	
	>59 & ≤66	61 (32.3)	571 (25.5)	**1.8 (1.2–2.8)**	
	>66	55 (29.5)	508 (22.7)	**1.8 (1.2–2.8)**	
Gender	Male	74 (39.3)	1038 (46.3)	Ref	0.086[e]
	Female	112 (60.2)	1202 (53.7)	1.3 (1.0–1.8)	
Smoking habits	Never smoked	65 (35.0)	959 (42.8)	Ref	0.114[e]
	Ex-smoker	101 (54.3)	1076 (48.0)	1.4 (1.0–1.9)	
	Smoker	20 (10.5)	205 (9.2)	1.4 (0.9–2.4)	
Education level	High	41 (22.0	681 (30.4	Ref	<0.001[d]
	Middle	76 (40.9	999 (44.6	1.3 (0.9–1.9	
	Low	69 (37.1	560 (25.0	**2.1 (1.4–3.1**	
Body mass index (BMI)	Normal	51 (27.4	725 (32.4	Ref	0.010[d]
	Obese	51 (27.4	446 (19.9	**1.6 (1.1–2.4**	
	Overweight	73 (39.3	1001 (44.7	1.0 (0.7–1.5	
	Underweight	11 (5.9	68 (3.0	**2.3 (1.2–4.6**	
Kept goats as hobby over past 5 years (46 missing)	No	173 (96.7	2141 (97.3	Ref	0.624[e]
	Yes	6 (3.6	60 (2.7)	1.2 (0.5–2.9)	
Kept poultry as hobby over past 5 years (93 missing)	No	149 (85.2	1848 (86.6	1.2 (0.7–1.7)	0.598[e]
	Yes	26 (14.9)	290 (13.4)		
Childhood on farm (24 missing)	No	112 (61.5	1472 (66.3	Ref	0.193[e]
	Yes	70 (38.5)	748 (33.7)	1.2 (0.9–1.7)	
Grew up outside study area (26 missing)	No	141 (77.5	1684 (75.9	Ref	0.638[e]
	Yes	41 (22.5)	534 (24.1)	0.9 (0.6–1.3)	
Chronic lung diseases[b]	No	138 (74.2	2051 (91.6	Ref	<0.001[d]
	Yes	48 (25.8)	189 (8.4)	**3.8 (2.6–5.4)**	
Comorbidities[c]	No	136 (73.1	1854 (82.8	Ref	0.001[d]
	Yes	50 (26.9)	386 (17.2)	**1.8 (1.3–2.5)**	
Reception of yearly influenza vaccination (19 missing)	No	98 (55.9	1537 (69.1	Ref	<0.001
	Yes	84 (46.2)	688 (30.9)	**1.9 (1.4–2.6)**	

[a] Based on Wald chi square test statistic. Variables with a p-value lower than 0.15 were included in the multivariable models in addition to livestock exposure variables

[b] Variable consists of at least one episode of chronic bronchitis, COPD or asthma recorded in the electronic medical record during the three years preceding the medical examination

[c] Comorbidity '1' means that participant had at least one (maximum 5) episodes of any hereafter mentioned comorbidity groups recorded in the electronic medical record during the three years preceding the medical examination: chronic cardiovascular disease, cerebrovascular disease, chronic liver disease, chronic nephropathy, malignancies, auto-immune diseases or neurological comorbidities.

[d] Significant or [e] non-significant p-value when adjusting for multiple testing using the Benjamini-Hochberg correction with a false discovery rate of 10%

Neither the presence, nor the number of children in the household (age categories: <4, 4–17, >17 years) were associated with pneumonia (data not shown). Adding 'Living within 2000m of a Q-fever positive farm' to the model did not change the OR between presence of goat farm within 500 to 2000m meters and pneumonia substantially and was therefore omitted from the model [range: OR 4.4, 95% CI 2.0–9.9 (within 500m) to OR 2.0, 95% CI 1.4–2.8 (within 2000m)].

3.2 General findings regarding associations between livestock farm exposure and pneumonia

The main study focus was on goat and poultry exposure, hence, results presented here are primarily focused on these two species.

Overall, different goat farm exposures were consistently associated with pneumonia in the univariate analysis (Table 3), whereas we found no significant associations for poultry or other animals/ animal farms (Table 3). After adjusting for risk factors, the significant positive associations between goats/ goat farms and pneumonia remained (Table 3). Overall, adjusted odds ratios (aOR) did not vary substantially between the three adjusted models (S3 Table).

In multivariable logistic regression analyses, none of the other determinants adjusted for remained an independent risk factor with exception of chronic lung diseases.

3.2.1 Association between seropositivity against Coxiella burnetii and pneumonia.

For 2,358 of the 2,426 participants (97%), serological results for *Coxiella burnetii* were available. Of these, 146 were seropositive (6.2%). Among those seropositive, 13 (8.9%) also have had pneumonia in the previous three years. However, no statistically significant association between having had pneumonia and being seropositive for *Coxiella burnetii* (dichotomous variable) was found (Table 4). Results did not change when using the three-category serostatus for *C. burnetii* (negative, borderline or positive; p-value>0.7).

Exposure to Q fever-positive farms. When investigating associations between goat farms that tested positive for Q fever during the epidemic and pneumonia, we found that no Q fever-positive farm was located within 500 or 1000m of the residence of a case. The number of pneumonia cases that lived within 1500 or 2000m to Q fever-positive farms was low (Table 4). Neither living within 1500m, nor 2000m of a Q fever positive farm was associated with pneumonia (Table 4).

Distance to farms with Q fever-positive bulk milk samples in 2010 was similar for cases and non-cases (median distance 3777m, IQR: 2924–5611 versus 4023m, IQR 2860–6108; p-value 0.133). Median distance between residence and farms with Q fever-induced abortions was significantly shorter for cases (median distance 7042m, IQR: 3233–12,550 versus 9760, IQR 3770–12,748; p-value 0.0013). However, distances to Q fever-positive farms were still substantially larger compared to distances to farms with ≥50 goats, as described above.

Table 3. Exposure status and associations from univariate (crude OR) and multiple logistic regression analyses (adjusted OR) between pneumonia and livestock exposure variables. The proportions of cases attributable to respective exposure is indicated by the population attributable fraction (PAF).

		Pneumonia cases (%) n = 186	Non-cases (%) n = 2240	Crude OR (95% CI)	Adjusted OR[a] (95% CI)	PAF[9] (%)
Number of farms (any type) within 1000m of residence[1]						
	<6	36 (19.4)	596 (26.6)	Ref	Ref	
	>6 and <11	88 (47.3)	842 (37.6)	**1.7 (1.2–2.6)**	**1.8 (1.2–2.6)**	20.4
	>11 (max 32)	62 (33.3)	802 (35.8)	1.3 (0.8–2.0)	1.36 (0.9–2.1)	
Presence of any type of farm within a certain distance of residence[2]						
	500m	125 (67.2)	1445 (64.5)	1.1 (0.8–1.6)	1.14 (0.8–1.6)	
	1000m	176 (94.6)	2151 (96.0)	0.7 (0.4–1.4)	0.73 (0.4–1.4)	
Presence of farm with minimum amount of animals within 500m-intervals of residence[3]						
500m[4]	Goat	11 (5.9)	31 (1.4)	**4.5 (2.2–9.1)**	**4.4 (2.0–9.8)**	4.6
	Poultry	32 (17.2)	322 (14.4)	1.2 (0.8–1.8)	0.95 (0.6–1.5)	
	Pig	60 (32.3)	626 (27.9)	1.2 (1.0–1.7)	1.21 (0.8–1.8)	
	Cattle	100 (53.8)	1176 (52.5)	1.1 (0.8–1.4)	0.91 (0.7–1.3)	
	Horse	50 (26.9)	521 (23.3)	1.2 (0.9–1.7)	1.03 (0.7–1.5)	
	Sheep	12 (6.5)	169 (7.5)	0.9 (0.5–1.6)	0.91 (0.5–1.7)	
1000m[5]	Goat	35 (18.8)	229 (10.2)	**2.0 (1.4–3.0)**	**2.0 (1.3–3.1)**	9.5
	Poultry	112 (60.2)	1226 (54.7)	1.3 (0.9–1.7)	1.10 (0.8–1.6)	
	Pig	152 (81.7)	1773 (79.2)	1.2 (0.8–1.7)	1.02 (0.7–1.6)	
	Cattle	174 (93.5)	2110 (94.2)	0.9 (0.5–1.7)	0.64 (0.3–1.3)	
	Horse	143 (76.9)	1599 (71.4)	1.3 (0.9–1.9)	1.28 (0.8–2.0)	
	Sheep	58 (31.2)	701 (31.3)	1.0 (0.7–1.4)	0.93 (0.7–1.3)	
1500m[6]	Goat	62 (33.3)	485 (21.7)	**1.8 (1.3–2.5)**	**1.9 (1.4–2.7)**	15.9
	Poultry	156 (83.9)	1890 (84.4)	0.96 (0.6–1.5)	0.74 (0.5–1.2)	
	Pig	183 (98.4)	2172 (97.0)	1.91 (0.6–6.1)	1.65 (0.5–5.6)	
	Cattle	186 (100)	2233 (99.7)	/	/	
	Horse	172 (92.5)	2032 (90.7)	1.26 (0.7–2.2)	1.34 (0.7–2.6)	
	Sheep	102 (54.8)	1357 (60.6)	0.79 (0.6–1.1)	0.78 (0.6–1.1)	
2000m[7]	Goat	90 (48.4)	742 (33.1)	**1.9 (1.4–2.6)**	**1.9 (1.4–2.6)**	23.1
	Poultry	175 (94.1)	2040 (91.1)	1.56 (0.8–2.9)	1.35 (0.7–2.7)	
	Pig	186 (100)	2228 (99.5)	/	/	
	Cattle	186 (100)	2240 (100)	/	/	
	Horse	180 (96.8)	2138 (95.4)	1.43 (0.6–3.3)	1.15 (0.5–2.9)	
	Sheep	145 (78)	1776 (79.3)	0.92 (0.6–1.3)	0.88 (0.6–1.3)	
Distance (quartiles expressed in meters) between residence and closest farm with minimum number of animals						
50 goats	>3490 to 11477	39 (21.0)	568 (25.4)	Ref	Ref	
	>2478 to ≤3490	41 (22.0)	566 (25.3)	1.06 (0.7–1.7)	1.05 (0.7–1.7)	
	>1629 to ≤2478	41 (22.0)	565 (25.2)	1.06 (0.7–1.7)	1.02 (0.7–1.6)	
	99 to ≤ 1629	65 (34.9)	541 (24.2)	**1.8 (1.2–2.7)**	**1.8 (1.2–2.7)**	15.3
250 poultry	>1296 to 4145	46 (24.7)	561 (25.0)	Ref	Ref	
	>923to ≤1296	33 (17.7)	573 (25.6)	0.70 (0.4–1.1)	0.69 (0.4–1.1)	
	>644 to ≤923	47(25.3)	557 (24.9)	1.03 (0.7–1.6)	1.03 (0.7–1.6)	
	39 to <644	60(32.2)	549 (24.5)	1.33 (0.9–2.0)	1.38 (0.9–2.1)	
Number of animals within 1000m of the residence[8]						
Goats (no tertiles)	0	133 (71.5)	1857 (82.9)	Ref	Ref	
	>0 to ≤50	18 (9.7)	155 (6.9)	1.62 (0.97–2.72)	1.54 (0.88–2.69)	3.4
	>50	35 (18.8)	228 (10.2)	**2.14 (1.44–3.19)**	**1.67 (1.06–2.63)**	7.5

		Pneumonia cases (%)	Non-cases (%)	Crude OR	Adjusted OR[a]	PAF[9]
		n = 186	n = 2240	(95% CI)	(95% CI)	(%)
Poultry (tertiles)	0	55 (29.6)	794 (35.4)	Ref	Ref	
	>0 to ≤28250	64 (34.4)	689 (30.8)	1.34 (0.92–1.95)	1.12 (0.74–1.69)	
	>28250	67 (36.0)	757 (33.8)	1.28 (0.88–1.85)	1.09 (0.70–1.69)	

[a] Adjusted for age and gender
[1] Any type of animal farm (main farming category, expressed in tertiles)
[2] Main farm category (any type of animal farm) as registered in the livestock database
[3] Minimum amount of animals: 50 goats, 250 poultry, 25 pigs, 5 cattle, 5 horses, 50 sheep
[4] Also adjusted for presence of other farms within 500m with a minimum number of animals
[5] Also adjusted for presence of other farms within 1000m with a minimum number of animals
[6] Also adjusted for presence of other farms within 1500m with a minimum number of animals
[7] Also adjusted for presence of other farms within 2000m with a minimum number of animals
[8] Also adjusted for number of other animals within 1000m
[9] Population attributable fraction based on model A

Table 4. Characteristics of the study population for Q fever related exposures.

Characteristic		Pneumonia cases	Non-cases	Crude odds ratio	p-value[a]
		n (%)	n (%)	(95% CI)	
Serostatus for Coxiella burnetii (Q-fever) (68 missing	Negative	165 (92.7)	2047 (93.9)	Ref	>0.5[d]
	Positive & borderline combined	13 (7.3)	133 (6.1)	1.2 (0.7–2.2)	
Serostatus for *Coxiella burnetii*[b]	Negative	165 (94.8)	2047 (96.1)	Ref	>0.4[d]
	Only positive	9 (5.2)	84 (3.9)	1.3 (0.7–2.7)	
Living within 1500m of Q-fever positive farm[c]	No	177 (95.2)	2120 (94.6)	Ref	>0.7[d]
	Yes	9 (4.8)	120 (5.4)	0.9 (0.5–1.8)	
Living within 2000m of Q-fever positive farm	No	159 (85.5)	1996 (89.1)	Ref	>0.1[d]
	Yes	27 (14.5)	244 (10.9)	1.4 (0.9–2.1)	

[a] Based on Wald chi square test statistic. Variables with a p-value lower than 0.15 were included in the multivariable models in addition to livestock exposure variables
[b] For the sensitivity analysis we excluded participants with borderline serological results.
[c] No pneumonia cases lived within 500 or 1000m of a Q-fever positive goat farm (defined as testing positive for *Coxiella burnetii* in bulk milk or having had abortion storms during the Q fever epidemic)
[d] non-significant when adjusting for multiple testing using the Benjamini-Hochberg correction with a false discovery rate of 10%

Presence of farms within 500m increments to residence. The presence of at least one goat farm (with a minimum of 50 goats) within 500m increments of the home address (ranging from 500m to 2000m), was significantly positively associated with pneumonia, whereas such associations were not, or not consistently, found for other types of livestock farms (Table 3). Adjusted ORs for presence of goat farm increased in magnitude the closer to the residence goat farms were located (Table 3); a trend analysis confirmed the observed dose-response relationship (p-value <0.001). This was also observed when an adjustment was made for the presence of other animal farms than goat farms.

Distance between farms and residence. Pneumonia cases lived significantly closer to farms with ≥50 goats, compared to non-cases (median distance 2090m, IQR 1222–3291 versus 2501m, IQR 1663-3511m; p-value 0.0015). A shorter distance to goat farms with at least 50 goats (Q4, ~100 to 1,600m) compared to a longer one (Q1, ~3,500 to 11,500m) was significantly associated with pneumonia (Table 3).

Poultry farms with a minimum amount of 250 birds were located closer to cases than non-cases (median distance 794m, IQR 560–1291 versus 936m, 647–1298; p-value 0.041). A shorter distance to poultry farms with at least 250 birds, was also positively associated with pneumonia compared to a larger distance (Q4 vs. Q1), but associations were not statistically significant (Table 3).

Number of animals within 1000m of the residence. The presence of more than 50 goats within 1000m of the residence was significantly associated with pneumonia (aOR 1.7, 95% CI 1.1–2.6), compared to having no goats close to the home (Table 3). Although number of poultry within 1000m of the residence was substantially higher, no association with pneumonia was found (Table 3).

Population attributable fraction. The number of pneumonia cases exposed to goat farms ranged from 11 (living within 500m) to 90 (living within 2000m of a goat farm; Table 3). The population attributable fraction (PAF; based on model A), i.e. the fraction of cases that could be avoided if the exposure to goat farms were to be removed, was low (4.6%) for those living within 500m of a goat farm. For a 1000m to 2000m-perimeter, the PAF gradually increased from 9.6% to 23.1%, respectively (Table 3).

3.3 Sensitivity analyses using EMR recorded pneumonia as outcome variable

When repeating the analyses using pneumonia based on the EMR only as outcome, i.e. using 'Having had a pneumonia episode registered in the EMR in the three years preceding the medical examination', the associations between goat farm exposure and pneumonia remained robust and the magnitude was comparable to when the combined pneumonia outcome was used (S4 Table). When using the EMR-based pneumonia variable, associations between pneumonia and presence of a poultry farm (with at least 250 birds) within 1000m of the residence reached statistical significance (OR adjusted for age and sex: 1.7, 95% CI 1.1–2.7). For other 500m-intervals (500m, 1500m, 2000m), no statistically significant association with poultry farm presence was found (S4 Table).

3.4 Non-response analyses

Participants who were older, female and lived closer to a livestock farm were more likely to fill in the short questionnaire, as well as more likely to participate in the medical examination (data not shown). Also, people with a goat farm present within 1000m of their home (irrespective of minimum number of goats) were more likely to participate in the medical examination. However, in all subpopulations (total population, individuals who responded to the short questionnaire, all individuals invited to the medical examination, individuals who took part in the medical examination) [11], we found consistent associations between pneumonia (as registered in the EMR) and presence of a goat farm within 1000m of the home. ORs adjusted for age and gender ranged from 2.1 (95% CI 1.6–2.8) for the total population to 2.4 (95% CI 1.3–4.5) for those who participated in the medical examination.

4. DISCUSSION

In this study in non-farming adults from the general population, we found that the presence of goat farms near the home was strongly positively associated with pneumonia, with increasing odds the closer the residence was located to the farm. Remarkably, the present study investigated the occurrence of pneumonia between 2012 and 2015, whereas similar results were found in a study conducted during a Q fever outbreak in 2009 [8]. A positive test for antibodies against *Coxiella burnetii* was not associated with pneumonia. The magnitude of the association between the presence of a poultry farm within 1000m and pneumonia was comparable to previous findings when using EMR-recorded pneumonia as an outcome. No associations between other animal farms/ types and pneumonia was found.

The finding that goat farms still pose a risk factor for pneumonia in a period where the Q-fever incidence among humans had dropped to pre-epidemic levels–i.e. 63 notifications with onset of illness in 2012, 20 in 2013, and 26 in 2014 –was unexpected and we explored several explanations for the association between pneumonia and proximity to goat farms. One explanation is that people who lived closer to goat farms were more likely to participate in this study compared to those living further away, which could possibly introduce selection bias. However, as the non-response analyses resulted in comparable associations between presence of goat farms and pneumonia in all three subpopulations, self-selection bias seems unlikely.

Bias in self-reported pneumonia is unlikely to be the explanation for the association as they were also found with EMR-registered pneumonia. For data obtained through the questionnaire, we attempted to avoid misclassification bias–meaning that, for instance similar clinical presentations, such as acute bronchitis/ bronchiolitis might falsely be classified as pneumonia–by only including cases who reported pneumonia and who reported having had their diagnosis confirmed by a physician. When comparing reports of self-reported, physician diagnosed pneumonia with pneumonia records from the EMR, 121 (65%) participants had a record of pneumonia in both sources. For 65 (35%) participants pneumonia was only recorded in the questionnaire but not in the EMR. The data from the EMR represented data from a GP primary care network, hence, we could not check whether the 35% that stated having had pneumonia in the questionnaire were diagnosed by a specialist or in a hospital, which would explain why no pneumonia was recorded in the EMR. In the Netherlands, about 80% of pneumonia cases are managed in primary care [16]. This would suggest that 15% of the pneumonia cases in our study could be misclassified, which could overestimate odds ratios by introducing differential misclassification bias, if cases who lived closer to goat farms falsely reported having had pneumonia. Among the 65 cases who did not have an EMR entry for pneumonia in the three years preceding the medical examination, 13 had an entry for acute bronchitis/bronchiolitis (20%), a condition that can present with similar symptoms as pneumonia [17].

A previous study conducted between April 2008 and March 2009 showed that cases with Q fever pneumonia were more likely to live in a region with a high goat density or close to sheep [18]. However, in the present study, we found no positive association between pneumonia and being seropositive for *Coxiella burnetii*, or living within 500 or 1000m of a Q fever-positive farm, respectively, which makes it unlikely that the associations presented here are primarily attributable to this zoonosis. A limitation of our study was that no information was available concerning causative pathogens or season in which pneumonia occurred, neither did we conduct serological analyses for other pathogens. A previous study in a large hospital in the province of North-Brabant, investigated which pathogens are most frequently detectable in patients with community-acquired pneumonia (CAP) [19]. This prospective observational study conducted between November 2007 and January 2010, coinciding with the Q fever epidemic, showed, that *Streptococcus pneumoniae* was most frequently identified bacterium (22%), followed by *Coxiella burnetii* (14%), *Mycoplasma pneumonia* (6%) and *Haemophilus influenzae* (6%). Although these results give interesting insights in the etiology of CAP, the authors stressed that continuous microbiological surveillance in combination with clinical symptoms is needed to be able to monitor seasonal variations and allow extrapolation to other years [19]. More recently, another group studied CAP in four Dutch hospital cohorts covering the periods 1998–2000 and 2004–2010, and found that atypical microorganisms, such as *Legionella* species, *Coxiella burnetii*, *Mycoplasma pneumoniae* and *Chlamydia* species were predominantly detected during the non-respiratory season defined as week 20 to week 39 (40.4%), as compared to the respiratory season (12.4%; week 40 to week 19) [20]. Awareness of the importance of seasonal patterns and including exposure to animals in medical history can help to guide clinicians in targeted testing for atypical pathogens outside the respiratory season. Whether atypical/ other zoonotic pathogens might also play a role in people living in a goat-dense area presenting with CAP could be addressed in future studies.

Besides infectious causes, non-infectious causes, such as mold, inhalable dust and endotoxins, might also play a role in explaining the associations between animal farms and pneumonia. Various experiments demonstrated that inhalation of fine dusts and endotoxins can increase susceptibility for infection with common human pathogens [21, 22] and that inhalation of (urban) particulate matter can lead to pneumonia in humans, through increased adhesion of *S. pneumoniae* to human airway epithelium [23]. Exposure to dust was also identified as a risk factor for CAP among professionals exposed to different working conditions [24], however measurements of such non-infectious exposures were not yet available in this study which constitutes a limitation. To analyze the relation between CAP and the composition of dust emission from goat farms, air filter techniques, as described previously, could be utilized in future research [25]. Dairy goats in the Netherlands are kept in deep litter stables with partly open roofs or walls. The deep litter husbandry system is characterized by topping up soiled litter, such as straw or hay, with fresh litter such as straw or hay every few days. When the layer of litter becomes too high, manure is removed and stored on a dunghill before being spread on farmland as a fertilizer [26]. A small study investigating the contribution of different dust sources to dust mass transmitted via air showed, that straw used for bedding contributed more than 50% to fine and coarse dust emissions. However, compared to cow farms, goat farms emit twelve times less total dust per kg metabolic weight [25], suggesting that dust emitted from goat farms might not be the most evident explanation of our results.

Although dust and endotoxins might not contribute substantially to explaining the observed association between goat farms and pneumonia, it might be relevant in explaining part of the association between pneumonia and poultry farms (S3 and S4 Tables). A comprehensive study from Denmark studied emissions of inhalable dust and endotoxins between different farming types and seasons and found, that compared to other farming types, poultry and pig farmers were exposed to the highest levels of fine dust and endotoxins [27]. That poultry farm emissions might also be relevant for the health of neighboring residents was shown in a Dutch study, which found elevated endotoxin levels 250m downwind from poultry farms [7].

Another interesting future research avenue is to investigate the indirect effects of non-infectious agents on the risk of CAP by studying the composition of the human upper respiratory tract microbiome in residents living close to goat farms. A recent study hypothesized that exposure to farm emissions may result in changes of the composition of the upper airway microbiome, which might lead to commensals, such as *S. pneumoniae*, to become pathogens. An association was found between living close to poultry farms and CAP that possibly resulted from alterations of the oropharyngeal microbiota composition. As this was the first study that showed such an association, the researchers stressed that these findings need to be replicated in larger studies [28].

The relatively small sample size constitutes a limitation of our study. In the study by Smit et al. [8], the association between presence of a poultry farm within 1km of the home address and pneumonia was relatively low (OR 1.25, 95% CI 1.06–1.47). It is conceivable that the current study was underpowered to detect an association with poultry exposure when using the combined pneumonia outcome, although results were close to statistical significance. When using EMR-based pneumonia as outcome, a significant association with poultry exposure within 1000m was found. Similarly, to above, zoonotic pathogens might play a role in those presenting with CAP and living close to poultry. For instance, *Chlamydia psittaci*, a zoonotic bacterium associated with poultry causes psittacosis in humans. In CAP-etiological studies, this bacterium is often not considered or only incidentally isolated. However, a study from the Netherlands identified 7/147 (4.8%) CAP patients with psittacosis [18, 19, 29], suggesting that the role of *C. psittaci* is more important than often assumed.

The research center Heeswijk-Dinther had the highest prevalence of pneumonia (16.3%, Table 1), which also coincided with the highest seroprevalence against *Coxiella burnetii* (10%; average seroprevalence: 6%) [10]. We examined goat farm density per research center and found that exposure to farms with ≥50 goats (defined as presence within 500m increments of the residence) for individuals examined at Heeswijk-Dinther was highest compared to other study centers. However, in the current study, no association between Q fever and pneumonia was found (Table 4). As older age is as a risk factor for pneumonia, we examined whether a higher percentage of older people lived in Heeswijk-Dinther compared to other study centers. However, although we found slight, nevertheless significant, differences between the centers (Kruskall Wallis test, Chi^2 value 21.1, p-value = 0.03), the median score for age was generally lower for Heeswijk-Dinther, compared to other study centers (data not shown), thereby dismissing this hypothesis. As there is currently no good explanation why the proportion of pneumonia cases was higher in Heeswijk-Dinther compared to other study centers, this question could be addressed in future research.

In conclusion, in this study we showed that living in the vicinity of goat farms still constitutes a risk factor for pneumonia. As it was surprising that the association between goat exposure and pneumonia was still found in a period where Q fever in humans has only occurred sporadically, however, Q fever itself seemed an unlikely explanation of the findings, future research should be directed to studying the role of alternative infectious and non-infectious causes to be able to assess possible implications for public health and provide evidence-based recommendations. To shed light on the contribution of atypical pathogens other than *Coxiella burnetii [20]* to the CAP burden near goat and poultry farms, molecular diagnostics for *Legionella* species, *Mycoplasma pneumoniae* and *Chlamydia* species should be considered in CAP patients from these regions. In addition, the role of the composition of the human upper airway microbiome [28] in people with CAP living close to goat farms should be investigated in future research.

REFERENCES

1. Van Os J and Gies T, Grootschalige veehouderij in Nederland—Aantal bedrijven, locaties en milieuvergunningen, in Alterra-rapport. 2011: Wageningen.
2. Roest H, Tilburg J, Van der Hoek W, Vellema P, van Zijderveld F, Klaassen C, et al., The Q fever epidemic in The Netherlands: history, onset, response and reflection. Epidemiol Infect, 2011. 139(1): p. 1–12. pmid:20920383
3. Van der Hoek W, Morroy G, Renders NH, Wever PC, Hermans MH, Leenders AC, et al., Epidemic Q fever in humans in the Netherlands. Adv Exp Med Biol, 2012. 984: p. 329–64. pmid:22711640
4. Dijkstra F, van der Hoek W, Wijers N, Schimmer B, Rietveld A, Wijkmans CJ, et al., The 2007–2010 Q fever epidemic in The Netherlands: characteristics of notified acute Q fever patients and the association with dairy goat farming. FEMS Immunol Med Microbiol, 2012. 64(1): p. 3–12. pmid:22066649
5. Slok EN, Dijkstra F, de Vries E, Rietveld A, Wong A, Notermans DW, et al., Estimation of acute and chronic Q fever incidence in children during a three-year outbreak in the Netherlands and a comparison with international literature. BMC Res Notes, 2015. 8: p. 456. pmid:26384483
6. EFSA Panel on Animal Health Welfare, Scientific Opinion on Q fever. EFSA Journal, 2010. 8(5).
7. Heederik D and IJzermans C, Mogelijke effecten van intensieve-veehouderij op de gezondheid van omwonenden: onderzoek naar potentiële blootstelling en gezondheidsproblemen. 2011, IRAS, NIVEL, RIVM,: Utrecht.
8. Smit LA, van der Sman-de Beer F, Opstal-van Winden AW, Hooiveld M, Beekhuizen J, Wouters IM, et al., Q fever and pneumonia in an area with a high livestock density: a large population-based study. PLoS One, 2012. 7(6): p. e38843. pmid:22685612
9. Wouters I, Samadi S, De Rooij M, Beekhuizen J, Hooiveld M, Van der Sman F, et al., Animal confinement buildings as a source of ambient PM10 and endotoxin PM10 [abstract]. Epidemiology, 2011.

10. Maassen K, Smit L, Wouters I, van Duijkeren E, Janse I, Hagenaars T, et al. Veehouderij en gezondheid omwonenden (Livestock farming and the health of local residents). 2016; 136]. Available from: http://www.rivm.nl/Documenten_en_publicaties/ Wetenschappelijk/Rapporten/2016/juli/Veehouderij_en_gezondheid_omwonenden.
11. Borlee F, Yzermans CJ, van Dijk CE, Heederik D, and Smit LA, Increased respiratory symptoms in COPD patients living in the vicinity of livestock farms. Eur Respir J, 2015.
12. Schnoor M, Klante T, Beckmann M, Robra BP, Welte T, Raspe H, et al., Risk factors for community-acquired pneumonia in German adults: the impact of children in the household. Epidemiol Infect, 2007. 135(8): p. 1389–97. pmid:17291378
13. Teepe J, Grigoryan L, and Verheij TJ, Determinants of community-acquired pneumonia in children and young adults in primary care. Eur Respir J, 2010. 35(5): p. 1113–7. pmid:20436174
14. Borlee F., Yzermans J., Krop E., Aalders B., Zock J-P., Van Dijk C., et al. Assessment of Asthma and COPD Prevalence: A Comparison Between Three Different Data Sources. in B48. Asthma: Insights from the bench, genetics, and epidemiology. 2016. American Thoracic Society.
15. Benjamini Y and Hochberg Y, Controlling the False Discovery Rate: A Practical and Powerful Approach to Multiple Testing. Journal of the Royal Statistical Society. Series B (Methodological), 1995. 57(1): p. 289–300.
16. Snijders B, van der Hoek W, Stirbu I, van der Sande MA, and van Gageldonk-Lafeber AB, General practitioners' contribution to the management of community-acquired pneumonia in the Netherlands: a retrospective analysis of primary care, hospital, and national mortality databases with individual data linkage. Prim Care Respir J, 2013. 22(4): p. 400–5. pmid:24042173
17. Scott JA, Wonodi C, Moisi JC, Deloria-Knoll M, DeLuca AN, Karron RA, et al., The definition of pneumonia, the assessment of severity, and clinical standardization in the Pneumonia Etiology Research for Child Health study. Clin Infect Dis, 2012. 54 Suppl 2: p. S109–16.
18. Huijskens EG, Smit LA, Rossen JW, Heederik D, and Koopmans M, Evaluation of Patients with Community-Acquired Pneumonia Caused by Zoonotic Pathogens in an Area with a High Density of Animal Farms. Zoonoses Public Health, 2016. 63(2): p. 160–6. pmid:26214299
19. van Gageldonk-Lafeber AB, Wever PC, van der Lubben IM, de Jager CP, Meijer A, de Vries MC, et al., The aetiology of community-acquired pneumonia and implications for patient management. Neth J Med, 2013. 71(8): p. 418–25. pmid:24127502
20. Raeven VM, Spoorenberg SM, Boersma WG, van de Garde EM, Cannegieter SC, Voorn GP, et al., Atypical aetiology in patients hospitalised with community-acquired pneumonia is associated with age, gender and season; a data-analysis on four Dutch cohorts. BMC Infect Dis, 2016. 16: p. 299. pmid:27317257
21. Becker S, Soukup JM, Sioutas C, and Cassee FR, Response of human alveolar macrophages to ultrafine, fine, and coarse urban air pollution particles. Exp Lung Res, 2003. 29(1): p. 29–44. pmid:12652814
22. Rivas-Santiago CE, Sarkar S, Cantarella Pt, Osornio-Vargas A, Quintana-Belmares R, Meng Q, et al., Air pollution particulate matter alters antimycobacterial respiratory epithelium innate immunity. Infect Immun, 2015. 83(6): p. 2507–17. pmid:25847963
23. Mushtaq N, Ezzati M, Hall L, Dickson I, Kirwan M, Png KM, et al., Adhesion of Streptococcus pneumoniae to human airway epithelial cells exposed to urban particulate matter. J Allergy Clin Immunol, 2011. 127(5): p. 1236–42 e2. pmid:21247619
24. Almirall J, Serra-Prat M, Bolibar I, Palomera E, Roig J, Boixeda R, et al., Professions and Working Conditions Associated With Community-Acquired Pneumonia. Arch Bronconeumol, 2015. 51(12): p. 627–31. pmid:25544548
25. Aarnink A. RHJ, Cambra-Lopez M., Zhao Y., Mosquera J. and Ogink N.W.M., Emissions and Concentrations of Dust and Pathogens from Goat Houses, in Ninth International Livestock Environment Symposium. 2012.
26. Hermans T, Jeurissen L, Hackert V, and Hoebe C, Land-applied goat manure as a source of human Q-fever in the Netherlands, 2006–2010. PLoS One, 2014. 9(5): p. e96607. pmid:24788538

27. Basinas I, Sigsgaard T, Heederik D, Takai H, Omland O, Andersen NT, et al., Exposure to inhalable dust and endotoxin among Danish livestock farmers: results from the SUS cohort study. J Environ Monit, 2012. 14(2): p. 604–14. pmid:22159073
28. Smit L, Boender G, de Steenhuijsen Piters W, Hagenaars T, Huijskens E, Rossen J, et al., Increased risk of pneumonia in residents living near poultry farms: does the upper respiratory tract microbiota play a role? Pneumonia, 2017. 9(3).
29. Spoorenberg SM, Bos WJ, van Hannen EJ, Dijkstra F, Heddema ER, van Velzen-Blad H, et al., Chlamydia psittaci: a relevant cause of community-acquired pneumonia in two Dutch hospitals. Neth J Med, 2016. 74(2): p. 75–81. pmid:26951352

CHAPTER 17

Participatory Evaluation of the Relative Livestock Populations and the Assessment of the Status and Impacts of Newcastle Disease in Rural Communities of Two Northeastern States, Nigeria

Waziri Ibrahim Musa[1*], Sa'idu Lawal[2], Mohammed Bello[3], Paul Ayuba Abdu[1]

[1]*Department of Veterinary Medicine, Ahmadu Bello University, Zaria, Nigeria*

[2]*Veterinary Teaching Hospital, Ahmadu Bello University, Zaria, Nigeria*

[3]*Department of Veterinary Public Health and Preventive Medicine, Ahmadu Bello University, Zaria, Nigeria*

ABSTRACT

Livestock is an important component of food security in many developing nations. However, animal diseases continue to undermine animal production and public health efforts thus widening poverty gap. Unfortunately, the conventional and what seems to be inefficient "top to bottom" disease surveillance and control policies are heavily relied upon such that negative impacts of diseases are underestimated. Participatory disease surveillance (PDS) techniques using semi-struc- tured questioning assisted by key informants targeting focus groups were employed in 60 randomly selected villages of two Nigerian States during the years 2012 to 2014. Haemagglutination and ELISA tests to detect antibodies to Newcastle disease (ND) were conducted on 950 poultry sera. The status and economic burden of Newcastle disease and the relative livestock populations and some ethno-veterinary practices of these livestock farming communities were brought to the fore. Poultry, goats, cattle, sheep and pigs in descending order were the major livestock species kept in the study areas to which Kendalls Coefficient of Concordance (W = 0.9) agreed strongly. Accordingly, ND, lousiness, fowl pox and coccidiosis with percentage scores of ND-52%, Fowl Pox- 31%, lousiness-17% and Salmonellosis-10% were important causes of poultry morbidity and mortality. ND sero-prevalence was 39%, relative morbidity; mortality and case fatality rates of 95%, 78% and 82% respectively were appraised. Again ND, coccidiosis, ectoparasitism and fowl pox were reported as seasonal poultry diseases (W = 0.6). Solanum nodiflorum and Momordica balsalmina were used to treat ND. Major livestock kept, and the ND status and effects in poultry in these livestock farming communities are here reported. Institutionalization of PDS would better inform strategic livestock policy reforms, and improve national food security and diseases surveillance and reporting system in Nigeria.

Keywords: Participatory, Evaluation, Newcastle Disease, Rural, Communities

1. INTRODUCTION

Participatory epidemiology (PE) is an emerging branch of veterinary epidemiology which is based on participatory rural appraisal methods that involve key stakeholders to improve both the understanding as well as developing the best and sustainable options for livestock disease control [1] -[3]. For many years, conventional research methodologies and disease control initiatives that utilized the "top to bottom" approach were faced with limitations to the quality and general acceptability of research outputs in many communities. In many instances human and financial resources available to support conventional disease surveillance efforts are becoming scarce in many developing nations. It has become necessary therefore that identified disease risks should deserve priority for a cost effective and sustainable surveillance which seems to be the basic concept upon which participatory epidemiology is established [1] [2] [4]. Farmers' participation in research is important to understand their priorities for interventions and resource allocations so that solutions developed through this approach would be appropriate to farmers' technical level and culture thus increasing their acceptability and sustainability [5]. With the increasing global focus on one health, emerging and re-emerging zoonoses are better approached by integrating veterinary and public health surveillance programmes offered by participatory concept of disease control.

Local communities are knowledgeable of the principal diseases that affect their livestock, have rich terminology for distinguishing between them, and possess traditional methods of handling them [6] -[8]. For diseases with significant economic impacts, communities have traditional medicaments and often attempt to actively track the evolution of outbreaks and take actions to mitigate risk of further spread. Participatory surveillance taps into these information networks to find out the past and present situations of diseases in order to obtain a better understanding of livestock-keepers knowledge and practices necessary for disease control [9].

Livestock population in a nation is known through census which appears not to be given needed attention in many places. The last time livestock census was conducted in Nigeria, for example, was over 23 years ago and often livestock population is needed in research and planning. Qualitative estimates of livestock populations can easily be obtained at a given time using participator epidemiology tools [9].

Newcastle disease is an infectious virus disease of over 250 bird species that is caused by avian Paramyxovirus seroptype-1 (APMV-1) virus belonging to the genus Avulavirus and considered as one of the major contagious disease of poultry worldwide. ND economic impacts take effects from its worldwide distribution that restricts poultry trade, drop in productivity, mass mortality and financial drain in control measures [10]. In most developing countries, the greatest impact of ND is on rural or backyard chicken production and generally referred to as the most important infectious disease affecting village chickens causing great economic losses [11] [12].

2. MATERIALS AND METHODS

2.1. Sites Selection

Bauchi State (HPAI reported) and Gombe State (HPAI not reported) as of December 2014. These states are located in north eastern Nigeria, share boarders and trade in live bird and thus were selected for this study. Muti-stage sampling technique was used to select one LGA from each of the six senatorial districts and simple random sampling technique was also used to select one district from each of the selected LGA of the two states.

2.2. Collection of Samples

Ethical clearance to conduct this research was approved by the Ahmadu Bello University Ethical Committee on Animal Use. Samples were collected in the months of January to June 2012 and November 2013 to January 2014. Collection of blood from birds was done with birds held laterally with the wing extended to expose the brachial vein and a few small feathers were removed to facilitate viewing of the vein. The vein was wiped with a swab containing 70% alcohol and a 21 gauge needle attached to a 5 ml syringe inserted into the vein and a maximum of 2 to 3 ml of blood collected. Pressure with a piece of cotton wool was then applied to the vein at the insertion site of the needle until the bleeding stopped. The blood was then transferred to a labeled plain blood collection tube. Samples were allowed to clot at room temperature. The clotted (coagulated) blood was then fractionated by centrifuging at 449 g for 5 minutes at room temperature (27°C). After centrifugation, the sera were transferred to labeled two mill screw cap containers and stored in a freezer at -20°C until used.

2.3. Sample Testing Procedure

2.3.1 Preparation of 1% Chicken Red Blood Cell

Two millilitres of blood was collected from 5-day old chicks and pooled in an equal volume of Alsever's solution. Cells were washed three times in physiological buffered solution (pH 7.2) by centrifuging at 447.2 g for 5 minutes [13]. One per cent RBC suspension was made by adding 0.9 ml of PBS to 0.1 ml of washed RBC. Antibody titre was determined by haemagglutination inhibition (HI) test using a U bottom microtitre plate [13]. The HI was the highest dilution of serum causing complete inhibition of agglutination of RBC by 4 HAU antigens. The agglutination was assessed by tilting the plates. Only wells in which the RBCs stream at the same rate as the control wells (containing 0.25 ml 1% RBCs and 0.05 ml PBS only) were considered to show inhibition.

All sera samples were tested at the Virology and Public Health Laboratories of the Faculty of Veterinary Me- dicine, Ahmadu Bello University, Zaria using the haemagglutination inhibition test following standard protocol and competitive enzyme linked immunosorbent assay (cELISA) procedures as described by the manufacturers. The Newcastle disease virus antibody ELISA test kits were kindly supplied b AFFINITECH, LTD (USA) and IDVET (France) respectively

3. PARTICIPATORY DISEASE INVESTIGATION

3.1. Conduct of Interview

Participation by farmers was made voluntarily but restricted to livestock farmers only. Interview was conducted using semi-structured questioning that was based on a check list/interview guide of written points that explored livestock diseases [14]. The semi-structured interview is a generally accepted interviewing method in PE and rural appraisal that is usually flexible to accommodate the type of information being sought [14]. A check list of steps in the interview was: 1. Self introduction of all participants with a brief for the purpose of the study. 2. What livestock species were kept in the visited communities? 3. What where the major constraints of livestock. 4. What diseases affected poultry? 5. How were poultry disease managed traditionally? 6. Was vaccination carried out against most important poultry diseases? 7. What was the impact of vaccination against major poultry disease? 8. Any other relevant issues. Careful observations were made to ensure respondents were relaxed, confident and actively participated. An average of 15 person's focus group discussions was conducted with participants comprising of elders and youths of both genders. Tools and techniques used in this study were as described by Catley, 1997 and AFENET, 2011 [14] [15]. These were generally or specifically used according to particular needs but not predetermined by prescribed PE methodology but were selected based on the information sought.

3.2. Simple Ranking

Simple ranking was conducted to get an estimate of livestock specie population. Open ended question (what are the livestock species kept in this area?) requested farmers to mention all livestock species found in the community. Livestock species mentioned were symbolized on pieces of cardboard paper. Farmers were requested to arrange or rank the papers in order in terms of livestock specie population beginning with the highest to the lowest.

3.3. Proportional Piling

Proportional piling (PP) was used to relatively estimate livestock species by numbers and their relative importance in the communities and to further estimate ND impact following outbreaks [14] [15]. Participants were asked to list the livestock species kept in the community. Circles were drawn on cardboard papers with each circle representing livestock species mentioned within the communities. Participants then allocated 100 counters (beans) (assuming that the livestock population in the community made up a 100%) to each circles in piles according to the relative numbers of each species with the highest score indicating the most abundant livestock specie in the area. The exercise was repeated to assess livelihood means in the communities, relative incidences of the major livestock diseases and case fatality rates of ND in vaccinated and unvaccinated flocks using the same procedure.

3.4. Participatory Mapping

This was drawn before interview began in most communities surveyed [14]. Participants were requested to identify and draw physical features of their community starting with the boundaries and coordinates. Important social (livestock markets, schools, health and veterinary centres, and forests, etc.) and physical features (grazing areas, rivers, centres, and forests, etc.) were identified and drawn in the map.

3.5. Disease Impact Matrix Scoring

For the important livestock species mentioned in the community, a matrix was constructed on a cardboard, with benefits derived from those species along the y axis and the important diseases that affect them on the x axis. Participants were given 100 counters and asked to allocate them among the livestock-derived benefits according to the relative importance of each benefit mentioned, with the most important benefit receiving the highest number of counters. The counters for each benefit were then sub-allocated to each disease to show the relative negative impact of each disease on a family's ability to achieve that benefit. In this way the disease having the greatest impact received the highest number of counters. The number of counters allocated for each disease was counted to give a measure of the overall impact of that disease on livestock-derived livelihood means [1] [14] [15].

3.6. Matrix Scoring for Disease Definition

The association between diseases based on easily recognizable clinical signs and risk factors was also assessed using matrix scoring method. Simple matrices were constructed on the ground. In the first for poultry, various clinical signs or risk factors formed the y axis and probable diseases, one of which was ND that formed the x axis. For the second and third, the same was done for cattle and then sheep and goats. In all, risk factors or consistent disease clinical signs formed the y axis and probable diseases formed the x axis. For each clinical sign or risk factor, participants allocated 30 counters in proportion to their relative importance for the different diseases [9] [14] [15].

3.7. Seasonal Calendar

A horizontal line was drawn and divided into 12 months. Participants were made to divide the year into different seasons paying attention to the characteristics of each season. Where seasonal diseases existed, their relative occurrences were indicated using counters at different points along the line of seasons mentioned. Key risk factors such as rainfall, humidity, vector populations, etc. in relation to disease incidence were also scored. Disease onset, prevalence and seasonality were determined by scoring (PP) using counters. This exercise was continuously debated until all the farmers agreed upon the scores and the actual situation /data were arrived at [9] [14] [15].

3.8. Traditional/Indigenous Veterinary Knowledge

The community traditional veterinary knowledge is often related to clinical, epidemiological, gross manifestations of diseases conditions as well as remedies or solutions to livestock production and health. Most times these are known, given literary meanings that may correspond to specific disease entities in various languages and transferred from generations to generations in such communities. Information on these was extracted from livestock farming communities [9] [14] [15 .

3.9. Data Analyses

Data obtained using PE tools were exported to SPSS 2002 version 17.0 where Kendall's coefficient of concordance (W) was used to measure levels of agreement between different informant groups. Kendall's coefficient of concordance measures associations between sets of ranks assigned to objects using values between 0 and 1. W ranges from 0 - 1 with 0 denoting no agreement and 1 complete agreement. Agreement may be weak ($W < 0.26$), moderate (W—between 0.26 and 0.38) and good ($W > 0.38$) [9] [14].

4. RESULTS

4.1. Use of Ranking and Scoring Techniques

4.1.1. Proportional Piling and Simple Ranking

Proportional piling and ranking application as the most popular PE scoring tools indicated crop farming, livestock farming and petit trading to be the most important livelihood activities that the communities were involved. It further identified indigenous poultry, goats, sheep cattle and pigs in order of population as the abundant livestock species kept in the communities as indicated in Table 1. Poultry farmers used PPP and SR to identify Newcastle disease (ND), fowl pox (FP), lousiness (L), and salmonellosis (SM) as important causes of poultry morbidity and mortality (Table 2). Proportional piling for disease impact indicated a mortality estimate of ND outbreaks to be over 70% in unvaccinated poultry flocks. With vaccination in place, ND mortality estimate was quantified at 30% (Figure 1 and Figure 2). Proportional piling exercise also revealed the relative occurrence of poultry diseases with ND as the most prevalent (52%) poultry disease in the study areas (Figure 3).

Table 1. Median scores (95% CI) of the relative proportions of livestock species kept in selected villages of Bauchi and Gombe States, Nigeria determined through proportional piling exercise.

State	Local Government Area	Village	Relative livestock populations				
			Poultry	Sheep/goats	Cattle	Pigs	Others
Bauchi	Bauchi	Fadaman	57 (54 - 60)	40 (37 - 43)	1 (0 - 4)	0 (0)	3 (0 - 6)
		Wunti	52 (49 - 55)	33 (30 - 36)	1 (0 - 4)	0 (0)	14 (11 - 18)
	Misau	Luggudi	58 (55 - 61)	35 (32 - 38)	5 (2 - 8)	0 (0)	2 (0 - 5)
		legal	50 (47 - 53)	41 (36 - 42)	8 (5 - 11)	0 (0)	1 (0 - 4)
	katagum	Madangal	52 (49 - 55)	35 (32 - 38)	11 (8 - 14)	0 (0)	2 (0 - 5)
		Kakimari	49 (46 - 52)	48 (45 - 51)	2 (0 - 5)	0 (0)	1 (0 - 4)
		Mean	53.0	38.7	4.7	0.0	3.8
Gombe	Gombe	Pantami	56 (53 - 59)	42 (39 - 45)	1 (0 - 4)	0 (0)	1 (0 - 4)
		Malam inna	53 (50 - 56)	36 (33 - 39)	8 (5 - 11)	0 (0)	3 (0 - 6)
	Yamaltu deba	Deba	49 (46 - 52)	33 (30 - 36)	9 (6 - 12)	2 (0 - 5)	7 (4 - 10)
		Kwadon	60 (57 - 63)	33 (30 - 36)	5 (2 - 8)	0 (0)	2 (0 - 5)
	Kaltungo	Kalaring	48 (45 - 51)	32 (29 - 35)	8 (5 - 11)	3 (0 - 6)	9 (6 - 12)
		Ture-mai	49 (46 - 52)	36 (33 - 39)	10 (7 - 13)	1 (0 - 4)	4 (1 - 7)
		Mean	52.5	35.3	6.8	1.0	4.3
		Overall mean	52.8	37.0	5.75	0.5	5.9

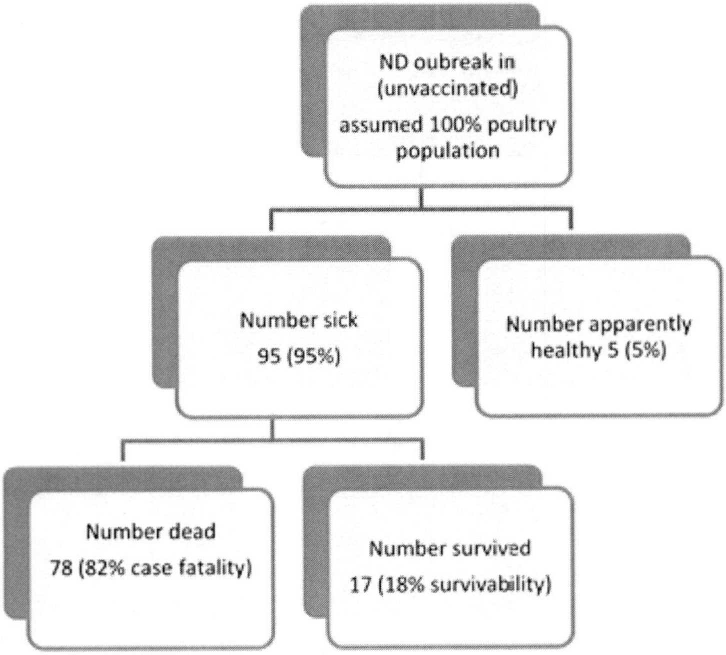

Figure 1. Farmers' perception of the relative impact of the outbreak of Newcastle disease in unvaccinated naïve poultry flock.

4.1.2. Matrix Scoring for Disease Definition

Farmers recognized ND in chickens to be associated with high morbidity and mortality, was characterized with nervous signs, depression, whitish and sometimes greenish diarrhea. It occurred throughout the year but had higher frequencies of occurrence during the cold windy hamattan.

4.1.3. Disease Impact Matrix Scoring

This was studied based on the economic values of various livestock species in one of the visited communities. It clearly indicated poultry as the major source of egg, meat and emergency source of fund.

4.1.4. Semi-Structured Interview

The use of semi-structured interview (SSI) showed that most livestock and especially poultry were predomi- nantly kept on free range with little or no veterinary care and feed supplementations.

Table 2. A summary of matrix scoring (by proportional piling of 30 counters each) of major poultry diseases versus clinical manifestations as reported by rural poultry farmers in seven selected villages of Bauchi and Gombe States, Nigeria.

Clinical attribute	Disease			
	Newcastle disease	coccidiosis	Fowl pox	Ectoparasitism
High morbidity (W = 0.83)	18 (15 - 21)	3 (0 - 6)	3 (0 - 6)	8 (5 - 11)
High mortality (W = 0.90)	20 (17 - 23)	8 (5 - 11)	1 (0 - 4)	1 (0 - 4)
Nervous signs (W = 0.83)	27 (24 - 30)	0 (0 - 0)	0 (0 - 0)	3 (0 - 6)
Respiratory distress (W = 0.90)	27 (24 - 30)	0 (0 - 0)	2 (0 - 2)	1 (0 - 1)
Greenish diarrhea (W = 0.51)	28 (25 - 30)	1 (0 - 1)	1 (0 - 1)	0 (0 - 0)
Bloody diarrhea (W = 0.54)	2 (0 - 2)	28 (25 - 30)	0 (0 - 0)	0 (0 - 0)
Whitish diarrhea (W 0.90)	27 (24 - 30)	0 (0 - 0)	0 (0 - 0)	2 (0 - 2)
Papules/nodules (W = 0.54)	0 (0 - 0)	0 (0 - 0)	27 (24 - 30)	2 (0 - 2)
Insects on the body (W = 0.83)	0 (0 - 0)	0 (0 - 0)	2 (0 - 2)	2 (0 - 2)
Decrease productivity (W = 0.90)	15 (12 - 18)	6 (3 - 9)	7 (4 - 10)	2 (0 - 2)
Vector associated (W = 0.83)	2 (0 - 2)	0 (0 - 0)	6 (3 - 9)	22 (19 - 25)

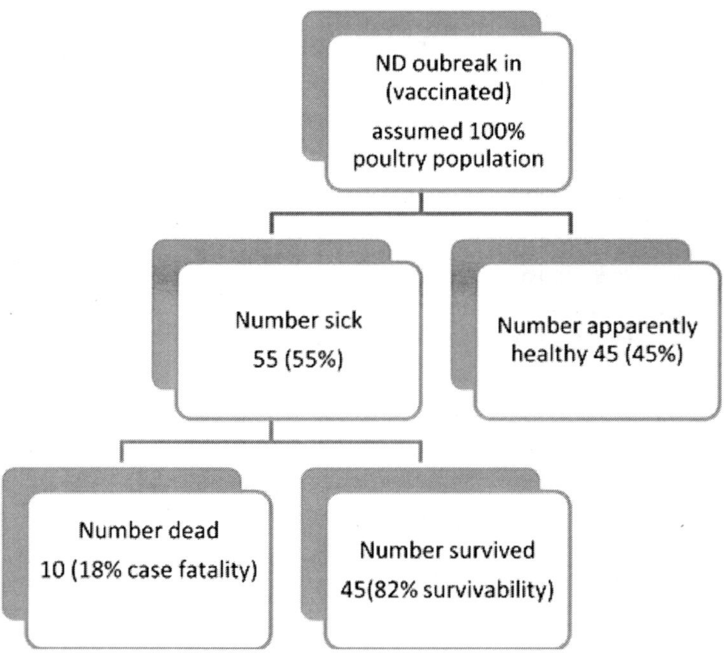

Figure 2. Farmers' perception of the relative impact of the outbreak of Newcastle disease in vaccinated poultry flock.

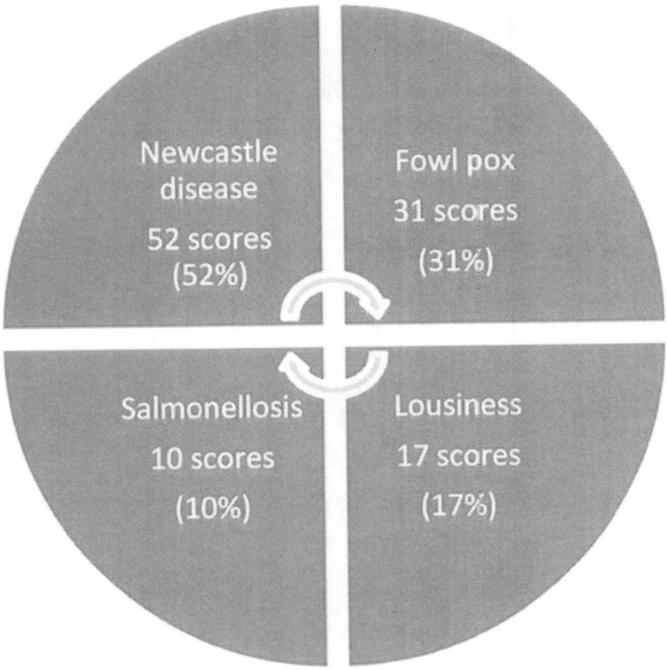

Figure 3. Proportional piling (scores) to indicate the relative occurrence of various poultry diseases.

4.2. Use of Visualization Technique

4.2.1. Seasonal Calendar

Seasonal calendar puts the year into 4 recognizable local seasons kaka (harvest period/early dry season), bazara/hamattan (cold dry windy season), zafi (hot season) and damina (rainy season). ND, though occurred all year round, was most common during the cold hamattan period. Lousiness was most common just before the onset of rains while pox was most common with early rains.

4.2.2. Participatory Mapping

This exercise involved both the literate and illiterate farmers by identifying on the community map the positioning of social amenities as well as natural features of the communities.

4.3. Ethnoveterinary Knowledge

Traditionally, gautankaji (Solanum nodiflorum), gunan bera (Momordica balsalmina) were used as remedies for ND in the villages visited. Palm and engine oils were commonly used to treat lousiness and fowl pox.

4.4. Sero-Prevalence Study

An overall sero-prevalence of ND in both States was 39% with the highest prevalence of 60% in northern Gombe (Table 3).

5. DISCUSSION

Extensive researches have been conducted using PE tools in recent years; also reviews focusing on the usage of PE by Veterinarians that led to the reliability and validity of information obtained [2] [3] [15]. Participatory rural appraisal is now known for its problem solving methodology that allows experts to learn about traditional knowledge systems and community development priorities [16] in many rural communities of the world, priority livestock diseases were established and practical control measures instituted through PDS [17] -[21]. In this study, common poultry diseases with the most important being ND were appraised in various communities. Traditional management of ND was emphasized in all visited communities. As a trial in 2009/2010, Nigeria chose to use PE practitioners trained by the EDRSAIA project to investigate the status of priority livestock diseases in five geopolitical regions of the country. The PS team in addition to the targeted diseases diagnosed PPR in Nigeria, which led to an effective emergency PPR disease control program [22]. PDS therefore has an additional advantage of unveiling other diseases not initially targeted.

Of all the livestock populations' poultry was ranked first within all the visited communities. It was reasoned that there was hardly any house without at least a chicken in the rural areas. Poultry was easy to start and maintained, it was a source of quick money to solve pressing needs and was frequently used for religious and spiritual celebrations, and as gifts to friends and visitors so it became a community pride to own chickens.

During ranking and scoring, ND was placed the most economically significant poultry disease in these farming communities. It is well established that ND is a pandemic and that its impact is felt by every poultry farmer worldwide, thus it was given local names in traditional settlements worldwide, disease impact matrix scoring strongly showed the likelihood of ND outbreaks to invariably affect egg and meat consumption and ultimately quick source of money to solving emergency family needs in many rural communities of Nigeria. [10] -[12]. The high sero-prevalence (38% - 60%) of ND in all communities studied indicates clearly the endemic nature as well as the wide spread of ND. In Indonesia for instance, participatory disease surveillance and response (PDSR) was successfully used targeting village poultry production system to report evidence of ND virus circulation and apparent absence of HPAI. This information was used to target ND control [23]. Furthermore, participatory disease investigation makes use of traditional knowledge, historical information and extensive oral communication network making it a more reliable and sensitive clinical surveillance [3] [9] [14] - [16] [24]. The traditional veterinary knowledge is concerned with history and patterns of diseases in an area. The local communities often recognize the principal diseases that affect their livestock and have a rich terminology for distinguishing between them [3]. Though this study targeted ND, fowl pox, lousiness and salmonellosis were found to be important poultry diseases in the rural communities studied (Figure 3) which opens a new horizon to holistic control of poultry diseases in rural communities who seem to hold over 65% of total poultry populations in many developing nations [8] [11]. Some of the matrix scoring findings in this study showed ND to be associated with high morbidity, mortality and occurred year round with higher incidences in the cold windy hamattan. The information that ND occurred with high incidence during hamattan may assist farmers as well as livestock policy makers to initiate appropriate ND control measures especially vaccinating birds and avoiding introduction of new birds before the onset of hamattan. It further elucidates the fact that ND outbreaks will directly affect most needed animal protein intake as well as settlement of emergency fees such as school fees in many rural communities in northeastern Nigeria. The results of matrix scoring (Table 2) showed that farmers usually recognize ND by nervous signs, whitish diarrhoea, high morbidity and mortality which agree with some established literatures [8] [10] [11]. In many literatures however, greenish diarrhoea especially in acute ND infections supersedes the white diarrhoea reported by farmers [10] [11]. This means that farmers may lack the ability to recognize acute ND that may not be associated with the normal nervous signs and high mortality and most especially may lack the ability to recognize HPAI therefore limiting their ability to report suspected cases. Matrix scoring further made it possible to evaluate certain disease syndromes like "makere (somnolence), fakat (massive death), and farin kasha (white diarrhea)" to mean ND as understood by the Hausas and this seems to agree with the clinical presentations of ND as existed in many scientific findings on ND [8] [10] [11].

Table 3. Newcastle disease sero-prevalence in poultry and wild birds from six senatorial districts of Bauchi and Gombe States using ELISA and haemagglutination inhibition (HI) test.

	State					
	Bauchi			Gombe		
	Region					
	North	Central	South	North	Central	South
Sera samplesa tested (n = 950)	164	170	166	150	150	150
ELISA positive (n = 170)	34	60	36	100	40	70
HI positive	58	66	56	90	40	56
ELISA seroprevalence (%/region)	20.7	35.3	21.6	66.7	26.6	46.7
HI seroprevalence (%/region)	35.4	38.8	33.7	60.0	26.6	37.3
ELISA seroprevalence (%/state)		26.0			46.7	
HI seroprevalence (%/state)		36.0			41.3	
Overall ELISA both states seroprevalence (%)			35.8			
Overall HI both states serophrevalence (%)			38.5			

Key: ELISA-enzyme linked immunosorbent assay, n-number of birds sampled, HI-haemagglutination inhibition

The major limitations in the conduct of participatory research especially in the livestock subsector and particularly in Nigeria are willingness of target farmers to participate in the absence of incentives and also participants tend to look at the study from a political perspective. These may likely affect the validity of results obtained.

6. CONCLUSION

A sustainable and workable livestock disease surveillance system should have a perceived benefit to data providers. Participatory epidemiology in view of this ensured animal health researchers and famers benefited from oral information exchange and arrived at reliable and sustainable intervention strategies.

REFERENCES

1. Jost, C.C., Mariner, J.C., Roeder, P.L., Sawitri, E. and Macgregor-Skinner, G.J. (2007) Participatory Epidemiology in Disease Surveillance and Research. Review Science Technical Office International des Epizootics, 26, 537-549.
2. Catley, A. (2005) Participator Epidemiology: A Guide for Trainers. African Union/ Interafrican Bureau for Animal Resources, Nairobi, 1-18.

3. Catley, A., Alders, R.G. and Wood, J.L.N. (2012) Participatory Epidemiology: Approaches, Methods, Experiences. The Veterinary Journal, 191, 151-160. http://dx.doi.org/10.1016/j.tvjl.2011.03.010
4. Mariner, J.C., Hendrickx, S., Pfeiffer, D.U., Costard, S., Knopf, I., Okuthe, D., Chibeu, D., Parmley, J., Musenero, M., Pisang, C.J., Zingeser., B.A., Jones, S.N., Ali, B., Bett, M.M., Araba, F.A., Mehta, P. and Jost, C.C. (2011) Integration of Participatory Approaches into Surveillance Systems. Review Science Technical Office International des Epizootics, 3, 653-659.
5. Apantaku, S.O. (2006) Analysis of Participation of Farmers in Participatory Poultry Production Research in Lagos State, Nigeria. Livestock Research for Rural Development, 18, 1-12.
6. Catley, A., Okoth, S., Osman, J., Fison, T., Njiru, Z., Mwangi, J., Jones, B.A. and Leyland, T.J. (2001) Participatory Diagnosis of a Chronic Wasting Disease in Cattle in Southern Sudan. Preventive Veterinary Medicine, 51, 161-181. http://dx.doi.org/10.1016/S0167-5877(01)00240-9
7. Musa, I.W., Ndahi, M., Pam, E.G., Okike, A., Cyrile, P.C., Saskia, H. and Jost, C. (2013) Using Participatory Epidemiological Techniques to Establish Rural Based Poultry Disease Profile; Practical Field Experience in the Jos Plateau, Nigeria. World Poultry Science Journal, 69, 387-400. http://dx.doi.org/10.1017/S0043933913000378
8. Musa, I.W. and Kaltungo, B.Y. (2014) Participatory Methods in the Profiling of Livestock Diseases in the Jos Plateau, Nigeria. Animal and Veterinary Sciences, 2, 154-160. http://dx.doi.org/10.11648/j.avs.20140205.15
9. Mariner, J.C. and Paskin, R. (2000) Food and Agricultural Organisation. Manual on Participatory Epidemiology, Methods for the Collection of Action-Oriented Epidemiological Intelligence, 5-81.
10. Alexander, D.J. (1995) The Epidemiology and Control of Avian Influenza and Newcastle Disease. Comparative Pathology, 112, 105-126. http://dx.doi.org/10.1016/S0021-9975(05)80054-4
11. Spradbrow, P.B. (1993) Newcastle Disease in Village Chickens. Poultry Science Reviews, 5, 57-96.
12. Aini, I., Ibrahim, A.L. and Spradbrow, P.B. (1990) Vaccination of Chickens against Newcastle Disease with a Food Pellet Vaccine. Avian Pathology, 19, 371-384. http://dx.doi.org/10.1080/03079459008418687
13. Collee, J.G., Duguid, J.P., Fraser, A.G. and Marmion, B.P. (1982) Practical Medical Microbiology. 13th Edition, Mackie and McCartey, Eds., Churchill Livingstone, 37-39.
14. African field epidemiology network (AFFENET) (2011) Public Health Participatory Epidemiology. Introductory Training Module, Manual for Trainers, 120-210.
15. Catley, A.P. (1997) Adapting Participatory Appraisal for the Veterinary Epidemiologist: Participatory Appraisal Tools for Use in Livestock Disease Data Collection. Proceedings of the Society for Veterinary Epidemiology and Preventive Medicine, Chester, United Kingdom, 4-9.
16. Chambers, R. (2007) From PRA to PLA and Pluralism: Practice and Theory. IDS Working Paper 286, July 2007.
17. Mariner, J.C., Manzoor Hussain, Rodder, P.L. and Catley, A. (2003) Use of Participatory Disease Searching in Pakistan for Rinderpest and More (Vina del Mar, Chile). Proceedings of the 10th International Symposium on Veterinary Epidemiology and Economics, Rome, November 2003, 23.
18. Admassu, B. (2005) The Participatory Epidemiological Investigation of FMD in Erzurum Province. http://www.participatoryepidemiology.info/userfiles/Admassu-FAO-Turkey-PE.pdf
19. Catley, A., Chibunda, R.T., Ranga, E., Makungu, S., Magayane, F.T., Magoma, G., Madege, M.J. and Vosloo, W. (2004) Participatory Diagnosis of a Heat-Intolerance Syndrome in Cattle in Tanzania and Association with Foot-and-Mouth Disease. Preventive Veterinary Medicine, 65, 17-30. http://dx.doi.org/10.1016/j.prevetmed.2004.06.007

20. Hussain, M., Afzal, M., Ali, Q., Taylor, W., Mariner, J. and Roeder, P. (2008) The Epidemiology of Peste Des Petits Ruminants in Pakistan and Possible Control Policies. Revue Scientifique et Technique, 27, 869-876.
21. Jost, C.C., Nzietchueng, S., Kihu, S., Bett, B., Njogu, G., Swai, E.S. and Mariner, J.C. (2010) Epidemiological Assessment of the Rift Valley Fever Outbreak in Kenya and Tanzania in 2006 and 2007. American Journal of Tropical Medicine and Hygiene, 83, 65-72. http://dx.doi.org/10.4269/ajtmh.2010.09-0290
22. Mohamadou, F., Jost, C. and Ihedioha, J. (2011) Financial Costs of Disease Burden, Morbidity and Mortality from priority Livestock Diseases in Nigeria. Nigeria Integrated Animal and Human Health Project. International Livestock Research Institute, Nairobi, 6-20.
23. Muhammad, J.W. and Resti, Y. (2010) The Clinical Profiles of Avian Influenza in Endemic and Non-Endemic Regions of Indonesia. Hospital-Based Studies and Its Implication on Clinical Management in the Future. Indonesian Journal of Tropical and Infectious Diseases, 1, 114-117.
24. Mariner, J.C. and Roeder, P.L. (2003) Use of Participatory Epidemiology in Studies of the Persistence of Lineage 2 Rinderpest Virus in East Africa. Veterinary Record, 152, 641-647. http://dx.doi.org/10.1136/vr.152.21.641

CHAPTER 18

Dynamics of Ruminant Livestock Management in the Context of the Nigerian Agricultural System

O. A. Lawal-Adebowale

Department of Agricultural Extension and rural Development, Federal University of Agriculture, Abeokuta Ogun State, Nigeria

1. INTRODUCTION

Among all the livestock that makes up the farm animals in Nigeria, ruminants, comprising sheep, goats and cattle, constitute the farm animals largely reared by farm families in the country's agricultural system. Nigeria has population of 34.5million goats, 22.1million sheep and 13.9million cattle. The larger proportion of these animals' population are however largely concentrated in the northern region of the country than the southern region. Specifically about 90 percent of the country's cattle population and 70 percent of the sheep and goat populations are concentrated in northern region of the country. Concentration of Nigeria's livestock-base in the northern region is most likely to have been influenced by the ecological condition of the region which is characterised by low rainfall duration, lighter sandy soils and longer dry season. This submission is predicated by the fact that drier tropics or semi-arid regions are more favourable to the ruminants, Notwithstanding this situation, certain breeds of sheep and goats, particularly the West African Dwarf (WAD) species, are peculiarly adapted to the southern (humid) region of the country and are commonly reared by rural households in the region. Although, no breed of cattle is peculiar to the southern humid region of Nigeria, the available cattle in the region was largely due to settlement of the Hausa/Fulani pastoralists, who constitute the main cattle rearers, in the region.

The option of settled lifestyle of the Fualani pastoralists in the southern region of Nigeria was largely informed by a number of changes in the ecological condition of the region. One of the changing conditions that made the southern/humid region of the country habitable for cattle rearing was the drastic reduction in the incidence of tsetse fly (Glossina spp) infestation- a vector of the cattle disease known as trypanosomoses or sleeping sickness, in the region. The reduced incidence of tsetse fly the reduced incidence of tsetse flies was brought about by considerable transformation of the southern region's forest-base to derived savanna arising from continuous and expanded land clearing for agriculture and human habitation; and the emerging incidence and severity of bush burning. These actions respectively lowered the region's humidity and heightened its heat intensity, thereby making the environment less conducive for the tsetse flies' survival or lifecycle completion. In the same vein, the successful settlement of the pastoralists in the southern region to the animal's development of a level of tolerance or resistance to the trypanosomosis or sleeping sickness as a result of prolonged exposure to tsetse flies. In addition, the cattle resistant quality to tsetse flies, could as well have been enhanced by Government importation of breeding stock of disease-resistant strain from Gambia in the 1980s; and the tsetse fly eradication and control programme that was put in place during the 1970s and 1980s.

2. BREEDS OF RUMINANTS' CHARACTERISTICS AND DISTRIBUTION IN NIGERIA

With the changing ecological condition of the southern Nigeria and its conduciveness to cattle survival, the animal have become common in the region, though with the Fulani and Hausa tribes that have chosen to settle in the southern region with their herds of cattle. Based on this cattle, sheep and goats, as commonly found in the northern region of Nigeria, are as well found in the southern part of the country, though in less proportion to that of the northern region. Most of the available ruminants in the country are however of indigenous breeds.

Cattle breeds: breeds of locally available cattle in Nigeria are basically indigenous and are grouped as the Zebu and Taurine. The zebus as locally recognised by the cattle rearers in northern part of Nigeria include Bunaji,

Rahaji, Sokoto Gudali, Adamawa Gudali, Azawak and Wadara. The Taurines on other hand include Keteku, N'dama and Kuri [11, 12]. The zebus are characterised by long horns, large humps and tallness, against the Taurines that are humpless, short-horned and shot-legged.

Although, there are varying estimations of cattle population in Nigeria ranging between 10 and 15million [2,3,14] the mean average of the nation's cattle population was put at 13.9 million as at 1990 [12]. While about 11.5 million of the cattle population was kept in pastoral systems, the remaining 2.4 million were kept in villages. Country-wide distribution of the cattle population however showed that the sub-humid region of Nigeria has about 4.5million heads[1] - [13], with the mean cattle density of about 15 per km2 or 6.6 hectare per head; and approximately 45% of the national herd could be readily found in the sub-humid zone of the country on year [12].

Production characteristics of surveyed cattle in the Kaduna plain of Nigeria, entails an average of 45.9 head, out of which 64.4% were females; 60months (5years) as first age of calving, 25months (about 2years) of calving intervals and calving percentage of 48%. Calf life-weight and mortality to 1 year of age averaged 103 kg and 22.4% respectively. Milk production by the cattle, after adjusting for length of calving intervals, for humans and calves averaged 112 and 169 litres/cow/year respectively.

Sheep: Nigeria has a population of about 8 to13.2million sheep out of which about 3.4million are found the southern/humic region and the larger proportion of the animal in the northern region of the country. Available breeds of sheep in the country are mainly indigenous and these are the West African Dwarf (WAD) sheep, Balami, Uda and Yankasa. Out of these four major of breeds of sheep in the country, the WAD breed is common to southern region against the widespread of Balami, Uda and Yakansa breeds in the northern region of the country. Characteristics analysis of sheep in the country, especially among the Fulani pastoralists showed that ewes had approximately 120% fertility rate, 12% rate of twinning and 25% lamb mortality rate at 3months old. Sheep productivity index puts lamb weight at 0.327 kg at a weaning age of 90 days, and 0.490 kg at a weaning age of 180days per ewe per year. Mature males of the local breeds of sheep have a live weight of about 30 to 65kg and their female counter parts often weigh between 30 and 45kg.

Goats: on the other hand has a population of about 22 to 26million in Nigeria with rough estimates of 6.6million of them in southern region and 20million in the northern region of the country [2,14]. The breeds of goats in Nigeria are largely indigenous; and the common ones [19] include the West African Dwarf (WAD) goat, Sahel/desert goat- known as West African Long-Legged goat; and Sokoto Red/Maradi. The Kalahari goat breed, which is of South Africa origin is gradually being adapted to the Nigeria's ecological zones on experimental efforts. Distribution of the goat breeds in the country showed that the West African Dwarf (WAD) goat is common to southern Nigeria while the Sahel or desert goat and Sokoto Red are common to the northern region of the country. Production characteristic of the small ruminant showed that breeds of goats in the country had low fertility rate (below 100%), 40% twins and triplets birth rates, and low mortality rates of 22% for kids and 14.4% for adults. The productivity indices for 90 and 180 days weaning age were 0.259 kg and 0.437 kg kid/kg doe respectively. Milk production characteristic of the goats varies from breed to breed. [20,21] The Sokoto Red produces a daily milk yield of about 0.5 to 1.5kg and 100days of lactation; Sahel goats produce between 0.8 and 1.0kg of milk daily with lactation period of 120days; and the WAD breeds produce about 0.4kg milk per day on a lactation period of 126 days. It was further indicated that these local breeds of goat usually weighs between 18kg and 37kg.

3. SOCIAL AND ECONOMIC VALUES OF RUMINANTS IN NIGERIA

The economic values: the ruminants play significant roles in the social and economic wellbeing of the Nigerians in various ways. Economically the animals serve as source of income earning to major ruminants' dealers-sellers of live animals and butchers/meat sellers; generates employments and creates markets for larger number of people who explore the animals' product and by-products for economic gains. Meat constitutes the foremost animal product that is highly explored by the Nigerian households, particularly for direct consumption and as such, the ruminants, especially cattle, constitute the major and cheapest source of meat consumption for most households in the country [22] about 1million cattle are annually slaughtered for meat in the country. This suggests heavy dependence on cattle for meat consumption by households in the country. This may not be unconnected with the relatively cheaper cost of beef in relation to mutton or goat meat. For instance, while a Kilogramme beef might cost about N400 (US$2.5)[2], the equivalent is about N1000 (US$6.25) for mutton or goat meat. In addition, the large size of cattle also makes it possible for daily meat demands of the Nigerians to be readily met. Although, the small ruminants, especially goat, are as well slaughtered for meat sale, the small size of the animals and high market price of their meats makes the animals less demanded for regular meat consumption.

However, live goats and sheep are much more easily acquired by individuals in relation to cattle owing to market price differentials between the small and large ruminants. For instance, a sizeable cow or bull sells for about ₦70, 000 (US$437.5) in most open cattle markets in the southwestern part of Nigeria, against the average market price of ₦10, 000 (US$62.5) for WAD sheep and goats, ₦18, 000 (US$112.5) for Sahel goats; and ₦20, 000 (US$125) for sheep (Uda and Balami)[3] - . This situation thus accounted for the widespread of sheep and goats among individuals in Nigeria either for consumption, though mostly on events celebration, or rearing for widespread sheep and goats as important animals of trade within humid West Africa though with different demand and consumption patterns in the region. For instance, while sheep are largely consumed during Muslim religious holidays, goats are used for all ceremonies throughout the year, especially for ceremonies such as births, deaths, marriages and festivals; thereby making the demand for goats consistently high. As a result of this, there is a clear price premium for male sheep during the festival period, and some early purchasing for fattening and re-sale takes place.

The market value of the ruminants not only creates employment and generates income for those that directly owned the animals, but indirectly for the butchers, foragers and government. For instance, cattle slaughtering and dressing cost N3, 000 per head per cow and the same services on sheep and goat cost N1, 000 per head per the animal. And to a lesser extent, the animals indirectly generate income for the Nigerian Government through licensing of abattoirs and taxation on every slaughtered animal at the registered abattoirs. Although, ruminants are generally kept on free range management system, conscientious feeding is provided the farm animals primarily kept for commercial purpose. Based on this, forages, either fresh or dry, are sought from the foragers for feeding the commercially-oriented farm animals. In the light of this, crop debris such as dried cowpea shafts and ground vines and husks becomes additional source of income for farmers that cultivate cowpea and groundnuts. Valuation of the Nigerian livestock resources [23] puts the total livestock value at N60billion, based on mid-1991 market prices and as indicated by [22], account for as much as one third of the country's agricultural gross domestic product (GDP).

Social values: socio-cultural value of the ruminants varied across the country. While the sheep and goats are highly prized for cultural heritage in the southwest Nigeria, cattle is of much significance among the Hausa/Fulani in the northern region. Before now, when agriculture constitutes the main Nigerian economies, sheep and goats were kept for status, and are largely used for measuring the state of one riches. But with the relegation of agriculture from the economic fore, use of the number or size of farm animals as measuring tool of social status is no longer tenable, especially at rural level in southwest Nigeria where subsistence agriculture is the main practice. However, the small ruminants still found value in sacrificial offerings among the traditional worshippers in southwest Nigeria. Up till now, goat, specifically doe, constitutes traditional requirement as part of bride price and the animals are kept in memory of the enacted marital relationship between in-laws. Unlike beef and mutton, goat meat are generally considered and consumed as delicacy.

Unlike the devalued state of the socio-cultural value of the small ruminants in southwest Nigeria, cattle, sheep and goats remained relevant as measuring tools of social status and economic strength among the rural households in the northern region of the country. The size of cattle herds and flock of sheep owned by a particular individual or household determines the economic strength of such ones. In addition, a herder's stock of animals constitutes his financial base thereby disposing the animals for income generation whenever it is necessary [24]. Cattle also serve as good means of transportation and animal traction among the livestock farmers in the northern region of the country, whereby the animals are used for land cultivation in preparation for crop cultivation, transportation of farm families to and from the farms and transportation of farm produce between farms and storage points. Given the volume and nature of excreta produce by cattle, the large ruminant have served as valuable source for manure for soil fertility and development of organic agriculture. [22] cattle produces manure outputs of 1368 kg DM/head/year and 248 kg DM/head/year by sheep for soil fertility.

4. DYNAMICS OF RUMINANT LIVESTOCK MANAGEMENT SYSTEM IN NIGERIA

In general, farm animals are poorly managed in Nigeria's agricultural system owing to the fact that the animals are mostly managed on free range/extensive system and semi-intensive system. These management systems are basically influenced by cheap means of feeding the stock all year round. Based on this, the animals are thus allowed to roam the streets and neighbourhood to fend for themselves with little or no special or conscientious provision of supplements for the animals. Although, commonly raised farm animals under the free range and semi-intensive systems include the monogastrics and ruminants, sheep and goats, alongside chicken constitutes the major farm animals largely raised in these systems of livestock management by the Nigerian rural households or livestock farmers.

Small ruminant management system: the small ruminants are however intensively managed on the free range/extensive system, especially in the southern part of Nigeria where crop farming dominates the agricultural practices and with farmers keeping an average of 10 sheep and/or goats. Under the free range system, the animals move about freely to feed on forages/grasses, which are abundantly available during the raining season, and on other feed source such as left over foods/ kitchen wastes and refuse dumps. Hardly are the animals provided supplementary feeds and even shelter by their keepers. The animals thus squat around corridors or available shades in the compounds. Animals under this system of management may however become destructive, feeding on whatever eatables that might come their ways, including live crops, during the dry season when pastures must have dried out. This implies that the free range system may be a healthy practice for ruminant management during the rainy seasons, at least for abundance of forage availability, and but unhealthy during the dry seasons as livestock infringement on the neighbourhood property often lead to conflicts.

Some households or livestock keepers on the other hand maintain semi-intensive management system whereby the animals are provided shelter and kept indoors for security purpose. The animals somehow have their movements regulated and as such are released to fend for themselves in the early and late hours of the day, after which they are kept indoors over the night. As it were in the extensive or free range system, the animals feed on natural pasture and kitchen wastes or by-products of processed foods/farm produce, especially during the rainy season. Although, hardly are the animals under semi-intensive management provided supplements or essential ration for consumption, efforts are made by their keepers to feed them with by-products from farm produce, especially during dry season when pasture are hardly available for free grazing.

The extensive management system is however largely applied for the WAD sheep and goats than for other breeds such as Balami, Uda and Yakansa breeds of sheep; and Sokoto Red goats in southwest Nigeria. This may not be unconnected with economic value of these breeds of small ruminants arising from their bigger body size and better market prices than the WAD breeds. In addition, these breeds of small ruminants are highly prized for social ceremonies and prestige; and are more tempting to be stolen than the WADs. The Balami, Uda, Yankasa; and Sokoto Red breeds of the small ruminants are thus kept on a 'modified intensive management system' whereby the animals are mostly tethered or kept in a guarded enclosure and fed on cut-and-feed forages and by-products of farm produce.

Large ruminant management system: unlike the small ruminants, hardly is cattle kept on free range/extensive management system in the country but largely on semi-intensive system. A level of modification is however applied to the semi-intensive management for cattle. Unlike the small ruminants that could be left to freely range about all alone, cattle are never left all alone to freely graze about or scavenge, but are conscientiously guided by the rearers in the search for pasture and water; and thereafter, are securely checked into the provided shelter. This may not unconnected with the social and economic value of the large ruminant, as the loss of a cattle, either in death or getting misplaced, is at great cost to the herder(s) and as such, the animals are jealously guided for survival, productivity and profitability. Socially, the size of the animal is highly intimidating to humans as appearance of unguided cattle in the public is known to cause commotions whereby people run helter-skelter. This farm animal is never neared as one would near sheep and goats. This situation thus accounted for the need to guide the cattle on grazing over a wide range of vegetations.

Nomadism/Exclusive pastoral system: in addition to the modified semi-intensive management of the cattle by herders, [12] other pastoral management systems commonly practised by cattle herders in the country include the exclusive, transhumant and agro pastoral systems. The exclusive pastoral practice or nomadism entails sole management of the ruminants, especially cattle for the socioeconomic wellbeing of the pastoral farmers. [12] The exclusive pastoralists do not grow crops but simply depend on sales of their ruminants and dairy products to meet their food needs. As a feeding practice, the exclusive pastoralists usually move their animals over long distances, usually through a set migration routes, in search of pasture for their animals or by going into advance arrangement with crop farmers for collection of crop residue for their animals.

Transhumance pastoral system: this entails rearing of ruminants in settlements with a low level of crop cultivation. The transhumant pastoralists [25], often have a permanent homestead and base at where the older members of the community remain throughout the year. The herds are however regularly moved in response to seasonal changes in the quality of grazing and the tsetse-fly challenge, or in an attempt to exploit seasonal availability of pasture. The herds are however regularly moved in response to seasonal changes in the quality of pasture and the tsetse-fly challenge, or in an attempt to exploit seasonal the availability of pasture. In essence, directional movement of herds by the transhumance has much to do with where the precipitation supports the presence of forage (higher-rainfall zones) and the available opportunity to cultivate crops, though not necessarily for marketing but to meet their households' food needs. They however meet their other basic needs through the proceeds from sales of milk and other dairy products. While the women take care of the production and marketing of the dairy products in the local markets, the men take away majority of the herds in search of grazing, leaving the older members of the community with a nucleus of lactating females. The male herders however return at the start of the wet season to help with crop cultivation and where necessary, household income is supplemented with the sales of surplus male sheep or cattle.

Agro pastoral system: the agro-pastoralist practice entails conscious crop cultivation for both home consumption and marketing purposes alongside their reared cattle.[25] Agro-pastoralists hold land rights and cultivate acquired land for crops such as maize, sorghum, millet, yams and cassava, using family or hired labours. While cattle are still valued property, the size of herds are averagely smaller than that of other pastoral systems, usually about 30 head per household in southwest Nigeria [26], possibly because they no longer solely rely on cattle for their livelihood sustenance. In this case, the large ruminants are guided on grazing within a short distance range from their permanent place of abode while the women explored the lactating animals for milk and having it processed into local cheese (wara) and skimmed sour milk (nono) for consumption and local marketing. The Agropastoralists, [25], invest more in housing and other local infrastructure, and where their herds become large, they often send them away with more nomadic pastoralists. In addition, the agropastoralists often act as brokers in establishing cattle tracks and negotiation of "camping" of herds on farms, whereby crop residues can be exchanged for valuable manure, and as well for rearing of work animals, all of which add value to overall agricultural production.

Implications of the extensive and semi-intensive management systems: widespread adoption of extensive and semi-intensive systems of management for livestock in general. Ruminants, is believed to have been highly influenced by its relatively low cost of feeding the animals. Although, the animals may feed on freely available pasture and forages, these systems exposed the livestock to environmental dangers, ranging across stealing and death of the animals [27]. In addition, these systems of livestock management accounted for the generally observed poor production performance of the local breeds of ruminants in terms of meat, milk and litter production in Nigeria, and does not allow for proper recording keeping of the animals production performance [28]. On the same note, [24,29] stress that farm animals kept under the extensive and semi-intensive management systems are burdened with high incidence of diseases, parasites, low productivity and small contribution to household's earnings.

5. RUMINANTS' PESTS AND DISEASES AND DYNAMICS OF MANAGEMENT

Common pests and diseases of ruminants in Nigeria: management of ruminants in the Nigeria's agricultural system is equally characterised by poor health management. As a matter of fact, [30] maintenance and sustenance of healthiness of farm animals constitutes a major challenge to efficient livestock production among the Nigerian livestock entrepreneurs. Several surveys of ruminants kept by the rural farmers, and even in the markets, across the country revealed that the animals are mostly infected with one form of diseases/pests or the other [30-32]. According to Dipeolu (2010), most of the diagnosed livestock diseases in the country were identified to be bacteria, viral, fungi and parasitic-caused diseases. Specifically, the diseases include rinderpest, foot-and-mouth disease, and contagious bovine pleuropneumonia to be the common diseases of cattle in Nigeria. In addition to these are small number of cases of dermatophilosis, lumpy skin disease, papillomatosis and keratoconjunctivitis. [33] on the other hand, indicated that infections such as pneumonia, helminthiasis, peste des petits and enterotoxaemia as common diseases of sheep and goats in Nigeria. Assessment of seasonal pattern of tick load on Bunaji cattle under the traditional management by [31] revealed the dominant tick species as Amblyomma variegatum; Boophilus decoloratus, Rhipicephalus (simus) senegalensis, R. tricuspis and Hyalomma spp.

Although, the incidence and intensity of pests and diseases infestation in the ruminant farm animals may vary between the rainy and dry seasons, and Across Nigeria's ecological zones, the infestation portends a great danger for healthiness and productivity of the animals. According to Dipeolu (2010), livestock farmers may experience total loss of stock in death, or partial losses (through morbidity) in which the productivity of the animals becomes greatly reduced. Disease such as pneumonia, especially PPR, as the major causes of deaths in of ruminants; diarrhoea is mostly caused by parasitic gastroenteritis and PPR; and abortions and neonatal deaths are associated with starvation. In order to overcome these gruesome effects of pests and diseases on the ruminants, it becomes essential for the livestock farmers to either prevent or control the incidence of the diseases.

Pests and diseases control: in terms of control of livestock diseases, the livestock farmers hardly take up veterinary treatment of the affected animal(s), especially the small ruminant farmers, as they considered the veterinary treatment as too expensive to bear (Fabusoro, Lawal-Adebowale & Akinloye, 2007; Oluwafemi, 2009). In as much as the small ruminant farmers may which to save any diseased animals by taking to ethno-veterinary treatment, they may afford to lose the animal in death rather than expending their hard earned income on veterinary treatment of a diseased animal. The losses may be marginal in case of one or two of the animals are lost in death, but will be a great economic loss where about five or more of the animals are lost in quick succession as result of disease infestation (Dipeolu, 2010; Aina, 2012). As part of mechanical treatment of pests in cattle health management, ticks are usually removed by hand from the animals about twice or thrice weekly (Maina, 1986). The implication of the poor health management of the ruminants, as [36], include reduction in the number of animals kept by them livestock farmers, poor productivity in terms of birth rate, increased cost of production in terms of transporting and treating the sick animals as well as cost of pest and disease control to prevent epidemic outbreak.

On another note, ante-mortem and post-mortem inspection of the ruminants, particularly cattle, meant for slaughtering across the major abattoirs in the country further underline the poor state of ruminant, especially cattle, management in the country. The ante-mortem inspection of cattle to be slaughtered at a major abattoir in Ibadan, southwest Nigeria, between 1990 and 1994 showed that between 2.4% and 6.3% of the slaughtered cattle were pregnant (Dipeolu (2010). The implication of this, [37], was a tremendous loss of potential cattle offspring that would have contributed to the cattle population growth and meat supply profile of the country. A post-mortem study of another major abattoir based in Lagos, Nigeria, between 2004 and 2007 showed that the slaughtered cattle portends a health risk to beef consumers as about 1.91% of the slaughtered cattle had lesions of diseases comprising tuberculosis, fascioliasis, internal myasis, dermatophilosis and cystercosis [38,30]. Laboratory examination of some of the meat-borne diseases showed that the meats are tainted with bacteria pathogens such as Campylobacter spp., Clostridium spp., Escherichia coli, Salmonella serotypes, and other enteric bacteria which may not cause clinical diseases in the animals but a potential threat to public health (Dipeolu, 2010).

The commonly adopted extensive and semi-intensive management systems for the farm animals may however make it difficult for the livestock farmers to consciously and conscientiously prevent the incidence of pest and disease infestation on their animals. This is based on the fact that, as the animals are allowed to freely roam the neighbourhood they readily contact infectious diseases or pests from other infected animals they mixed with in the course of fending for themselves, and may as well sustain injuries which in turn may eventually impair their health status and probably lead to their deaths (Lawal-Adebowale & Alarima, 2011).

6. RUMINANT FEEDS AND DYNAMICS OF UTILIZATION

Given the distinct nature of the ruminant's stomach, the farm animals heavily depend on forage or raoughage as major feeds. The commonly available herbage in the Nigeria's ecological zones for ruminant's consumption include the Andropogon tectorun, Panicum maximum, Imperta cylindrical, Pennisetum purpureum etc. These grasses, which are fibrous in nature, are rich in cellulose and provide the ruminants a high level carbohydrate and some measures of vitamins and minerals. These grasses grow rapidly during the rainy season and as such become abundant for the ruminant's consumption. The ruminant kept on free range thus feed freely on the naturally occurring forages. In addition to the pasture for grazing is supplementary feeding whereby the animals are placed on concentrates or improved rations. Supplementary feeding of cows significantly improve weights of the calves at birth (20.1kg) and at one year of age (107.9 kg) when compared with other animals not placed on supplements (with birth weight of 18.6 and 99.3 kg at one year). At 365 days of age, viability of calves from supplemented dams averaged 88% against 67% in calves from non-supplemented dams. Milk for calves and humans from dams on supplements averaged 128 and 179 litres/cow/year. Although, supplementary feeding did not improve calving intervals, it suggests that it every essential to place the ruminants on supplements for better productivity in term of milk and meat production.

Cost of supplementary feeding and non-availability of forage during the dry season greatly challenged efficient livestock feeding and management in Nigeria. Based on the need for adequate feeding, it is believed that about 85% of cost of livestock production is feeding, and given the poverty status of most livestock farmers and poor marketing system of farm animals, hardly could they take up supplementary feeding. This accounted for preference of extensive and semi-intensive systems of management. Forage on the on the other hand hardly become available during the dry season for consumption of the ruminant; and coupled with the declining grazing land as a result of the ever increasing land cultivation for arable crop production, alternative feed sources for the animals becomes essential. Utilisation of fodder from crop residues compensates for non-availability of grasses during the off-season. Other alternative to mitigate the effect of dry season feeding was the establishment of fodder bank whereby legumes are established and properly managed in a concentrated unit [41]. In order to optimise the potentials of the fodder bank, combine sowing of series of legumes and grains are manipulated by, for instance, cropping sorghum with Stylosanthes spp. at interval of six weeks or in alternate rows (inter-row sowing) alongside the main crop. A study of the grazing behaviour of cattle among the settled Fulani pastoralists showed that the farm animals utilized a wide range of different feed resources, notably sorghum and millet residues, during the dry season. As further indicated, the residue accounted for 12.6% of annual grazing time in Abet- a farming area, and for 6.6% in Kurmin-Biri- a grazing reserve. Browsing accounted for 1.4% annual grazing time in Abet, and 11.2% in Kurmin-Biri.

The fodder bank alternative however mainly benefits selected animals as not all animals are allowed to graze the bank. Fodder banks are designed not to supply forage year-round for an entire herd but rather to be used strategically for limited periods with selected animals, thus only pregnant and lactating animals are allowed to graze the bank. This suggests that, only a few ruminants had access to grazing or foraging during the dry season, and thus portends that dry season feeding constitutes a major challenge to livestock production in Nigeria. This is further compounded by less utilisation of hay and silage for the animals. Since the reared animals cannot survive without food, the implications of dry season feed problem include straying or deliberate guiding of the animals into farms for grazing thus leading to conflicts and violent clash between the crop and livestock farmers.

7. FUTURE OF RUMINANT LIVESTOCK DEVELOPMENT IN NIGERIA

Although, ecological categorisation of the Nigeria has varied over time arising from changing trends of the commonly used natural factors [12], critical examination of the country's ecozones in relation to livestock distribution revealed that the ruminants are distributed throughout the three major ecozones in the country, namely the semi-arid, sub-humid and humid zones. The semi-arid region, characterised by average rainfall of 500 – 1000mm, prolonged dry season and sparsely distributed vegetations, is known to have greatly favoured livestock management in the country over the years. But with the changing climatic trends in the country, the sub-humid zone and its characteristics rainfall distribution range of 1000 – 1500mm, vegetative cover and moderately dry periods, now enclaves about 45% of the cattle in the country. In the same vein, studies have affirmed that the changing situation of tsetse flies infestation in the region, coupled with the prolonged rainfall period and good rainfall distribution range of more than 1500; has equally made the environment favourable to cattle and other small ruminants' management. This observation suggests that Nigeria's agro-ecologies have the potentials to favourably support livestock development in the country. This notwithstanding, there is need to consciously harness the environment to enhance the country's livestock development through the following:

Efficient livestock feeding: exploration of the environment and the country's breeds of ruminant potentials for livestock industry development are yet to be fully harnessed. The larger proportion of the ruminant livestock in Nigeria lies in the hands of herders who keep them under extensive and semi-intensive management systems, whereby the animals only rely on natural pasture and crop residue for survival. The ruminants may though have access to enough forage during the rainy season; it becomes a great deal of challenge to efficiently feed the animals during the dry season. In order to sustain the animals and ensure better productivity, there is need to explore the available natural pasture for silage and hay making such that the animals could be adequately fed during the dry season. In addition, there is need for paddock establishment, especially in the rural communities or reserved areas, for grazing by the ruminants. Although, forage constitutes the bulk of food needed by the ruminants, supplementary feeding is equally essential, especially for the lactating animals. In view of this, the farm animals' diet needs to be supplemented with meals such as cottonseed cake, wheat bran, molasses, drugs and mineral salt licks etc. In view of the fact that the indigenous cattle can gain an average of 0.9 to 1.2 kg per day on silage and concentrate rations [22], it suggests that the local breeds of cattle have the potentials for efficient utilisation of feed for better production performance.

Veterinary services: pests and diseases portend a major risk to livestock development in Nigeria, as incidence of pests and diseases are common in the country's livestock system. Although, prevention is known to be better than cure, it is invariably impossible to out rightly prevent the farm animals from being infested with either pests or diseases. This premise thus calls for establishment of sound veterinary services where infected animals could be taken care of. This requirement has been a great challenge in the Nigeria's livestock management system. Apart from inadequate veterinary services in the country, current veterinary therapy in Nigeria is suffering from both scarcity and the high cost of drugs thereby making it impossible to save the livestock industry as it were in the country. Although, the livestock herders may take to ethno-veterinary treatment of their animals, this becomes possible only when the symptoms become manifested, and by then a serious internal damage or impairment of the animals' health might have taken place. The implication of this is that, it may be impossible to adequately treat the animals or ensure proper clinical remedy. This situation thus calls for government and non-government organisations intervention for development of the veterinary services such that it becomes affordable to be patronised by the stock herders. The easiest and most rational solution to the problem of livestock health is to develop acceptably effective drugs from reasonably inexpensive sources for use as supplements to commercial drugs. The veterinary traditional medicine practices may still be of value in the animal health care, but should be subjected to scientific investigation for efficacy. In the light of this, it becomes important to have baseline data about traditional ethno-veterinary practices for ethno-veterinary medical information generation. Combination of the orthodox and ethno-veterinary care could thus save the animals of impaired health and enhance productivity.

Livestock breeding: livestock breeding is crucial to livestock development globally. Good system of management of the resulting breeds/offspring from the crosses – in terms of intensive keeping, good health care and feeding, is however crucial to better performance of the animals. Adopted poor management systems for farm animals in Nigeria and most other developing countries certainly accounted for the poor production performance of the local ruminant breeds. The same poor management system accounted for poor performance of the exotic breeds imported into the country in the 70 (Blench, 1999). Just as the exotic breeds are known to have performed excellently well in their countries of origin under good management practice, results from experimental stations results from stations and universities farms across Africa showed that productivity of the animals could be improved under more intensive management. Similarly, where crossing has been successful under good management practice, dairy cattle dairy cattle portrayed a linear increase in milk yield as the exotic gene is increased up to the 7/8 level. The F1 Friesian x Bunaji cow (50%) gives 1684 kg, the 3/4 (75%) gives 1850 kg and the 7/8 gives 2051 kg of milk in a lactation of about 260 days. This suggests that good practices and cross breeding with the exotic breeds of desirable quality stand the chance of enhancing the country's livestock development.

Profitable livestock marketing system: among all other agricultural enterprise production, livestock management remains a delicate and expensive venture; it however has the potentials of profitable returns. The livestock is delicate in the sense that the animals need to be adequately fed, not just with any ration, but a balanced ration for productive performance. In the same vein, the health of the animals cannot be forgone as healthiness of the animals is not only a vital for production performance, but survival and sustenance of the livestock venture. Placement of the ruminant on a good ration is certainly at a great deal of cost or financial incurment, the poor economic status of the ruminant keepers in the country however makes it extremely difficult to build the livestock industry. This situation may however be reverted through efficient marketing system of livestock and its products and by-products. Poor marketing system is one of the bane livestock development in the country, whereby the animals are locally sold either directly as live animal or meat.

Livestock research development: development of the Nigeria's livestock industry will not magically occur, but through conscientious efforts in livestock research. This calls for baseline data generation about the breeds of ruminants in the country, their production performance and marketing. Other information-base that must be established include the common livestock feeds (pasture and feed meal supplements) and common pests and diseases of livestock and their effects on the animals. This will harm the livestock research institutes with the salient information as bench mark for research work and generation of livestock innovation. Social scientists inclusion in livestock research development is crucial as this disciplines helps to ascertain the psychology of the ruminant keepers and their economic status to adopt and adapt generated livestock innovation. Similarly, the social scientist, especially the economists, will help to ascertain the economic implications of the innovations and the market driving force for ensuring efficient production and marketing of livestock and its products.

8. LIVESTOCK DEVELOPMENT IN NIGERIAN: POLICY RECOMMENDATION

Over 90 percent of the ruminant livestock lies in the hands of rural livestock farmers, especially the pastoralists, in Nigeria. The animals though, are of considerable economic importance in Nigeria's economy, poor management system of the stock has greatly hindered the development of the livestock. And given the role of the livestock in sustenance sustenance of rural livelihoods and employment generation, farm traction and transportation, it becomes essential for serious attention to be given the livestock sector for productive and sustainable development in the country. In this regard, the livestock research institutes, comprising National Animal Production Research Institute (NAPRI), National Veterinary Research Institute (NVRI), and Nigerian Institute for Trypanosomiasis Research (NITR), need to be strengthened in terms of qualified and adequate research personnel and equipment for quality research on livestock related issues. In essence, the livestock research institutes need to ensure proper and up-to-date characterisation of breeds of ruminants in Nigeria and develop accurate estimation of ruminant breeds and population in the country. In essence, the livestock research institutes need to ensure proper and up-to-date characterisation of breeds of ruminants occurring pests and diseases in livestock, and the lethal effects of ill-health causative agents on the animals.

To effectively achieve this, research in livestock development should go beyond the traditional field visit to animal sheds for physical livestock condition monitoring and data collection. The country needs to harness the emerging information and communication technology (ICT) devices that allow for remote and continuous monitoring of livestock conditions and collection of data on the animals without physically being in the animals' sheds. With this, efficient data and information on farm animals' health status, productivity, feeding regime and feed conversion could be readily monitored. Similarly, documentation of particular livestock pedigree, characterisation of breeds of farm animals and simulation of the animals' characteristics and production performance could be enhanced for effective management and transformational development of the livestock sector. In addition to this is need for better development of better grazing system and management practices in the country's livestock sector. Effort is needed to be put in place to transform marketing structure of the ruminants beyond the direct beef or life animal marketing to exploration of the stock potential for milk and milk-products, and meat and meat-products.

APPENDIX 1

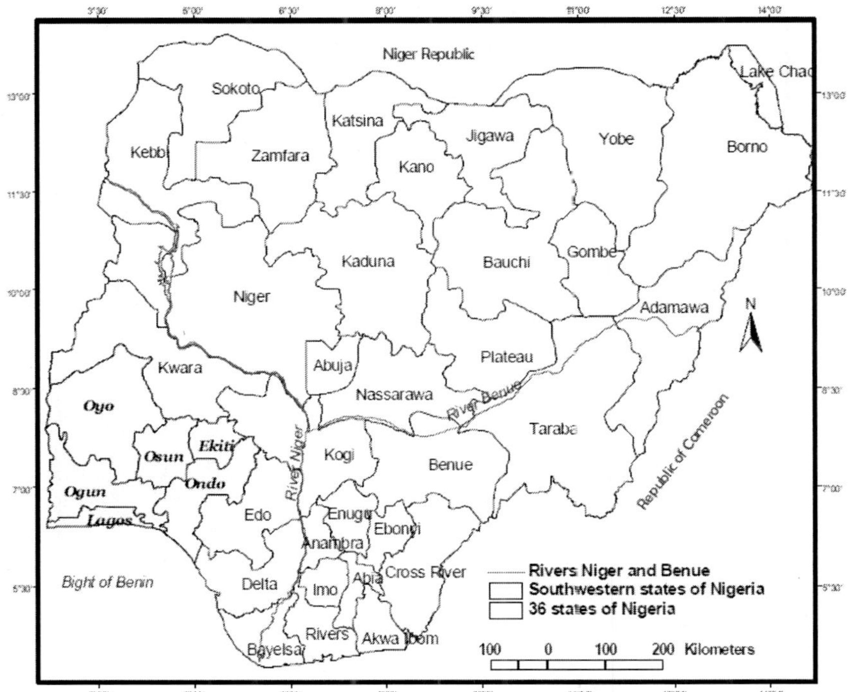

Map of Nigeria showing the natural division into three regions by rivers Niger and Benue

APPENDIX 2

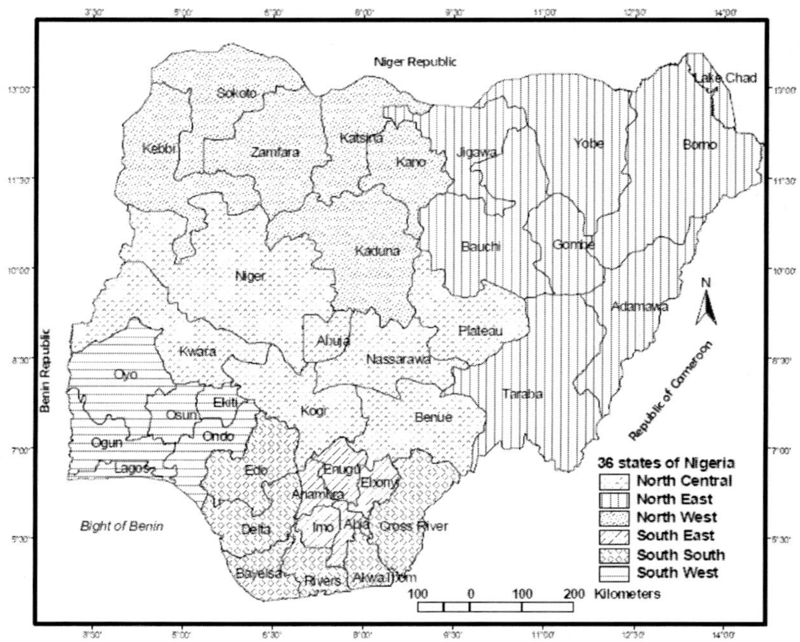

Map of Nigeria showing the six geopolitical zones in the country for political administration

APPENDIX 3

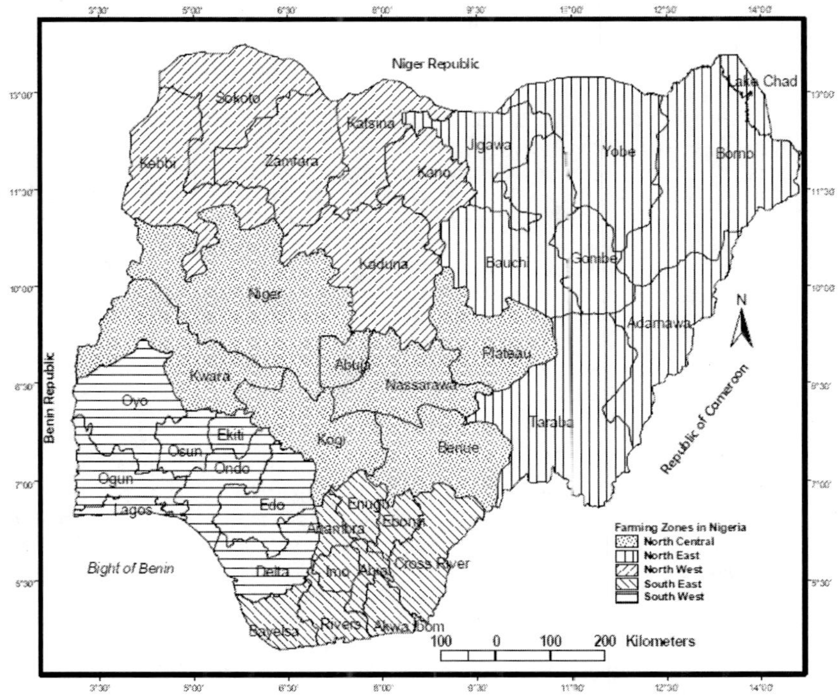

Map of Nigeria showing the five farming zones in the country for agricultural development administration

REFERENCES

1. Abiola, S. S., Ikeobi, C. O. N. & Dipeolu, M. A. (1999). Bovine wastages in abattoir and slaughter slabs of Oyo State, Nigerian: Pattern and ethical concerns. Tropical Journal of Animal Science, 1 (2): 143 – 148.
2. Ademosun, A. A. (1987). Appropriate management systems for West African Dwarf goats in humid tropics. In O. B. Smith & and Bosman, H. G. (Eds). Goat production in the tropics. Workshop proceeding, Obafemi Awolowo University, Ife.
3. Adesehinwa, A. O. K., Okunola, J. O. and Adewumi, M. K. (2004). Socio-economic characteristics of ruminant livestock farmers and their production constraints in some parts of South-western Nigerian. *Livestock Research for Rural Development*, 16 (8). Retrieved April 6, 2012, from http://www.lrrd.org/lrrd16/lrrd16.htm
4. Adu, I. F., Omotayo, A. M., Aina, A. B. J. & Iposu, S. O. (2000). Animal traction technology in Ogun State, Nigeria: potentials and constraints. Nigerian Journal of Animal Production, 27 (1): 95 – 98.

5. Adu, I. F. (1993). Some socio-economic aspects of settled Fulani in Ogun State. Biennial workshop of cattle research network (CARNET), Addis Ababa, ILCA.
6. Adu, I.F. & Ngere, L.O. (1979). The indigenous sheep of Nigeria. *World Review of Animal Production* 15(3): 51–62.
7. Aina, A. B. J. (2012). Goat (*Capra hircus*): A misunderstood animal. 35th Inaugural Lecture, Federal University of Agriculture, Abeokuta.
8. Bayer, W. (1986). Agropastoral herding practices and the grazing behaviour of cattle Supplementary. In R. von Kaufmann, Chater, S. & Blench, R. (Eds). Proceedings of ILCA/NAPRI Symposium, Kaduna, Nigerian. Retrieved February 2, 2012 from http://www.fao.org/Wairdocs/ILRI/x5463E/x5463e0b.htm#supplementarypaper2
9. Bayer, W. (1986a). Traditional small ruminant production in the subhumid zone of Nigeria. In R. von Kaufmann, Chater, S. & Blench, R. (Eds). Proceedings of ILCA/NAPRI Symposium, Kaduna, Nigerian. Retrieved February 2, 2012 from http://www.fao.org/Wairdocs/ILRI/x5463E/x5463e0b.htm#paper7
10. Bayer, W. (1986b). Utilization of fodder banks. In R. von Kaufmann, Chater, S. & Blench, R. (Eds). Proceedings of the Second ILCA/NAPRI Symposium, Kaduna, Nigerian. Retrieved February 2, 2012 from http://www.fao.org/Wairdocs/ILRI/x5463E/ x5463e0b.htm#paper17
11. Bourn, Wint, Blench & Woolley (2007). Identification and characterisation of West African shorthorn cattle. Nigerian Livestock Resource Survey, FAO cooperate document repository. pp 1 – 12.
12. Blench, R. (1998). The expansion and adaptation of Fulbe pastoralism to sub-humid and humid conditions in Nigeria. African studies Centre, Leiden.
13. Blench, R. (1999). Traditional livestock breeds: geographical distribution and dynamics in relation to the ecology of West Africa Overseas Development Institute, Portland House.
14. Bourn, D., Milligan, K. & Wint, W. (1986). Tsetse, trypanosomiasis and cattle in a changing environment. In R. von Kaufmann, Chater, S. & Blench, R. (Eds). Proceedings of ILCA/NAPRI Symposium, Kaduna, Nigerian. Retrieved February 2, 2012 from http://www.fao.org/Wairdocs/ILRI/x5463E/x5463e0b.htm#paper5
15. Bourn, D. (n.d.). Highlights of the Nigerian livestock resources reports. Retrieved April 11, 2012 from http://www.odi.org.uk/work/projects/pdn/papers/35d.pdf 78 Livestock Production
16. Cadmus, S. I. B., Thomas, J. O. & Oluwasola, O. A. (2003). Isolation of *Mycobacterium bovis* in human TBsuffereing from cervical adenitis in Ibadan, Nigerian. Proceeding of the International Conference on Emerging Zoonoses, Iowa, USA.
17. Dipeolu, M. A. (2010). Healthy meat for wealth. 29th Inaugural Lecture, Federal University of Agriculture, Abeokuta.
18. Ehoche, O. W., Buvanendran, V. (1983). The yield and composition of milk and pre-weaning growth rate of Red Sokoto goats in Nigeria. *World Review of Animal Production*, 19: 19 – 24.
19. Fabusoro, E., Lawal-Adebowale, O. A. & Akinloye, A. K. (2007). A study of rural livestock farmers' patronage of veterinary services for health care of small farm animals in Ogun State. *Nigerian Journal of Animal Production.* 34 (1): 132 – 138.
20. FAO (2001). Pastoralism in the new millennium. Animal production and health, paper150.
21. Food and Agriculture Organisation, Rome. Retrieved December 10, 2011 from http://www.fao.org/DOCREP/005/Y2647E/y2647e00.htm
22. Gall, C. (1996). *Goat breeds of the world.* Margraf Publishing, Weikersheim, Germany.Ibrahim, M. A. (1986). Veterinary traditional practice in Nigeria. In R. von Kaufmann, Chater, S. & Blench, R. (Eds). Proceedings ILCA/NAPRI Symposium, Kaduna, Nigerian. Retrieved February 2, 2012 from http://www.fao.org/ Wairdocs/ ILRI/ x5463E/x5463e0b.htm#paper9
23. International Technology Association (ITA, 2004). Nigerian livestock. The Library of Congress Country Studies; CIA World Factbook. Retrieved April 8, 2012 from http://www.photius.com/countries/nigeria/economy/nigeria_economy_livestock.html

24. Jabbar, M. A. (1994). Evolving crop-livestock farming systems in the humid zone of West Africa. *Journal of Farming Systems Research and Extension,* 4 (3): 47 – 60.
25. Jabbar, M. A., Reynolds, L. & Francis, P. A. (1995). Sedentarisation of cattle farmers in the derived savnna region southwest Nigeria: results of a survey. *Tropical Animal Health Production,* 27: 55 – 64.
26. Khan, B. B., Iqbal, A. & Mustafa, M. I. (2003). Sheep and goat production (part 1). University of Agriculture Faisalabad. Retrieved from Lawal-Adebowale, O. A. & Alarima, C. I. (2011). Challenges of Small Ruminants Production in Selected Urban Communities of Abeokuta, Ogun State. *Agriculturale Conspectus Scientificus* 76 (2): 129 – 134.
27. Lawal-Adebowale, O. A. (2012). Factors Influencing Small Ruminant Production in Selected Urban Communities of Abeokuta, Ogun State. *Nigerian Journal of Animal Production,* 39 (1): 218 – 228.
28. Maina, (1986). Animal health in subhumid Nigeria. In R. von Kaufmann, Chater, S. & Blench, R. (Eds). Proceedings of ILCA/NAPRI Symposium, Kaduna, Nigerian. Retrieved February 2, 2012 from http://www.fao.org/Wairdocs/ILRI/ x5463E/x5463e0b.htm#paper8
29. Mohamed-Saleem, M. A. & Fitzhugh, H. A. (1993). An overview of demographic and environmental issues in sustainable agriculture in sub-Saharan Africa. *In* J.M. Powell, S. Fernandez-Rivera, T.O. Williams, C. Renard (eds). Livestock and sustainable nutrient cycling in mixed farming systems of sub-Saharan Africa. Proceedings of an International Conference, ILCA, Addis Ababa, pp. 3 – 20. Dynamics of Ruminant Livestock Management in the Context of the Nigerian Agricultural System 79
30. Muhamed-Saleem, (1986a). The establishment and management of fodder banks. In R. von Kaufmann, Chater, S. & Blench, R. (Eds). Proceedings of ILCA/NAPRI Symposium, Kaduna, Nigerian. Retrieved February 2, 2012 from http://www.fao.org/ Wairdocs/ ILRI/x5463E/x5463e0b.htm#paper16
31. Muhamed-Saleem, (1986b). Integration of forage legumes into the cropping systems of nNigeria's subhumid zone. In R. von Kaufmann, Chater, S. & Blench, R. (Eds). Proceedings of ILCA/NAPRI Symposium, Kaduna, Nigerian. Retrieved February 2, 2012 from http://www.fao.org/Wairdocs/ILRI/x5463E/x5463e0b.htm#paper15
32. Ngere, L.O., Adu, I.F. & Okubanjo, I.O. (1984). The indigenous goats of Nigeria. *Animal Genetic Resources Information* 3: 1– 9.
33. Nuru, S. (1986). Livestock research in Nigerian. In R. von Kaufmann, Chater, S. & Blench, R. (Eds). Proceedings of ILCA/NAPRI Symposium, Kaduna, Nigerian. Retrieved February 2, 2012 from http://www.fao.org/Wairdocs/ILRI/x5463E/x5463e0b.htm#paper1
34. Oluwafemi, R. A. (2009). Cattle production and marketing in Nigerian: the impact of diseases. A case study of Maiakuya, Assakio and Shinge cattle Markets in Lafia Local Government Area of Nasarawa State, Nigerian. *The Internet Journal of Veterinary Medicine,* 6 (1). Retrieved April 12, 2012 from http://www.ispub.com/journal/ theinternet- journal-of-veterinary-medicine/volume-6-number-1
35. Omotayo A. M. (2003). Ecological implications of Fulbe pastoralism in southwestern Nigeria. *Land Degradation and Development,* 14: 445- 457
36. Omotayo, A. M., Adu, I. F. & Aina, A. B. (1999). The evolving sedentary lifestyle among nomadic pastoralists in the humid zone of Nigeria: implications for land-use policy. *Int. J. Sustain det. World Ecol.* 6: 220 -228.
37. Opasina, B. A. & David-West, K. B. (1988). Position paper on sheep and goat production. InV.M. Timon & Baber, R. P. (Eds), Sheep and goat meat production in the humid tropics of West Africa. Food and Agriculture Organisation Animal Production and Health Paper 70. Retrieved January 15, 2012 from http://www.fao.org/docrep/004/ s8374b/ S8374b00.htm#TOC
38. Oppong, E. N. W. (1988). Health control for sheep and goats in the humid tropics of West Africa. In V.M. Timon & Baber, R. P. (Eds), Sheep and goat meat production in the humid tropics of West Africa. Food and Agriculture Organisation Animal Production and Health Paper 70. Retrieved January 2012 from http://www.fao.org/docrep/ 004/s8374b/ S8374b00.htm#TOC

39. Otesile, E. B., Kasali, O. B. & and Nzeku, C. K. N. (1983). Mortality in goats on the University of Ibadan teaching and research farms, Ibadan, Nigerian. Bulletin of Animal Health and Production in Africa. 30: 235 – 239.
40. Otchere, E. O. (1986). The effects of supplementary feeding of traditionally managed Bunaji cows. In R. von Kaufmann, Chater, S. & Blench, R. (Eds). Proceedings of ILCA/NAPRI Symposium, Kaduna, Nigerian. Retrieved February 2, 2012 from http://www.fao.org/Wairdocs/ILRI/x5463E/x5463e0b.htm#paper10 80 Livestock Production
41. Otchere, E. O. (1986). Traditional cattle production in the subhumid zone of Nigeria. In R. von Kaufmann, Chater, S. & Blench, R. (Eds). Proceedings of ILCA/NAPRI Symposium, Kaduna, Nigerian. Retrieved February 2, 2012 from http://www.fao.org/Wairdocs/ILRI/x5463E/x5463e0b.htm#paper6
42. RIM, (1992). Nigerian National Livestock Resource Survey (vol 4). Report by Resource Inventory and Inventory and Management Limited (RIM) to FDL&PCS, Abuja, Nigeria. Sumberg, J. E. & Cassaday, K. (n.d.). Sheep and goats in humid West Africa. Retrieved April 4, 2012 from http://agtr.ilri.cgiar.org/Library/docs/x5555e/x5555e00.htm

END NOTES

1. Based the use of low-level systematic aerial surveys (Bourn, Milligan & Wint, 1986)
2. Exchange rate at ₦160 to US$1 as at March 2012
3. The indicated prices are based on personal market survey between February and March 2012 and off the festival periods. The market price of sheep goes for ₦50, 000 (US$) during the festival period, especially during the Muslim (Idi-el Kabir) celebration.

INDEX

A

adaptive capacities, 34, 45

Agritourism, 285, 286, 287, 302, 303

anaplasmosis, 219, 228, 238

animal system, 34

Arundo donax, 239, 240, 241, 254, 256

B

bacterial zoonoses, 219

borreliosis, 219, 237

C

Cattle, 58, 61, 67, 124, 151, 154, 157, 158, 159, 160, 373, 376, 380, 395, 400

CGE Model, 70, 84

crops systems, 34

E

ehrlichiosis, 219, 225, 228, 237

Evaluation, 68, 146, 184, 255, 256, 336, 357, 359, 360

F

fatty acids, 205, 213, 257, 258, 260, 266, 272, 273

feed resources, 121, 122, 123, 124, 125, 127, 128, 130, 143, 144, 145, 147, 319, 386

Feedstock Traits, 240

flexibility, 34, 35, 36, 37, 38, 39, 40, 42, 43, 45, 46, 47, 49, 50, 318

G

Gaseous emission, 258

GHG Emissions, 70

H

Hybrid, 240

I

Income, 52, 60, 67

indigenous enzymes, 203, 204, 212

Invasive Species, 240

L

lactation stages, 203, 204, 205, 206, 207, 208, 210, 211, 212

Land Use, 67, 69, 70, 75, 84, 146, 325

Livelihoods, 52, 67

Livestock, 1, 2, 3, 4, 27, 28, 29, 30, 31, 33, 50, 51, 52, 62, 64, 389, 390, 393, 394, 395, 396

M

Manure composition, 258, 270, 276

mediterranean, 306

milk, 2, 31, 40, 41, 42, 55, 60, 67, 72, 88, 89, 92, 95, 97, 98, 382, 383, 386, 389, 390, 394, 401

Mixed-Methods, 52

mortality, 167, 190, 228, 230, 231, 232, 233, 357, 360, 361, 366, 367, 371, 377, 378

N

Newcastle Disease, 359, 360, 373

P

parity, 42, 203, 209, 210

pedigree bulls, 190

Pork Production, 285, 286

R

resilience, 34, 35, 36, 37, 38, 39, 45, 46, 49, 50, 95

rickettsiosis, 219

S

Saccharum spp, 240

seasonality, 9, 54, 62, 190, 319, 365

semen evaluation, 190, 201

semiarid, 306, 307, 319, 320, 321

Shocks, 52

T

Thermochemical Conversion, 240

tick, 219, 220, 225, 227, 228, 229, 232, 233, 234, 235, 237, 238, 384

Tissue Components, 240

tularemia, 219, 233, 234, 235, 238

U

uncertainty, 24, 34, 35, 36, 42, 46, 47, 49

V

Volatile, 258, 273